Fungal Biology

Harry J. Hudson

Department of Botany
University of Cambridge

CAMBRIDGE
UNIVERSITY PRESS

CAMBRIDGE UNIVERSITY PRESS
Cambridge, New York, Melbourne, Madrid, Cape Town,
Singapore, São Paulo, Delhi, Tokyo, Mexico City

Cambridge University Press
The Edinburgh Building, Cambridge CB2 8RU, UK

Published in the United States of America by Cambridge University Press, New York

www.cambridge.org
Information on this title: www.cambridge.org/9780521427739

First published by Edward Arnold 1986
First published by Cambridge University Press 1992
Re-issued 2011

A catalogue record for this publication is available from the British Library

ISBN 978-0-521-42773-9 Paperback

Preface

In this text I have attempted to set out my own interests in and attitudes to fungal biology. For a variety of reasons, the aim has not been to give an all embracing definitive account but nevertheless to emphasize the ecological diversity and extreme versatility of the fungi. Omissions are vast, but selection and inclusion have not been entirely random. Some of the omissions stem from my own shortcomings as a mycologist. I plead guilty to ignoring, through my own ignorance and through conceptual difficulties, the vast conglomerate of fungi usually called 'soil fungi' but would argue, in mitigation, that many of their biological activities are displayed by litter-decomposing and wood decay fungi, which are dealt with in some detail. An immense range of texts have been published on fungi as plant pathogens; hence they get short shrift here, but some appraisal of their nutritional roles as necrotrophs and biotrophs is essential to add emphasis to their major role as parasitic symbionts – thus the final chapter. Fungi excel at living together with other organisms. The omnipresence of mycorrhiza, rather than just roots, and the ultimate in symbiosis, lichens – over one sixth of all the true fungi are lichenized – stress the ecological significance of mutualistic symbioses and justify Chapters 7 and 8. Symbioses with animals are less well understood but symbiosis between heterotroph and heterotroph extends the nutritional activities of the fungi. Travels in west, central and southern Africa and the fascination of finding (and subsequent devouring) *Termitomyces* irresistibly leads to a quest to understand in part at least the basis of the wide variety of fungus/insect associations. Leaves and wood must be the two major bulk substrates for decomposer heterotrophs on this planet and the role of fungi as their prime decomposers admirably illustrates the attributes of fungi as heterotrophs. The elegance of the reproductive structures of coprophilous fungi and the striking range in conidial forms in the aquatic hyphomycetes cannot fail to impress and may be sufficient justification for their inclusion. But again their roles as decomposer saprotrophs and their contribution to the re-cycling of carbon and mineral nutrients in their own particular spheres of the ecosystem as a whole is a highly significant one. Ecological diversity can in part be expressed by growth in extreme environments where the majority of other organisms do not flourish. Various excursions with research students into the study of fungal activity in composts have inevitably led to an attempt to understand thermophily in fungi and to seek comparison with psychrophily and xerophily. All such compounded form the basis of this text which is intended primarily for undergraduate students and others interested in fungi as organisms and the role that they play in their natural environments.

I gratefully acknowledge the help given by B.E.H. in preparing the manuscript and the artistic skills of L.J.H. in preparing some of the figures.

Cambridge   H.J.H.
1986

To B.E.H., L.J.H. and S.E.H.

Contents

Preface   v

1. **Fungi as organisms**   1
   Vegetative phases   2
   Hyphal walls   2
   Generalized life cycles and nuclear phases   7
   Hyphal growth   13
   Colony growth   21
   Chlamydospores and sclerotia   25
   Mycelial strands and rhizomorphs   29
   Spores   32
   Basidiocarp form and structure   45

2. **Fungi as decomposers of leaves**   57
   The leaf as a spore trap   57
   Phylloplane inhabitants   58
   Common primary saprotrophs   62
   Pathogens   63
   Exochthonous fungi   64
   The leaf surface as a habitat for fungi   64
   Distribution of the common primary saprotrophs   70
   Attributes of the common primary saprotrophs   71
   Decomposition of pine needles   77
   The role of the litter micro-fauna   79

3. **Fungi as decomposers of wood**   84
   The structure and components of wood   84
   Types of wood decay – white, brown and soft rots   85
   Lignin degradation   88
   Natural resistance of wood to fungal decay   93
   Other wood-inhabiting fungi   96
   Environmental factors and the decomposition of wood   99
   Habitat relations and specificity of wood-inhabiting fungi   100
   Ecological studies on decaying wood   103
   Decomposition of lignin and humus in the soil   106

4. Fungi as inhabitants of aquatic environments 110
The diversity of marine fungi 110
Substrates and hosts of marine fungi 114
The diversity of freshwater fungi 117
'Water moulds' 119
Parasitic Saprolegniales 122
Freshwater Ascomycotina 123
Aquatic Hyphomycetes 124
Aero-aquatic Hyphomycetes of stagnant waters 141
The fungi of stagnant waters 143

5. Fungi as inhabitants of animal faeces 146
Succession of fungi on herbivore dung 146
Adaptations to habitat 147
Herbivore dung as a substrate 148
Analysis of the fungal succession 150
Preferences for particular dung types 154
Basidiobolus and two-phased animal dispersal 155
Decomposition of the faecal pellets of arthropods 157

6. Fungi as inhabitants of extreme environments 159
Thermophily in fungi 159
Basis of thermophily 159
Variety and distribution of thermophilous fungi 161
Succession of fungi in wheat straw compost 164
Beneficial and detrimental activities of thermophilous fungi 166
Cultivation of *Agaricus bisporus* 167
Cultivation of *Volvariella volvacea* 171
Garden and municipal composts 172
Thermophilous fungi in soils 173
Nests as sources of thermophilous fungi 174
Psychrophily in fungi 175
Basis of psychrophily 175
Variety and distribution of psychrophilous fungi 176
'Snow moulds' 177
Fungi of frozen foods 177
Xerophily in fungi 178
Xerophilous and xero-tolerant fungi 179
The 'osmophilous' Aspergilli 180
Basis of xerophily 182

7. Fungi as mutualistic symbionts in ectomycorrhizas and lichens 183
Ectomycorrhizas 183
Structure 183
Fungi involved 184
Nutrition of fungi 185
Physiology, fungus and host benefits 187

Contents

Lichens   194
Fungi involved and structure   194
Physiology, carbon and mineral metabolism   199
Detrimental consequences of efficient absorption mechanisms   203
General biology and possible benefits to the autotroph   204
Lichen substances   209
Lichen chimeras   210

8. **Fungi as mutualistic symbionts in endomycorrhizas**   214
Vesicular-arbuscular mycorrhizas   214
Occurrence   214
Features of infection and fungi involved   215
Benefits to host   219
Benefits to fungus   225
Ericoid mycorrhizas   226
Occurrence, structure and fungi involved   226
Benefits to host and fungus   228
Ectendomycorrhizas   231
Monotropa hypopitys   231
Sequestration of heavy metals by ericoid mycorrhizas   232
Mycorrhiza of the Gentianaceae   233
Orchidaceous mycorrhizas   235

9. **Fungi as symbionts with insects**   242
Mutualistic associations between fungi and insects   244
Ectosymbiotic associations   244
Ambrosia fungi and wood-boring beetles   244
Plant galls and fungi   248
The *Sirex/Amylostereum* association   248
The *Septobasidium/Aspidiotus* association   252
The Attine ants and their fungi   254
Termites and their fungi   258
Endosymbiotic associations   262

10. **Fungi as parasitic symbionts of plants – an introduction**   264
Modes of nutrition   264
Obligate biotrophs   266
Necrotrophs   273

References   285

1

Fungi as organisms

Fungi are certainly not plants. They have amply sufficient distinctive features to set them well-apart from both plants and animals and most would now accept that they merit a kingdom of their own. Unlike plants, they cannot synthesize organic materials from carbon dioxide, mineral ions and water. They are heterotrophic for carbon. Unlike animals, they cannot ingest solids. They obtain nutrients by absorbing soluble inorganic and organic materials. Like plants, they have walled cells but the walls are never of true cellulose and usually contain some chitin. Their life styles are unique. Many may show two separate but contemporary phenotypes. The holomorph or whole organism in the fungi often consists of a teleomorphic state reproducing sexually by producing 'perfect' spores, as a result of a nuclear fusion followed by a meiosis, and an anamorphic state reproducing asexually by producing 'imperfect' spores following mitotic divisions. Each state thus produces a different type of spore.

   Some 63 500 species of fungi have been described with another 13 500 associating with algae as lichens (Table 1.1). The true fungi or Eu-mycota can be conveniently divided into four sub-divisions – the Mastigomycotina, Zygomycotina, Ascomycotina and Basidiomycotina. However most would recognize a fifth sub-division, the Deuteromycotina, with some 17 000 species. These are conidial fungi in which the anamorphic or imperfect state exists quite independently of the teleomorphic or perfect state. Hence they are also called Fungi Imperfecti and given separate Latin binomials. The classification and nomenclature of such fungi present a number of problems which are fully discussed by Webster (1980). Since by far the majority of these are anamorphic states of Ascomycotina and most of the remainder similar states of Basidiomycotina, they are best regarded as such and referred to merely as conidial fungi. There is as yet no universally accepted scheme of classification for the fungi. The scheme proposed by Ainsworth (1973) and adopted by Webster (1980) is the most widely used. A modification of this scheme is presented at the end of this chapter for reference. A further modification, removing the Mastigomycotina from the fungi, is given by Barnes (1984).

2 *Fungi as organisms*

Table 1.1 Distribution of fungi in 'Ainsworth and Bisby's Dictionary of the Fungi' (Hawksworth, Sutton and Ainsworth, 1983). Reproduced with permission of the Commonwealth Agricultural Bureau.

'Real Fungi'		
Mastigomycotina	1170	spp.
Zygomycotina	765	
Ascomycotina	28650	
Basidiomycotina	16000	
Deuteromycotina	17000	
Total	63585	
Lichens	13500	
Total for all fungi	77085	

Vegetative phases

The vegetative phase of fungi may consist of a single, globose or ellipsoidal cell as in many true yeasts, the yeast-like phases of other fungi and in some members of the Chytridiales (Fig. 1.1a,b,c). In many Chytrids and in the Blastocladiales, which possess a discrete vegetative form, there is, in addition, a specialized attaching and absorbing system composed of extremely narrow, very much branched and tapering aseptate, walled filaments (Fig. 1.1d,e). These are called rhizoids although they are far more delicate, branched and extensive than are the rhizoids of mosses, liverworts and fern prothalli. However, in by far the majority of fungi, the vegetative phase consists of branched, aseptate, or septate, cylindrical, walled tube-like filaments called hyphae (Fig. 1.2). It is partly the full exploitation of this relatively simple hyphal organization which has largely contributed to the ecological diversity and success of the fungi. One of the most striking features of the fungi is this morphological structural simplicity of their vegetative phase which contrasts so markedly with the range and complexity of structure seen in other filamentous groups, for example the Red Algae, and indeed with their own relatively complex reproductive structures.

Hyphal walls

Although the hyphae of different fungi appear superficially very similar, in being thin-walled tubes varying in overall range from 0.5–1000 μm diameter but with the majority 1–15 μm diameter, they differ greatly not only in wall composition and structure but also in internal structure and, when septate, in type of septum. This all indicates that the fungi are certainly polyphyletic in origin.

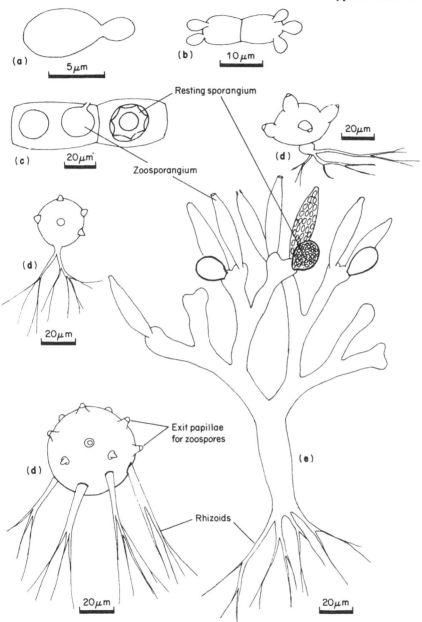

Fig. 1.1 (a) Budding in a yeast. (b) Budding in *Mucor rouxii*. (c) Single-celled zoosporangia and a resting sporangium of *Olpidium brassicae*. (d) Three chytrids with rhizoidal systems. (e) Tree-like vegetative phase of *Blastocladia* sp. with cylindrical zoosporangia and thick-walled resting sporangia.

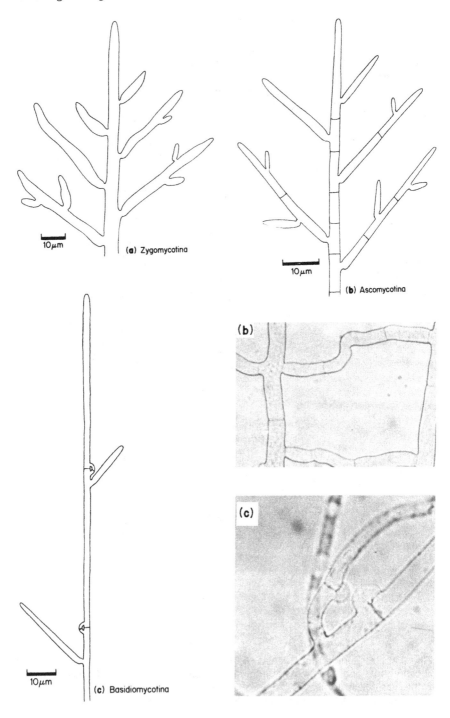

(a) Zygomycotina

(b) Ascomycotina

(b)

(c)

(c) Basidiomycotina

Hyphal wall chemistry

Hyphal walls contain 80–90% polysaccharides, 1–15% protein and 2–10% lipids. Two most notable features of fungal cell walls are their chemical hete-rogeneity and the fact that, in most cases, they are devoid of cellulose, which is characteristic of plant cell walls. The hyphal walls of most fungi contain some chitin, a polymer of β-1,4 linked 2-acetamido-2-deoxy-D-glucopyra-nose. Chitin makes up anything from 3–60% of the dry mass of the wall and is usually associated with non-cellulosic β-1,3 and β-1,6 linked glucans but also α-1,3 and α-1,4 linked ones. Bartnicki-Garcia (1968) has classified the hyphal walls of fungi on the dual combinations of the principal polysaccha-rides present. His groupings (Table 1.2) are of particular interest in that they indicate that there is a close correlation between the chemical composition of the hyphal wall and the taxonomic group to which the fungus belongs. The vast majority of fungi, including all those with typical septate hyphae, have a chitin-glucan wall. Three other significant groupings are cellulose-glucan, chitin-chitosan and mannan-glucan.

In the Mastigomycotina, the aseptate Oomycetes have been traditionally regarded as possessing cellulosic hyphal walls. In *Phytophthora*, for instance, glucans constitute about 90% of the dry mass of the walls and about one-quarter of this is believed to be cellulose, a β-1,4 linked glucose polymer. The remainder is a highly branched β-1,3 linked glucose polymer with β-1,6 linked branches. The cellulose in the hyphal walls is very poorly crystalline and it seems more likely that it is not pure cellulose but a complex branched chain polymer with mixed β-1,3 and β-1,4 linkages. Apart from the absence

Table 1.2 Cell wall chemistry and taxonomic groups. (After Bartnicki-Garcia, 1968.) Repro-duced, with permission, from the *Annual Review of Microbiology*, Vol. 22. © 1968 by Annual Reviews Inc.

	Composition	Taxonomic group
I	Cellulose-glycogen	Acrasiomycetes
II	Cellulose-glucan	Oomycetes
III	Cellulose-chitin	Hyphochytridiomycetes
IV	Chitin-chitosan	Zygomycetes
V	Chitin-glucan	Chytridiomycetes, Ascomycotina, Basidiomycotina, Deuteromycotina
VI	Mannan-glucan	Hemiascomycetes: Saccharomycetaceae and Cryptococcaceae
VII	Mannan-chitin	Basidiomycotina: Sporobolomycetaceae
VIII	Polygalactosamine-galactan	Trichomycetes

Fig. 1.2 (*Left*) (**a**) Aseptate hypha of a member of the Zygomycotina; (**b**) diagram of septate hypha of a member of the Ascomycotina – photograph, septate hyphae showing lateral fusions; and (**c**) dikaryophase hypha with clamp con-nections of a member of the Basidiomycotina – photograph, main hypha and branch, each with a clamp connection.

of any appreciable amounts of chitin in the majority, another distinguishing feature of the hyphal walls of Oomycetes is the presence of the amino acid hydroxyproline. This is not found in chitinous walled fungi but is characteristic of the cell walls of Green Algae and seed plants where it is thought to form an important link between the cellulose and proteins in the wall. Such characteristics, and quite a multiplicity of others including such facts that they are oogamous, possess biflagellate zoospores, have an ultrastructure with many resemblances to that of certain Green Algae, and synthesize lysine via α-aminoadipic acid, suggest that they are unique amongst fungi and that they could be considered, with some conviction, as heterotrophic aseptate Algae. They are classified with these in the Protista by Barnes (1984).

The other group of aseptate fungi, the Zygomycotina, also appear to possess their own distinctive wall components. In those few which have been investigated, including mucoraceous fungi such as *Mucor rouxii* and *Phycomyces blakesleeanus*, and species of the endomycorrhizal genus *Glomus,* the two major components are chitin and chitosan, the non-acetylated chitin-like polymer.

Polymers of mannose have been reported primarily from the cell walls of yeasts and the combination of mannan and glucan appears to be characteristic of true yeasts and the yeast-like phases of other fungi. Many fungi exhibit pleomorphism, a phenotype duality of form. They may develop in a typical filamentous manner or become yeast-like and bud depending upon the particular environmental conditions. This mould–yeast dimorphism is particularly true of a number of fungi which are parasitic on warm-blooded animals. The majority of such fungi are yeast-like in the parasitic phase. The walls of the yeast-like phase always contain more mannan than those of the hyphal phase. The presence of mannan-protein complex in the walls seems to be critical in the establishment of the yeast-like form. The walls, however varied their components, give rigidity, stability and shape to fungal hyphae. The polysaccharides are in the main responsible for this, although protein complexes may be involved.

Hyphal wall structure

Hyphal walls normally appear homogeneous in section under the light microscope; however ultra-thin sections under the electron microscope reveal a layering of the walls. In most fungi the hyphal wall has a distinct microfibrillar texture on the inner face and an amorphous appearance on the outer surface. Thus it is basically dual textured, but in many there is a multiplicity of layers. The microfibrillar components are composed of chitin in the majority of fungi but are cellulose-like in the Oomycetes. The amorphous components are the more varied glucans. One of the most elegant and informative studies reported to date on the ultrastructure of hyphal walls is that of Hunsley and Burnett (1970) and their findings may be used to emphasize some of the structural and chemical heterogeneity of hyphal walls. Hyphae of fungi from three different groups, the Pyrenomycete *Neurospora crassa* from the Ascomycotina, the Hymenomycete *Schizophyllum commune* from the Basidiomycotina and the Oomycete *Phytophthora parasitica* from the Mastigomycotina, were shadowed or sectioned for observation with the electron microscope before and after treatment with various single enzymes, each specific for a particular wall polymer, and various sequences or combinations of these enzymes. Hunsley and Burnett used chitinase, pronase (a commercial mixture of proteolytic enzymes), laminarase (which breaks β-1,3 and β-1,6 linkages) and cellulase for *P. parasitica*. They thus followed the progressive degradation of the wall from the outside inwards and were able to draw reconstruction diagrams for wall sections.

In *Neurospora crassa* the mature region of the wall is made up of four intergrading and co-axially distributed regions. There is an outer layer of amorphous glucan with mainly β-1,3 and β-1,6 linkages but with some α-1,3 linkages. Inside this is a coarse reticulum of a glycoprotein, like a coarse, stranded but very fine meshed wire netting. It is made up of a complex of glucan, peptides and galactosamine and lies on a heterogeneous matrix of an amorphous glucan enriched inwards with protein. Inside this is a layer of easily removable protein. Finally, innermost of all, is a layer of chitin microfibrils intermixed with protein (Fig. 1.3). As yet, no other fungus has been shown to possess such a complex wall structure. In *Schizophyllum commune* the wall differs in two important features. There is no reticulum of glycoprotein, and, on the outside of the amorphous β-1,3 and β-1,6 linked glucan, there is an additional layer of amorphous α-1,3 linked glucan (S-glucan). The walls of *Phytophthora parasitica* are structurally simpler, being basically two-layered, with an outer layer of amorphous β-1,3 and β-1,6 linked glucan and an inner microfibrillar layer of the cellulose-like polymer embedded in protein. Undoubtedly other fungi will be found to possess differences, in detail at least, in hyphal wall structure but the basic pattern seems to be one of an inner microfibrillar layer covered by an amorphous layer.

Generalized life cycles and nuclear phases

Fungal life cycles are extremely varied and often very complex. It is important to appreciate that nuclear phases differ amongst the fungi and from other organisms.

Mastigomycotina and Zygomycotina

The vegetative hyphae of many Mastigomycotina and all Zygomycotina contain haploid nuclei. The vegetative phase is thus haplophase and asexual spores are produced on this (Fig. 1.4). This contrasts with the vegetative diplophase of seed plants for example. At sexual reproduction, two haploid nuclei fuse to give the diplophase. This phase is restricted to a single cell. The fusion nucleus does not divide by mitosis before meiosis occurs. The diplophase becomes a thick-walled resting spore – in the case of the

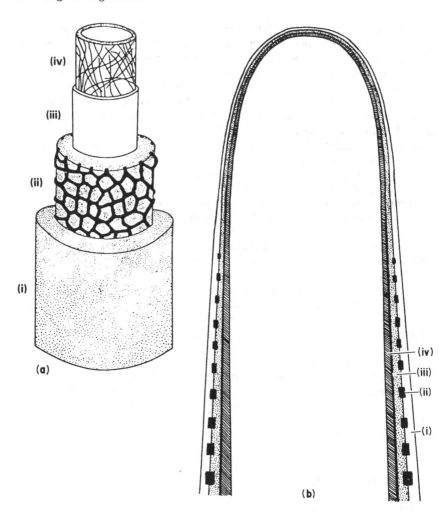

Fig. 1.3 (**a**) Diagram to illustrate the principal regions of the wall of *Neurospora*. Layers, from the base up: (**i**) outer mixed glucans; (**ii**) the reticulum, glucans merging into protein; (**iii**) principally protein; (**iv**) innermost proteinaceous region with embedded chitin microfibrils. (After Burnett, 1976.) (**b**) Diagrammatic representation of the wall structure at the apex of a hypha of *Neurospora*. Layers (**i**), (**ii**), (**iii**) and (**iv**) as in (**a**) (After Hunsley and Kay (1976). Reproduced by permission of the Society for General Microbiology.)

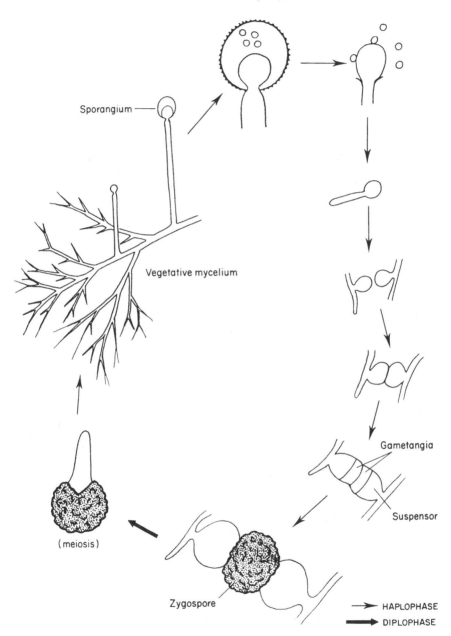

Fig. 1.4 Life cycle diagram of *Mucor*, a member of the Zygomycotina.

Zygomycotina, a zygospore. This type of life cycle is the simplest possible one which allows for nuclear fusion, meiosis and genetic recombination.

Ascomycotina

In the Ascomycotina (Fig. 1.5), there is again a vegetative haplophase. In many, asexual conidia are borne on this phase. The essential part of the sexual process is initiated by the fusion of two cells containing one or more nuclei. The two cells need not be recognizably distinct gametangia. Nuclear fusion does not occur immediately but the two nuclei form a pair of closely associated compatible nuclei called a dikaryon. The two nuclei then divide together and the products are separated into two daughter cells; by such repeated divisions a dikaryophase is established. In the Ascomycotina this is purely reproductive. It is the hyphae which bear the asci – the ascogenous hyphae – produced inside the ascocarp which itself is made up of haplophase hyphae. Nuclear fusion eventually occurs in the ascus initial. As in the Zygomycotina, this is followed immediately by meiosis. Thus the diplophase is again limited to a single cell. However, as any cell in the ascogenous hyphae can form an ascus, many meioses result from the initial fusion of the two haplophase cells. It is a haploid life cycle with a restricted and reproductive dikaryophase. The conidial Deuteromycotina exist as the independent haploid vegetative phase of this life cycle.

Basidiomycotina

In the Basidiomycotina, it is the haplophase which is restricted (Fig. 1.6). The hyphae produced from two compatible spores soon fuse to establish a dikaryophase. The mycelium has binucleate cells and it is the vegetative phase which is the dikaryophase. Basidiocarps are formed on this and they are wholly dikaryotic. Eventually in some cells, the basidium initials, the two nuclei fuse and immediately undergo meiosis. There is again no mitotic division of the diploid nuclei. The establishment and extensive multiplication of the dikaryon give rise to a far greater number of meioses and a resultant increase in possible genetic recombination per initial cell fusion.

The vegetative dikaryon is unique to the Basidiomycotina. In most Mastigomycotina, all Zygomycotina, Ascomycotina and Deuteromycotina, there is a prolonged independent haploid vegetative phase, unbuffered by heterozygosity or genetic complementation, and thus no system to mask genetic deficiencies so that the effects of natural selection are more vigorous and more immediate than in most diploid organisms. In contrast, the Basidiomycotina, with their dikaryophase, are functionally diploid in that the dikaryon can store recessive genes and complementation can occur. Furthermore, it has critical selective advantages over the diploid in terms of the establishment of dikaryotic mosaics. This envisages the direct exchange of nuclei between adjacent dikaryons after hyphal fusions, thus providing

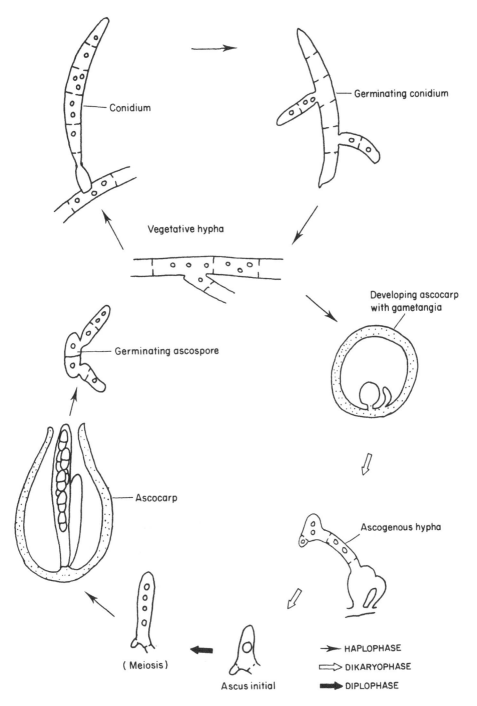

Fig. 1.5   Life cycle diagram of *Nectria*, a member of the Ascomycotina.

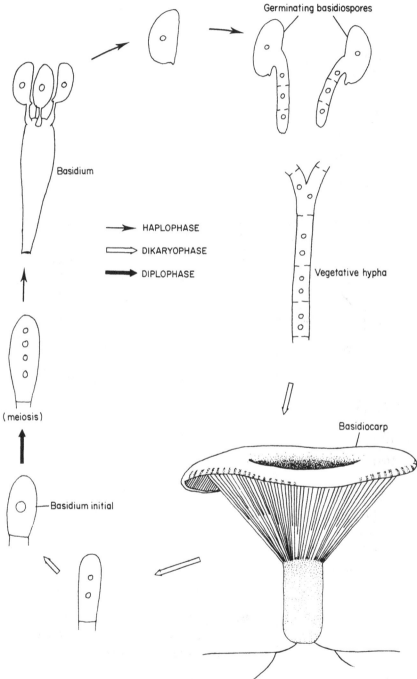

Germinating basidiospores

Basidium

HAPLOPHASE

DIKARYOPHASE

DIPLOPHASE

Vegetative hypha

(meiosis)

Basidiocarp

Basidium initial

Fig. 1.6 Life cycle diagram of *Clitocybe*, a member of the Basidiomycotina.

variation in genotype, without undergoing the disturbances and delays of meiosis and so permitting rapid adaptation to changing environments. This is advantageous in the immediate short term but needs a back-up system, which these fungi possess, of nuclear fusion and meiosis, to provide genetic variability. The heterokaryotic condition, where the cells or the mycelium contain a number of nuclei of more than one genotype, is widespread in the fungi. Differences between nuclei may arise by mutation or by fusions between hyphae, a widespread phenomenon in the septate fungi, followed by nuclear migration. Such a system is responsible for much of the genetic variability and adaptability shown by many purely conidial fungi, but it does lack the back-up of nuclear fusion and meiosis which a sexual system provides. The full significance of heterokaryosis to fungi is discussed by Webster (1980). True diploidy, with diploid nuclei multiplying by mitosis, is not at all common in the fungi but is found in *Saccharomyces*, which has a haploid and diploid budding phase, and more particularly in the Oomycetes, which have a diplophase vegetative mycelium.

Hyphal growth

Tip growth

It is a long established fact, yet a very significant one in understanding the mechanics of fungal growth, that hyphae extend only at the extreme apex. This can be demonstrated simply by measuring the distance from the hyphal tip to the first septum or branch and the distance between subsequent septa or branches at different time intervals. Only the apical segment will increase in length. There is no increase in length between septa or branches once they have been formed.

Hyphal growth is markedly polarized. Although increase in volume is achieved solely by extension of the hyphal tip, extension even there being limited to the curved portion of the apex, protoplasm for growth is synthesized by a considerable length of the hypha behind the tip. This is called the peripheral growth zone in contrast to the extension zone and it may extend back as far as 1–2 mm or even more from the hyphal tip. Several methods can be used to determine the length of the extension zone: for example by observing the relative displacement of markers, such as starch grains deposited on the hypha; by using the optical brightener Calcolfluor, which binds to the extension zone; or, more simply, by measuring the tapering portion of the tip. The extension zone varies in length with the rate of extension of the hypha. The faster the rate, the longer the zone. There is also a positive correlation between the length of the extension zone and hyphal diameter. The length of the extension zone is usually less than $20\mu m$, but even over this extremely short distance, there is a very rapid decline in extensibility from the very tip of the hypha to the base of the zone.

The wall laid down at the hyphal apex is considerably thinner and, in some, simpler in structure than the wall further back. In *Phytophthora parasitica*, there is a very thin layer of amorphous glucan through which the microfibrils can be seen and in *Neurospora crassa*, a very thin amorphous proteinaceous layer, with some small amounts of glucan on its surface, overlays randomly orientated chitin microfibrils. These components make up the extendable wall at the apex but behind rigidity is given by the addition of further components. In *N. crassa*, the reticulum of glycoprotein is laid down in the sub-apical region. It is first detectable about $3\mu m$ behind the actual apex. Further back still the reticulum increases in size and complexity and becomes covered by an increasing thickness of amorphous glucan. The protein content of the wall also increases further back and the microfibrils of chitin become wider and more closely packed. This can be seen if a comparison is made between those at the apex and those at about $10\mu m$ behind. The latter are about twice as wide and they are 15–20% more closely packed. This means that the space occupied by the microfibrillar components of the wall has virtually doubled. The microfibrillar components are often regarded as the major skeletal elements of the wall but it is abundantly clear from degradation with enzymes that it is often possible to remove any two of the three major components, the amorphous glucan, the protein or the microfibrillar polymers, without the wall disintegrating. All the varied components give strength and rigidity to the hyphal wall away from the apex, making it mechanically stronger and resistant to deformation both from within and without. Such addition must also mean that there is appreciable incorporation of materials behind the apex. Autoradiographic studies in which actively growing hyphae have been given a brief pulse of tritiated wall precursors, such as glucose, 2-acetamido-2-deoxy-D-glucopyranose and galactose, have demonstrated that incorporation is highest in the apical $1\mu m$ and falls off sharply after the first $5\mu m$ but that there is still appreciable incorporation from 5–$75\mu m$ behind the tip.

Turgor and extension

Turgor appears to be indispensable for, and at least a component of, the driving force for hyphal elongation. Due to the highly efficient absorptive powers of the hypha, the concentration of solutes within it is considerably higher than that of its environment and results in a substantial hydrostatic pressure, 0.4–0.8 MPa being exerted on the wall by the hyphal contents. This can readily be seen when the wall is punctured or damaged. The hyphal contents pour out. Furthermore, application of hypertonic solutions to hyphal apices brings about a rapid cessation of elongation which is resumed when turgor is restored. Thus, during extension growth, the production of the tubular hypha behind the apex is dependent upon a delicate balance between the rate of synthesis of new wall material and the hydrostatic pressure available for extending the wall. Turgor pressure appears to provide the driving force that extends the wall at the same time that the new wall material is added. The rigid nature of the wall behind the apex would also ensure that any mechanical pressure, which develops in these fine tubes as the result of the synthesis of new protoplasm, would be directed towards the apex.

The components needed for extension are produced in the peripheral growth zone in which the synthesis of RNA and protein proceed at a virtually constant rate throughout. This zone moves forward at the same rate as hyphal extension. Within young hyphae it is possible to distinguish three regions; the apical zone with an accumulation of small cytoplasmic vesicles, often to the exclusion of all other organelles; the sub-apical zone which is non-vacuolated and very rich in protoplasmic contents; and a zone of vacuolation. The first two zones correspond with the extension zone and the peripheral growth zone. As hyphae age, vacuolation increases, lipids accumulate and the proportion of cytoplasm is reduced accordingly. Vacuolation of the older parts of hyphae has often been suggested as a major factor influencing turgor pressure and forcing the protoplasm to move acropetally, but clearly it can be either a cause or a consequence of protoplasmic movement.

The apical vesicular complex

At the apex of all actively growing hyphae of septate fungi a dense cluster of cytoplasmic vesicles, the apical vesicular complex, can be seen. In the centre of this is a vesicle-free area or area of aggregated but very much smaller vesicles called the apical body or spitzenkörper (Fig. 1.7). In *Neurospora crassa*, about 80% of the volume of the apical 1μm of the hypha is occupied by vesicles, each some 75–150 nm diameter, as compared with under 5% by volume at the base of the extension zone. They disappear when growth is checked and re-form just before growth is resumed. The vesicles are formed in the sub-apical zone and are transported to the apex. It is thought that they contain materials and enzymes necessary for the growth of the wall and plasmalemma at the apex. The plasmalemma in this apical region is particularly crenulated and the crenulations are comparable in diameter with the vesicles. It is assumed that the crenulations represent vesicles whose membranes have fused with the plasmalemma and which are in the process of extruding their contents into the wall and so extending it. The vesicles contain amorphous substances and there is no evidence for the intracellular pre-polymerization of the microfibrillar components. Enzymes necessary for both the synthesis and lysis of chitin and various glucans have been shown to be present in the wall fabric.

Although the apical vesicular complex is of universal occurrence in the apices of growing hyphae, it differs in detail especially in aseptate fungi. In Zygomycetes, the vesicles are arranged as a cap lining the inside of the wall of the apical dome, whereas in Oomycetes they are scattered as in septate hyphae but in each group the apical body is absent (Fig. 1.7). Similar complexes of vesicles are involved in tip growth of other filamentous systems such as coenocytic algae, pollen tubes and root hairs.

The involvement of enzymes

Extension at the hyphal apex is not purely a physical process. Just as hyphae

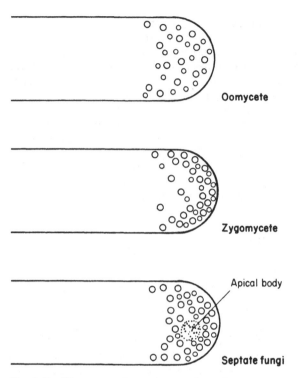

Fig. 1.7 Diagrammatic representation of the organization of apical vesicles in hyphal tips of different groups of fungi. (After Grove and Bracker (1970). Reproduced by permission of the American Society for Microbiology.)

may stop growing if placed in a hypertonic solution, many, unless they normally grow in water, will burst if placed in a hypotonic solution or water. This is not a simple physical osmotic effect as it is temperature dependent, with a temperature coefficient of 1.3–2.0, rather than 1.0, and has led to the conclusion that increased enzyme activity is involved in the weakening of the wall. Hyphal apices readily burst when disturbed by a number of treatments. Thus both the application of the antibiotic polyoxin-D, which inhibits chitin synthase, and the application of chitinase may cause hyphal apices to burst. This would suggest that both synthetic and lytic enzymes are involved in hyphal growth and that such apical growth is a very delicate and controlled balance between synthesis and lysis. Based on this sort of evidence, Bartnicki-Garcia (1973) has produced a useful hypothetical model for wall growth which is equally applicable to the apical growth of hyphae (Fig. 1.8). In his model he takes a wall of two layers – an inner microfibrillar layer covered on the outside by amorphous material. Lytic enzymes from cytoplasmic vesicles are secreted into the wall and these attack and break the microfibrils. The

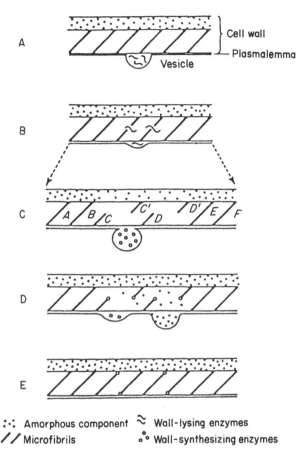

Fig. 1.8 Hypothetical representation of the events leading up to a unit of cell wall growth. (After Bartnicki-Garcia, 1973.)

wall in such an area can then no longer withstand the high turgor pressure from within and becomes stretched with a consequent increase in area. Other vesicles bring synthases which rebuild the microfibrils. It is assumed that the soluble precursors necessary for the synthesis of the microfibrils are transported directly across the plasmalemma. Other vesicles containing the components of the amorphous wall material, possibly in a preformed state, deposit their contents against the wall and these are forced through the microfibrillar framework and into place by turgor pressure. The wall has thus expanded without losing the differential layering of the wall polymers. Such a model can be applied to apical growth by assuming that such activities are concentrated around the dome-shaped apex of the hypha.

Hyphal branching and its significance

Such controlled and polarized tip growth enables fungal hyphae to extend very rapidly across solid surfaces, some at rates exceeding 1 mm h^{-1}, but two other features of hyphae are of particular importance in their growth through three dimensional substrates. They branch and they produce enzymes which act externally. Hyphae become extensively branched. Branches arise at some distance behind the hyphal tip in acropetal succession from parts of the wall which have not only ceased extending but which have also attained maximum rigidity. At the site of a branch, controlled partial dissolution of a minute area of the wall occurs, presumably by the action of lytic enzymes. The high turgor pressure from within causes the wall to bulge out and, what is to all intents and purposes, a new hyphal apex is created. The fact that it is a forced ballooning out of the wall is substantiated by the fact that the point of origin of a branch is often narrower in diameter than the branch itself. As in tip growth, cytoplasmic vesicles also concentrate at potential sites of branch initiation along the walls.

On a flat surface a monopodial system develops with a leading hypha and a series of alternate branches. The leading hypha extends at a more rapid rate than its branches. Such differences in linear growth rate reflect micro-environmental variations in the substrate supporting the various parts of the branching system. Only the leading hypha continuously extends into uncolonized substrate, whereas its branches grow out into regions which have already been colonized and changed to make them less favourable for growth.

As a series of leading hyphae diverge from a common point, such as when a spore germinates and grows into a colony, their dominance over their laterals weakens and some primary branches escape. These extend into the uncolonized parts of the substrate between the leading hyphae and so fill the gaps in the margin. Thus a colony, more or less circular in outline, develops. Such growth within a three-dimensional substrate would lead to a spherical colony. As branches are transformed into leading hyphae, they not only increase in linear growth rate but their extension zones increase in length, they become wider and the length of the apical compartment also increases. These morphological changes always occur in hyphae as they increase their rate of elongation. The production of branches has two very important effects on the ability of any fungus to colonize a particular substrate. Dense and regular branches endow the fungus with the potential to pervade any substrate thoroughly. The complex branching system and the rigid nature of the wall behind the apex both ensure that the older parts of the hypha are firmly anchored and enable the tip to exert considerable forward mechanical pressure as it extends. Some hyphae can penetrate metal films in this way. This, coupled with the production of substrate-hydrolytic enzymes which erode away, or at least soften, the substrate, greatly assists penetration and ensures complete permeation of even the hardest and toughest substrate. This is where the fungi have the advantage as heterotrophs over the bacteria. Both present a relatively large surface area in relation to their volume to the substrate: the fungi, because of the narrow dimensions of their hyphae and the bacteria, because of their minute unicellular form. In the fungi, this area is in the order of 4–5 $m^2 g^{-1}$ dry mass of hyphae. But the bacteria lack any powers of mechanical penetration and can accomplish the breakdown of their substrate only at an already exposed surface. Their activities are much more limited in bulky solid substrates with a low free surface area to volume ratio. The advantages of a hyphal organization are nowhere better illustrated than in the decay of wood (Chap. 3). The large surface area of a hyphal system is advantageous in the exploitation of any substrate, in terms of the production of hydrolytic enzymes and the absorption of the products and, in particular, in the absorption of materials in short supply, such as mineral ions in the soil, but it does have its disadvantages. It also increases its liability to injury by adverse environmental factors. The survival of such a structure depends upon the maintenance of a high degree of humidity and for some free moisture in the external environment. Most hyphae, having no cuticle, are very susceptible to desiccation. This is perhaps why most are immersed in the substrate in which they are growing and why in many, but certainly not all, which are exposed to the atmosphere the walls are melanized. Pigmented melanins are often deposited either in or as encrustations on hyphal and spore walls and the walls of survival structures, such as unicellular chlamydospores and multicellular sclerotia, organs of translocation, such as rhizomorphs, and some ascocarps, especially perithecia and pseudothecia. Melanins are phenolics related to lignin and form a resistant impervious binding between the microfibrillar components of the wall or an amorphous encrustation on the surface. In addition to preventing water loss and leakage out from the hyphae and other structures, they also protect from excessive ultra-violet irradiation from sunlight. This may be particularly important to air-borne spores and exposed hyphae, such as those of the phylloplane fungi (Chap. 2). Although the chemistry of fungal melanins is very poorly understood, they are different from those of animals. Depolymerization by alkali usually yields catechol rather than tyrosine as in the eumelanins of animals.

Septation

One of the major taxonomic dichotomies in the fungi is between the sub-divisions Mastigomycotina and Zygomycotina on the one hand and the Ascomycotina and Basidiomycotina on the other. Members of the two former groups do not produce any complex reproductive structures around their spores, whereas the majority of Ascomycotina and Basidiomycotina do, often forming macroscopic ascocarps and basidiocarps, loosely called fruit bodies. In addition, the filamentous Mastigomycotina and Zygomycotina have in the main actively growing hyphae which are typically aseptate. They are coenocytic with no cross walls. Complete septa do occur in these fungi but usually only to cut off reproductive structures and evacuated and injured parts of the hyphae. Fungi in the other two groups have septa dividing the hyphae into compartments but the septa are perforate. In the Ascomycotina, each septum grows inwards from the wall as a diaphragm and leaves a small pore in the centre. This central pore is about 0.05–0.5 μm diameter and allows cytoplasmic continuity between adjacent compartments and also nuclear migration. Such hyphae are thus not made up of independent cells but are really coenocytes with incomplete septa (Fig. 1.9a). In many Basidiomycotina, the pore is not a simple hole but may be flanged to produce an elongate channel and this may be loosely capped at each end by a crescent-shaped structure, the septal pore cap (Fig. 1.9b). The whole is visible only with the electron microscope and is restricted to dikaryotic hyphae. The channel is 0.1–0.2 μm diameter and again allows cytoplasmic continuity. This suggests that the distinction between the aseptate and septate hyphae is not a very profound one. The cross walls give rigidity to hyphae, for example by helping to prevent collapse under pressure. The fact that the pores can be blocked relatively easily also prevents excessive loss of contents when hyphae are damaged through walls being punctured or burst. However their main function may be to allow hyphae to undergo differentiation to produce a

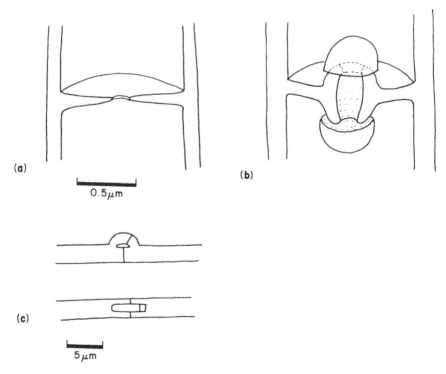

Fig. 1.9 (a) Perforate septum of a member of the Ascomycotina; (b) dolipore septum of a member of the Basidiomycotina; (c) clamp connections of a member of the Basidiomycotina seen from the side and above ((a) and (b) from longitudinal sections of hyphae as seen with the electron microscope; (c) intact hyphae as seen with the light microscope).

variety of different types of cell adjacent to each other, for example basidia and cystidia on a gill surface of an agaric and skeletal and binding hyphae in a basidiocarp of a polypore. The hyphae of many Basidiomycotina have clamp connections at the septa (Fig. 1.9c). These are narrow tubular bridges connecting one compartment of the hyphae with the adjacent one. The presence of these, together with the presence or absence of typical septa in the other, afford useful diagnostic criteria.

Colony growth

Very little is known about the actual size and shape of fungal colonies in their natural habitats. If they grow over a flat surface where they can be seen, the colony is often more or less circular in outline but the majority grow immersed in an opaque substrate such as soil, a decaying tree trunk or a rotting apple. In these cases, subject to the confines of their substrates and the physical factors of their environment, a colony spherical in form would develop.

Assessment of growth by linear spread

If a Petri dish containing nutrient agar is inoculated in the centre with a spore or mycelial inoculum, a colony with an almost circular outline will develop and growth can be assessed by measuring linear spread. The simplest method is to measure regularly along two or more marked diameters and take the mean. If this is done, it is possible to recognize a number of growth phases (Fig. 1.10). Initially there is a lag phase. This is the time taken for spores to germinate or, if a mycelial inoculum is used, the time taken for regeneration of damaged hyphae. This is followed by a further interval before the maximum growth rate is attained. After this there is a linear growth phase. The colony diameter increases linearly with time as a constant rate of growth is maintained at the colony margin. A colony may increase at a linear rate for an indefinite period. The rate varies widely between 0.1 and 6.0 mm h^{-1} and is characteristic of the fungus and the cultural conditions. The density of the hyphae at the colony margin remains more or less constant as the primary branches of the leading hyphae are periodically transformed into leading hyphae. This phase of linear growth may continue but usually the rate of spread declines as the margin of the dish is approached. This is called the staling phase and the decline in growth rate is usually associated with the harmful effects of the colony's own metabolic products – hence staling. It is frequently associated with unfavourable changes of pH, such as those caused by acid production from sugars in the medium. Staling may also be accompanied by autolysis of the mycelium in the centre. In Petri dish cultures the older parts of the hyphae regularly become empty, dead and sealed off by complete septa. The individual cells or compartments die as a result of the

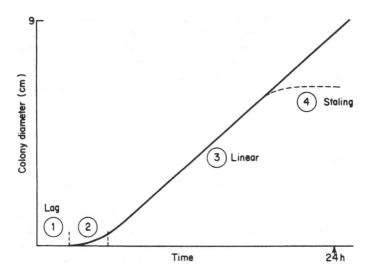

Fig. 1.10 Linear growth curve of *Neurospora crassa* at 22°C.

exhaustion of the food supply, coupled perhaps with the accumulation of toxic metabolites which they have produced.

Such measurements of linear spread have one fundamental limitation. They are not a measure of the actual amount of mycelium produced. A colony on a low nutrient may spread rapidly at near maximum radial growth rate but the hyphal network may be very sparse, whereas on a high nutrient medium it may spread more slowly but produce a dense network of hyphae. There is no necessary correlation between the spread of the mycelial front and the total amount of fungus produced. On plain agar without added nutrients, many fungi cover the surface of the Petri dish very rapidly. The hyphae are very sparsely branched and widely spread. In a habitat such as soil with local aggregations of rich organic matter, forming suitable substrates, separated by non-nutrient mineral matter, the advantages of such a growth habit are obvious. There would be much branched, densely packed hyphae where the substrate was plentiful and a sparse rapidly growing network in between. This ensures maximum utilization of any one substrate and rapid spread to the next.

If a fungus has an unlimited surface area to colonize, it would come to consist of a peripheral ring or annulus of actively growing hyphae producing more and more tips by branching, so ever increasing in diameter and tapping new sources of nutrients, but at the same time gradually ageing and dying off behind as a consequence, in particular, of the exhaustion of the food supply. Ageing is accompanied by marked vacuolation and often also by the production of staling substances, loss of viability as inocula and by pigment formation. There is a centrifugal flow of hyphal contents from the centre to the periphery. Finally, as in Petri dish cultures, the older rearward parts of the hyphae regularly become empty, dead and sealed off by complete septa as autolysis occurs.

'Fairy rings'

Many agarics are responsible for the production of 'Fairy rings' in grass. The best known of these are produced by *Marasmius oreades*, the Fairy-ring champignon, which usually produces circles of dead or dying grass bordered on the outside and inside by darker green, more vigorous grass (Fig. 1.11). Others, such as *Lepiota procera*, the parasol mushroom, and *Agaricus campestris*, the field mushroom, produce dark green rings in grass but usually with no dead or bare patches. In all these, and in many woodland litter decomposing agarics, the mycelium perennates by continually growing outwards and dying off behind. The rings of *M. oreades* are perhaps the most conspicuous because they often occur in lawns, golf courses, playing fields and parks disfiguring the turf. The rings increase in diameter up to 200 mm each year and vary from under a metre to well over 10 metres in diameter. The outward movement of the concentric rings is undoubtedly due to an intricate complex of factors. A mycelium becomes established in the soil and grows outwards in a circular manner in a very similar way to a colony growing in a Petri dish. The peripheral ring of actively growing hyphae in this case is invisible, submerged in the soil under the outer stimulated ring of grass and extends back through the dying and dead zone (Fig. 1.11). Basidiocarps appear on the inner side of this latter zone any time from June to November if there has been sufficient rain to keep the turf moist for about 7–10 days. Beneath the inner stimulated zone, the mycelium is dying off or dead. *M. oreades* is known to be able to degrade both the fulvic and humic acid fractions of humus and to mineralize the organic nitrogen and phosphorus associated with these fractions in the soil. As the hyphal front grows outwards utilizing this organic matter, it initially enzymatically releases inorganic nitrogen and phosphorus sources in excess of its requirements, or at least in such quantity that the grass roots can effectively compete for a share of them. They are absorbed by the roots in sufficient quantity to produce an abnormally luxuriant growth. As the hyphal front branches behind, the mycelium becomes much more dense and it consumes most of the available nitrogen and phosphorus, and the grass suffers from lack of these nutrients in the zone where it is dying or dead. Chronological sampling of expanding fairy rings has shown that nitrogen and phosphorus are mobilized during active mycelial extension. In the soil, under the centre of the outer stimulated zone of grass, the concentration of nitrogen extractable as ammonia has been found to be about three times higher than in the soil outside the ring. Extractable nitrate and phosphorus were also significantly higher in this zone. In the same position but some weeks later, when the fungus had extended further outwards and when the grass was beginning to die off, it was found that the concentration of ammonia nitrogen in the soil had been reduced by over 90% and that of nitrate nitrogen and phosphorus

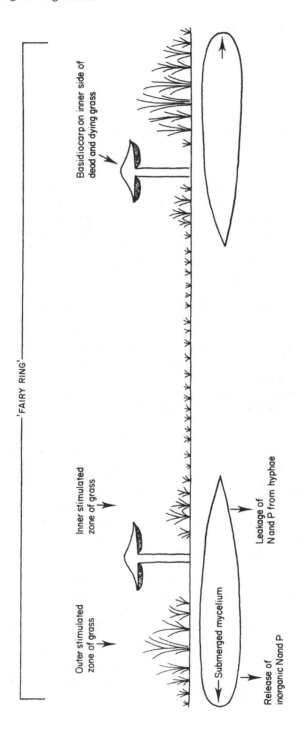

Fig. 1.11 Vertical section through a fairy ring of *Marasmius oreades*.

by over 70 and 72% respectively. Inside this zone the grass is again stimulated but this time by leakage of suitable sources of nitrogen and phosphorus from the autolysing and dead hyphae. The sequential fluctuation in the availability of mineral nutrients may be one of a complex of factors which helps to account for the concentric ring form. The depletion of its substrate, the soil humus, is an additional factor favouring the invasion of previously unoccupied soil by the fungus. The middle zone of dying and dead grass may also be partly due to lack of water and to hydrogen cyanide produced by the fungus. There is only a very small amount of mycelium in the top 20 mm or so of soil. This layer wets up easily as soon as it rains but further rain fails to percolate through the soil beneath. In this zone dense wefts of hyphae can be seen growing around the roots and completely permeating the soil. The fungus appears to make the soil water repellent, perhaps due to the air entangled between the meshes of the hyphae occupying the spaces between the particles of soil. As a consequence, the grass suffers directly from lack of water, there being such a dense growth of fungus that water cannot penetrate to the grass roots, or indirectly by the failure of mineral nutrients to diffuse back into the zone as they are absorbed because of the water repellent coat imparted by the fungus to the soil particles.

Marasmius oreades is known to produce hydrocyanic acid and other cyanogenic compounds as metabolites in culture and in fairy rings. These are also thought to be involved in killing the grass but opinions differ as to how they might do so. The hydrocyanic acid given off to the soil could predispose the root tips, root hairs and epidermal cells to colonization by *M. oreades*. The roots could take up cyanogenic compounds produced in the soil by the fungus and release hydrocyanic acid from these. The amount may not be sufficient to constitute a lethal dose but above a particular threshold level it would affect respiratory activity and ion uptake. The top 20 mm or so of soil could act as a water seal restricting diffusion of the hydrocyanic acid to the atmosphere, resulting in a build up to toxic concentrations that would kill all the vegetation, thus forming the dead zone of the fairy rings. After the soil in this zone eventually wets up, the hydrocyanic acid and any other cyanogenic compounds may be leached out making the area suitable for the growth of the grass once again. It would seem that cyanide production, lack of water and the poor fertility status of the soil all play their part in killing the grass.

Such rings afford easily observable examples of the over simplistic concept of a fungal colony as consisting of a peripheral growth zone, which continually advances centrifugally, ever extending into previously uncolonized substrates, and quitting areas depleted of nutrients.

Chlamydospores and sclerotia

The protoplasmic contents of the older parts of the hyphae of many fungi are not completely evacuated as the tips grow on. Chlamydospores may be produced (Fig. 1.12a) when the protoplasmic contents of a short length of a hypha accumulate at a particular point, round off and become surrounded by

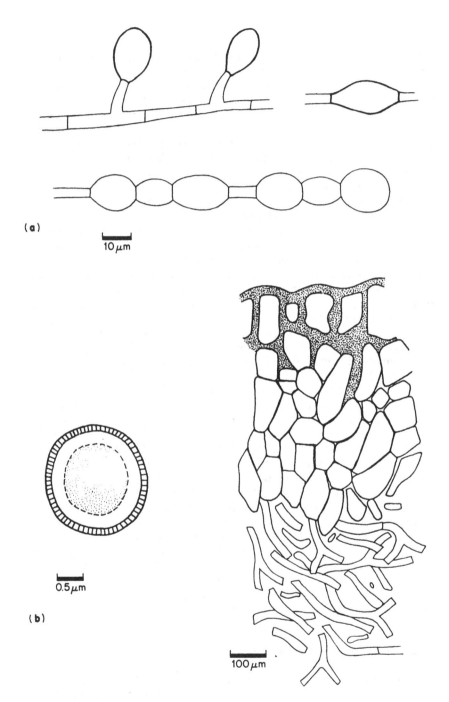

Fig. 1.12 (a) Terminal and intercalary chlamydospores. (b) Plan and detail of vertical section through a sclerotium of *Botrytis*.

a thick, often pigmented, wall. They are often rich in reserves of glycogen or oil. When the remainder of the hypha decays, rather than being dispersed to new environments, chlamydospores persist as survival spores and await the arrival of a suitable fresh substrate in the old one. They are thus dispersed in time. Chlamydospores are formed in particular by many soil fungi, such as Mucorales, which rely upon ephemeral substrates. Many soils possess a rich bank of these spores. They are particularly important survival units where substrates are essentially discontinuous in time and the environment is mostly in the solid phase.

As an alternative to chlamydospores, sclerotia may be formed. These are firm hyphal aggregates with determinate, rather than indeterminate, growth. They are exceedingly variable in size and form, ranging from small, irregular, loose clusters of cells less than 100 μm diameter to relatively massive, rounded, compact structures, 200 mm or more in diameter. The majority are less than 5 mm diameter. Sclerotia contain substantial quantities of reserve materials such as mannitol, trehalose and particularly glycogen and often lipids – they are both storage and survival structures. They can survive independently of the mycelium under more severe conditions, such as drought, intense cold and high temperatures, for longer periods than any other resistant bodies. They are important perennating structures for a large number of plant parasites, enabling the fungus to survive in the absence of a suitable living host. Sclerotia are characteristic of certain genera such as *Claviceps*, *Sclerotinia*, *Sclerotium* and *Typhula*, in which many species are parasitic. Some sclerotia are very long-lived. Those of *Sclerotium cepivorum* may show 70–80% germination after 4 years under field conditions. Those of *Verticillium dahliae* may persist in soil for 14 years. The sclerotia of *S. cepivorum* and those of a number of soil-borne root parasites do not normally germinate in natural soil and only do so as a direct response to the presence of the roots of members of its host genus. This ensures that they do not germinate in the absence of a suitable host. The sclerotia of other fungi may require activation by chilling. The significance of the low temperature requirement in the biology of temperate species of *Claviceps*, where autumn and winter chilling of the sclerotia (ergots) normally takes place, is quite clear. The successful infection of grasses by *Claviceps* depends upon the release of ascospores from ascocarps produced from the germinated sclerotia in the spring, being synchronized with anthesis and stigma extrusion by the host. Although some of the abundant endogenous reserves are used up in such resting periods, the bulk are utilized when favourable conditions return as the sclerotia germinate to form a mycelium or some form of reproductive structure from which spores are rapidly and widely dispersed.

Structure and development of sclerotia

The majority of mature sclerotia have a clearly defined outer rind consisting of swollen, almost globose, cells with very thick, melanized and agglutinated walls. The inside is often differentiated into an outer cortex and an inner medulla (Fig. 1.12b). The latter is prosenchymatous containing large, wide and thick-walled storage hyphae and smaller hyphae with dense contents containing numerous organelles. Both are often embedded in a mucilaginous matrix. The cortex usually consists of several rows of pseudoparenchymatous tissue, the hyphae being no longer recognizable as such and the cells having thin walls and dense contents.

Sclerotia may develop in a number of ways. Most arise as discrete initials among the vegetative hyphae. Three main types of development can be distinguished – the 'loose', the 'terminal' and the 'strand' types. In the 'loose' type, the initials develop by the increased and localized irregular branching and septation of adjacent hyphae. Loosely arranged barrel-shaped cells are formed to produce a relatively soft and homogeneous sclerotium as in *Rhizoctonia solani*. In the 'terminal' types, the initial is formed by the very prolific and often dichotomous branching of the tip of a hypha or the tips of several. Numerous septa are laid down producing short-celled branches, many of which fuse together forming a compact knot of hyphae as in *Botrytis cinerea*. In the 'strand' type, numerous small lateral branches arise in a localized region of a hyphal strand as in *Sclerotium rolfsii*. Frequent septation, fusion and interweaving of branches again take place to form a hyphal aggregate from which a mature sclerotium develops.

In developing sclerotia, hyphal branches appear to attract each other, as they do in mycelial strands, rather than repel each other, as they do in the vegetative mycelium. The numerous fusions which occur between hyphal branches provide permanent unions and not only facilitate the circulation of nutrients throughout the whole sclerotium but also increase the mechanical strength of the component tissues. Many fungi which form sclerotia produce abundant mucilage which accumulates as a layer on the outside of the hyphal walls. The hyphae become suspended in a mucilaginous matrix which may help to hold the sclerotial hyphae loosely together. The mucilage could also be important as a water reserve during germination. On soaking after a drought, the mucilage would rapidly take up water and retain it for a sufficient period for the sclerotia to germinate successfully. It may also act as a barrier to the movement of water out of the sclerotia and thus protect against desiccation or at least ensure that they dry out slowly. The high concentration of dissolved solutes, including trehalose and mannitol, within the cells helps to retain water. Like seeds, a slow rate of drying out is important if viability is to be retained. The close packing of the cells of the rind and their thick and melanized walls also help to prevent water loss. Melanins, as well as being impervious, have also been shown to reduce or prevent lysis. They physically protect chitin and glucans in the walls by covering them and also inhibit chitinase and glucanases produced by other micro-organisms. This may be important in any structures such as chlamydospores, sclerotia or rhizomorphs which either store, or, as in the last, conduct organic nutrients. Without such protection they would be a rich source of nutrients for any antagonistic micro-organisms.

Mycelial strands and rhizomorphs

It has already been discussed how the filamentous habit permits variation in colony density such that the energy sources available in the environment are used more efficiently. An alternative to this is for the hyphae to aggregate into either mycelial strands or rhizomorphs and for the fungus to extend via these. Although the two terms mycelial strands and rhizomorphs are very often used interchangeably for any cord- or strand-like structure produced by fungi, they are two types of fungal organs which develop in entirely different ways but are often virtually indistinguishable in mature form and thus superficially very similar. They are both linear aggregates of hyphae with the potential to extend unidirectionally. At one extreme, they may consist of a very few hyphae and at the other of several thousand, varying in diameter from about 50μm to several millimetres. Mycelial strands and rhizomorphs arise from the margin of a vegetative mycelium in a colonized substrate or sometimes from a sclerotium. Most are capable of unlimited extension, often over a surface, away from a nutrient substrate. They ultimately fan out and form a vegetative mycelium in a new substrate or produce reproductive structures. They are particularly characteristic of fungi which produce large reproductive structures and which colonize more bulky substrates such as wood or leaf-litter. Such structures are almost entirely confined to the Hymenomycete and Gasteromycete members of the Basidiomycotina.

Structure and development of mycelial strands

Mycelial strands are gradually built up acropetally around one or more leading hyphae. Silky, cotton-like strands are produced by *Collybia dryophila* growing through oak leaf litter and by *Hypholoma fasciculare*, the sulphur tuft fungus, growing through and from hardwood stumps. Many species of *Agaricus*, including *A. bisporus*, produce strands and they can be seen in abundance in commercial mushroom composts and spawn. *Helicobasidium purpureum*, which causes violet root rot of carrots and sugar beet, and a number of other root parasites also produce mycelial strands. The best known strands are those produced by *Serpula lacrimans*, the dry rot fungus. In this case the mycelial strands arise from a colonized woody substrate, such as a wooden joist in the wall of a house, when the fungus grows out over non-nutrient material such as brickwork or plaster. They arise as a result of three processes not associated with normal vegetative growth. These are (i) the formation of three different types of hyphae; (ii) a change in the orientation of hyphal branches; and (iii) the development of adhesiveness between hyphae. One, or sometimes several, recognizably wider leading hyphae, the strand initials, grow out from the mycelium. The strand initials become ensheathed by their own branches from behind. The branches, rather than spreading out at an angle from the leading hyphae as they would on an agar plate, are thigmotropically attracted to them and coil around them becoming 'tendril' hyphae. Lateral fusions or anastomoses also occur between the branches. In older parts away from the tip, some of the branches may become 'fibrous' hyphae. These are much narrower with thicker walls and consequently the cell lumen is reduced and sometimes occluded. They grow intrusively longitudinally through the strand and in many ways are similar to the skeletal hyphae of polypores (p. 47). The wider central main hypha or hyphae and their wide branches may have well-spaced septa which may be partially or completely broken down to form open tubes or 'vessel' hyphae. The whole is thus fashioned from behind and tapers towards the apex. Although there is some sort of differentiation into 'vessel', 'tendril' and 'fibrous' hyphae to provide both conductive and supportive elements, there is no precise ordered arrangement of these. They are all intricately intermixed and firmly stuck together by adhesive extrahyphal material.

Structure and development of rhizomorphs

A markedly different type of development is found in rhizomorphs. They are autonomous branching organs which grow fully-fashioned by the co-ordinated apical growth of the aggregated hyphae, behaving to all intents and purposes as an apical meristem comparable with that of the root apex. The best known rhizomorphs are those of *Armillaria mellea*, the honey fungus, which is particularly insidious and perhaps the most widespread of all the root parasites of trees. Undoubtedly the key to its success as a parasite lies in its ability to produce rhizomorphs. In *Armillaria*, although the rhizomorph is made up of several thousand closely associated, parallel, unbranched septate hyphae, which grow some five to six times faster than unorganized hyphae, individual hyphae are not easily discernible at the apex. Growth occurs from a point about 25μm behind the extreme tip, where a compact meristem-like apical centre composed of small, isodiametric apical initials is situated. Anteriorly, the apical centre gives rise to a relatively dense, intertwined layer of hyphae forming the rhizomorph's cap, which, with its gelatinous sheath, superficially resembles a root-cap. Posteriorly the apical centre produces a distinct region of inflated, loosely interwoven hyphae, the primary medulla. The cortex arises just lateral to the apex of the apical centre. Meristematic activity continues on the flanks of the apical centre, some way back from the apex giving rise to a lateral meristem (Fig. 1.13). Cells cut off from this eventually differentiate into components of the cortex on the outside and the medulla on the inside. The innermost region of the cortex remains meristematic, giving rise to endogenous lateral branches.

Fully mature rhizomorphs usually have a central hollow filled with air, surrounded by a longitudinal system of a loose network of wide, thin-walled, elongated cells. These are comparable with the 'vessel' hyphae of mycelial strands. Some of these collapse and become interspersed with 'fibrous' hyphae. They are also very irregular in both size and shape and are conspicuously laterally connected. Towards the outside, the walls become much thicker and black with deposits of melanin. These outer layers, the rind, clearly protect and support the rhizomorphs.

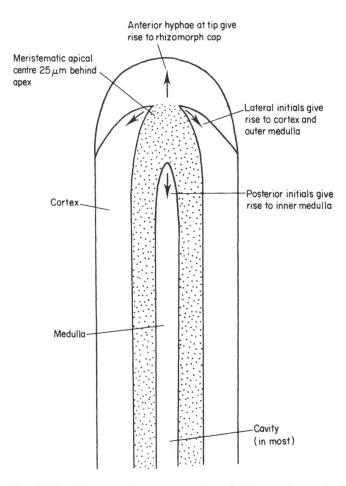

Meristematic apical centre 25 μm behind apex

Anterior hyphae at tip give rise to rhizomorph cap

Lateral initials give rise to cortex and outer medulla

Posterior initials give rise to inner medulla

Cortex

Medulla

Cavity (in most)

Fig. 1.13 Diagrammatic representation of rhizomorph apex in *Armillaria mellea*.

When growing from wood, into and through soil, they are cylindrical and very difficult to distinguish from roots. When growing under the bark of fallen trees which the fungus has killed, they are flattened and about the same width as bootlaces. Apart from the fact that they produce many small branches arising at very wide angles from the main rhizomorph, they so closely resemble these that the other common name for *Armillaria* is the 'bootlace fungus'.

Functions of mycelial strands and rhizomorphs

Both mycelial strands and rhizomorphs may extend considerable distances out from the substrate – in the case of *A. mellea*, several metres from an infected tree stump. In the context of growth, mycelial strands and rhizomorphs can be seen as organs which enable the fungus to spread over or through what is to it an unsuitable substrate in terms of available nutrients, to reach another, while remaining in contact with its original mycelium. Their major function, however, appears to be one of translocation of nutrients from an established colony to enable the fungus to colonize a new substrate at a distance. By using tracers, ample evidence has been obtained that mycelial strands and rhizomorphs translocate both organic and inorganic nutrients. Both may subtend basidiocarps, which during development are very powerful nutrient sinks, and in such cases they are the only connections between the basidiocarps and the food base on which the fungus is growing. In the mycelial strands of *Serpula lacrimans*, the longitudinal system of 'vessel' hyphae provides a suitable pathway for translocation but such a system is less apparent in the rhizomorphs of *Armillaria mellea*.

The advantages afforded by the possession of mycelial strands and rhizomorphs by root parasites of trees and saprotrophic wood-decay fungi are easy to see. They not only enable root parasites to reach other roots at some considerable distance from the original victim but they also provide an abundant supply of nutrients which can be used to increase the potential available for the penetration of mechanical barriers, such as root periderms, and also for the synthesis of the various enzyme complexes which are necessary if they are to invade such highly resistant tissues successfully. A single hypha or spore would not provide that potential. Furthermore, the central air space extending from the food base to almost the growing tip of the rhizomorph provides a diffusive pathway for oxygen to enable aerobic metabolism to occur at the site of attack, even at depth, in soil. Similarly, many wood-decay fungi may find that the wood which they are about to colonize is low in readily available carbon and nitrogen sources and so may depend upon translocation from behind to enable them to synthesize sufficient extracellular lytic enzymes, such as cellulase and hemicellulases, to initiate the process of degradation to such an extent that they become successfully established. This gives them an initial advantage over those fungi which do not produce such structures.

Spores

Fungi reproduce both asexually and sexually by spores and most fungi are dispersed by spores. The fungi present problems in terminology to the beginner mainly because there is such an absolute multitude of different kinds of spore. Although many spores are the unit of dispersal, other act as resting spores which tide the fungus over an unfavourable period. Some act efficiently as both. The various sub-divisions within the fungi are based on the nature of the spores formed, especially the sexually produced ones. In many Ascomycotina and Basidiomycotina, actual sex organs are not produced and the term sexual spores refers to spores produced after a nuclear fusion and a meiosis.

Asexually produced spores

There are basically two types of asexual spore – sporangiospores and conidia. Sporangiospores are again of two types – zoospores and aplano-spores. Conidia are exceedingly variable in both their development and form. They are best considered on the basis of their modes of development.

Sporangiospores – zoospores

Sporangiospores are enclosed during development within a sporangial wall, being released only at maturity when the wall fragments or when a distinct pore or pores develop in the wall. Zoospores possess one or two flagella and swim about by means of these. They are naked in the sense that they have no wall of their own and they are absolutely characteristic of the Mastigomy-cotina. No other true fungi possess motile spores. In the Oomycete Sapro-legniales, for example *Saprolegnia*, there are two distinct motile stages. Primary zoospores are produced inside long, cylindrical zoosporangia which are formed at the tips of hyphae. They become cut off from their subtending hypha by a septum. The zoospores are pip-shaped with two sub-apically attached flagella. The latter are equal in length and one is of the tinsel type and the other whiplash. The zoospores escape from the zoosporangium through a single distinct pore at the apex (Fig. 1.14). They swim about for a short time but soon withdraw their flagella, synthesize a wall and become a spherical cyst. Each cyst germinates to release a second type of zoospore. Secondary zoospores are bean-shaped, have two laterally attached flagella, and are far better swimmers than the primary zoospores. They in turn encyst and the secondary cysts may germinate to produce further secondary type zoospores but eventually each produces a germ-tube from which vegetative hyphae develop. Thus in *Saprolegnia*, in the dispersal phase, there are several periods of motility. Both the zoospores and the cysts are randomly dispersed. As far as the actual transport is concerned the motility of the zoospore gives them little advantage over the non-motile cyst. The actual movement of water in a stream or even a pond carries them far greater distances than they can achieve by flagellar movement and also determines the general direction in which they move. Zoospores are chemotactic and the major value of the zoospore seems to be in its ability to select the substrate on which it settles. Thus numerous temporally separated periods of motility may be an adaptive advantage in detecting a suitable substrate in an aqueous environment. There is a tendency in other members of the Saprolegniales to lose the primary type of zoospore altogether. In the terrestrial Peronosporales there is a further tendency to lose all motile stages and for the zoosporangia to become so modified that they are easily detached by wind and dispersed as a unit functioning as air-borne conidia (Fig 1.15).

Sporangiospores – aplanospores

Zoospores are sometimes called planospores. Non-motile sporangiospores, aplanospores, are found in the Zygomycotina and are characteristic of the Zygomycete Mucorales. In *Rhizopus*, for example, the sporangia are

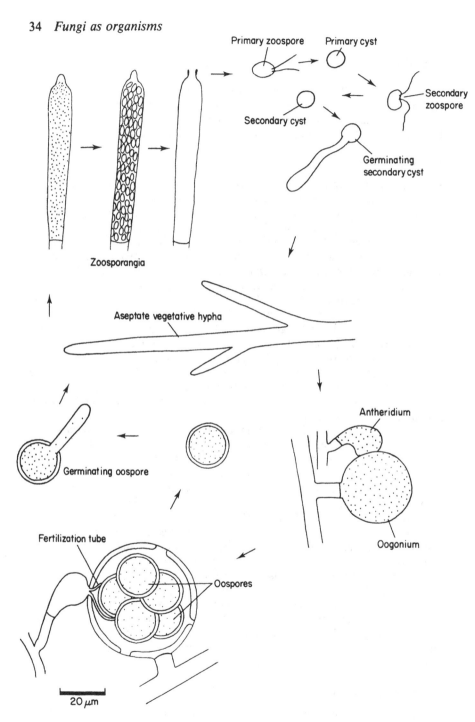

Fig. 1.14 Life cycle diagram of *Saprolegnia*, an Oomycete.

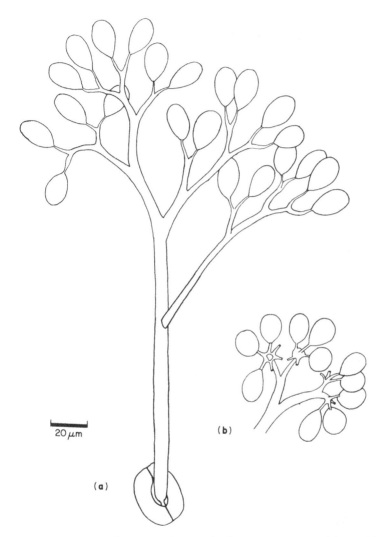

Fig. 1.15 (a) Sporangiophore of *Peronospora parasitica* with wind-borne sporangia. (b) Apex of sporangiophore of *Bremia lactucae*.

spherical structures borne at the ends of erect melanized hyphae, the sporangiophores, which arise in groups on the mycelium. Within the developing sporangium, a dome-shaped septum is laid down. This cuts off a distal spore-producing part from the sterile columella (Fig. 1.16). The latter is particularly large in *Rhizopus*. The sporangium dehisces on drying. The columella collapses, changing from being almost spherical to being like an inverted basin. This change in shape not only breaks the dry and very brittle

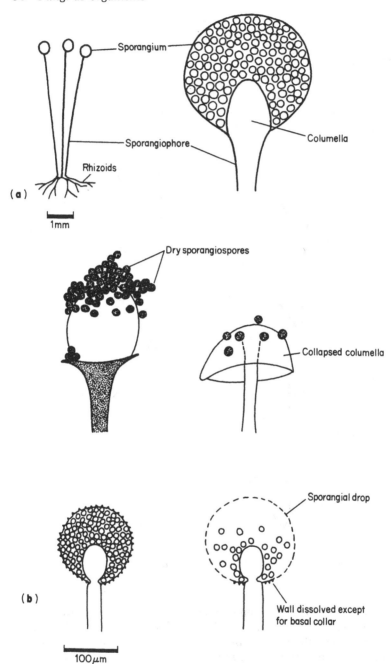

Fig. 1.16   (a) Sporangia of *Rhizopus* and dehiscence. (b) Sporangia of *Mucor* and dehiscence.

sporangial wall into massess of fragments but also leaves the mass of dry spores exposed to air movements so that they are easily blown away. In other Mucorales, especially many species of *Mucor*, the mature sporangium and sporangiophore becomes a stalked spore drop. The sporangial wall dissolves except for a basal collar-like region and the spores themselves are contained in mucilage (Fig. 1.16). Aplanospores from the sporangial drop are dispersed by adhering to passing insects or by rain splash. Ingold (1971) gives a very full and readable account of these and other modes of spore liberation in the Mucorales.

Conidia – blastoconidia

Conidia are distinguished from such non-motile sporangiospores by the fact that they are not enclosed by a separate sporangial wall. They are usually borne externally at the tips of hyphae. A large number of the Ascomycotina and somewhat fewer of the Basidiomycotina produce conidia. There are two basic modes of conidial development – blastic and thallic – but there is considerable variety in each type. In blastic types, there is a marked enlargement of the conidium initial before it is delimited by a septum and the conidium develops from part of a cell. In thallic types, if there is any enlargement of the conidium initial, it occurs only after the initial has been delimited by a septum or septa and the conidium develops from a whole cell. Conidial development in most is blastic. Blastic development is a form of budding where a limited area of the wall becomes plastic and balloons out to form a bud. Such simple budding is seen in the yeast, *Saccharomyces cerevisiae* (Fig. 1.1a, p. 3). Blastoconidia can be produced apically or laterally by budding from a cell of an undifferentiated hypha or conidiophore at one or more than one point. They may be produced singly but are often produced in straight or branched chains, the youngest being at the apex. The conidia become detached at a predetermined point on the conidiophore and leave behind an attachment scar or a small cylindrical or tapering stalk. Some common examples are illustrated in Fig. 1.17e–h.

Conidia – phialoconidia

Phialoconidia also show blastic development. They are produced from a special flask- or bottle-shaped cell called a phialide (Fig. 1.17a–d). A succession of conidia is produced from a fixed opening at the apex of the phialide. Each conidium reaches full size before being cut off by a septum and as soon as this happens another forms underneath. The conidia often hang together in long chains, with in this case the youngest at the base, but they frequently become aggregated into sticky heads at the apex of their conidiophores, forming stalked spore drops. Phialides are very variable in form. Some common examples are illustrated in Fig. 1.17.

Conidia – thalloconidia

One of the simplest ways to produce conidia is for the cells of an existing

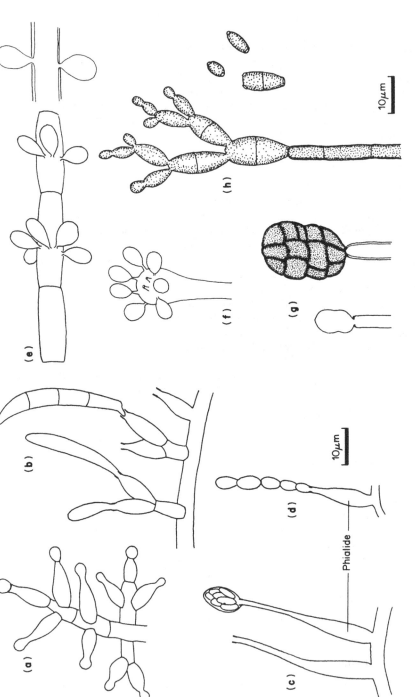

Fig. 1.17 (a)–(d) Phialoconidia: (a) *Trichoderma*; (b) *Fusarium* macroconidia; (c) and (d) *Fusarium* microconidia. (e)–(h) Blastoconidia: (e) *Aureobasidium*; (f) *Oedocephalum*; (g) *Stemphylium*; (h) *Cladosporium*.

hypha to disarticulate at the septa. This basic pattern is exhibited by a number of fungi (Fig. 1.18). This is thallic development and such conidia are called arthric thalloconidia. Again they may be dry or slimy, in short or long, and branched or unbranched chains. Thalloconidia may also be borne singly at the apex of a hypha which becomes the conidiophore (Fig. 1.18a–c). The conidium initial is delimited from the end of the conidiophore by a cross wall and, when fully developed and mature, it separates. Another conidium is not usually produced from the vacated conidiophore.

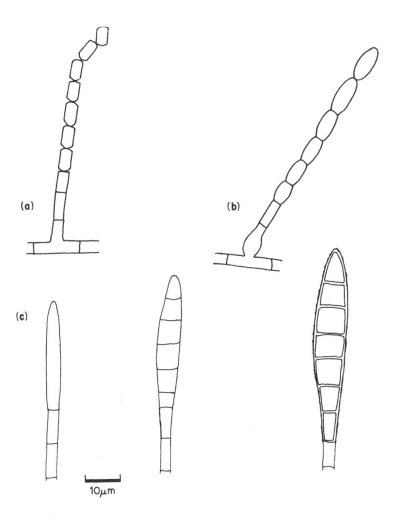

Fig. 1.18 Thalloconidia: (**a**) *Geotrichum*; (**b**) *Oidium*; and (**c**) *Microsporum*.

Sexually produced spores

It is the sexually produced spore which characterizes the members of any one group of fungi, for example, oospores in Oomycetes, zygospores in Zygomycetes, ascospores in the Ascomycotina and basidiospores in the Basidiomycotina.

Oospores

The Oomycetes are thus so-called because they produce oospores in an oogonium. In *Saprolegnia*, a spherical oogonium is delimited from a hyphal tip by a septum (Fig. 1.19a). The cytoplasm within cleaves into 5–10 oospheres or 'eggs' and an antheridium arises nearby, often on the same hypha, and becomes attached to the oogonium usually at a thin area in its wall (see Fig. 1.14). Fertilization tubes are formed which penetrate the oogonial wall and then the walls of the eggs. A male nucleus enters each oosphere and eventually fuses with its nucleus. The fertilized oospheres then develop into thick-walled oospores. They are then released when the oogonial wall breaks down. In Oomycetes, meiosis occurs in the antheridia and oogonia. The mycelium is unusual in being diploid.

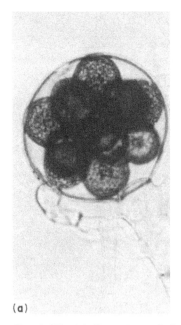

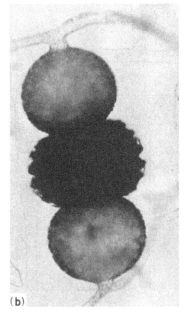

(a)   (b)

Fig. 1.19   (a) Oogonium of *Saprolegnia* with mature oospores. (b) A mature zygospore of *Rhizopus*.

Zygospores

In contrast to the Oomycetes, in Zygomycetes, such as *Rhizopus*, sexual reproduction results in the formation of zygospores (see Fig. 1.4; Fig. 1.19b). The gametangia arise as equal – unequal in some other genera – lateral protrusions from the hyphae. They become cut off from a supporting suspensor by a septum. The walls between the two applied gametangia break down and they fuse completely forming a single cell which develops into a thick-walled, warty, black zygospore. Like the oospore, it is a survival spore.

By far the majority of the Mastigomycotina and Zygomycotina do not produce complex reproductive structures. Their sexually produced spores are borne free and not surrounded by sterile hyphae. In contrast the Ascomycotina and Basidiomycotina usually, but not invariably, produce their sexual spores in, or on, microscopic or macroscopic reproductive structures called ascocarps and basidiocarps respectively.

Ascospores

The one feature common to all Ascomycotina is that they produce ascospores in a sac-like ascus. In a few the asci are borne unenclosed. For example, in *Schizosaccharomyces*, two haploid cells produced by fission fuse to form an ascus with eight ascospores (Fig. 1.20a). The asci are non-explosive. The ascospores are released by breakdown of the ascus wall.

Most asci are explosive, asci being turgid cells that eventually burst violently, liberating the contained ascospores. Basically there are two sorts of explosive asci – unitunicate and bitunicate ones. The former are found in Pyrenomycetes, such as *Sordaria*, and Discomycetes, such as *Coprobia* (Fig. 1.20b). In these the ascus is an elongated cylindrical cell, with a thin elastic wall and a very thin lining layer of cytoplasm, enclosing a relatively large volume of ascus sap, probably of an aqueous solution of sugars derived from glycogen. Eight ascospores are suspended in the sap. When mature, the ascus is in a state of tension, the thin, apparently one-layered wall being stretched by the osmotic forces (-1.0 to -1.5 MPa) of the sap. Finally the ascus ruptures in a definite manner by a distinct pore or slit developing at the apex (Fig. 1.20b). This is followed by a sudden contraction of the wall which squirts the contents, sap and ascospores, out to varying distances ranging from a few mm to 0.5 m. Thus they are discharged up into the turbulent air for dispersal.

Bitunicate or double-walled asci are characteristic of Loculoascomycetes, such as *Pleospora* and *Leptosphaeria*. The wall of the young ascus appears to be thicker than in unitunicate ones and really consists of two walls – a thin and inextensible outer wall and a thick extensible inner wall. The mature ascus is again in a state of tension. It normally discharges by a transverse slit in the outer rigid wall at the apex. The pressure of the wall on the contents is thus reduced and the elastic inner wall is free to expand under turgor. It elongates very rapidly to over twice its original length. The ascospores are then discharged through a pore in the apex of the inner wall (Fig. 1.20c,d). In

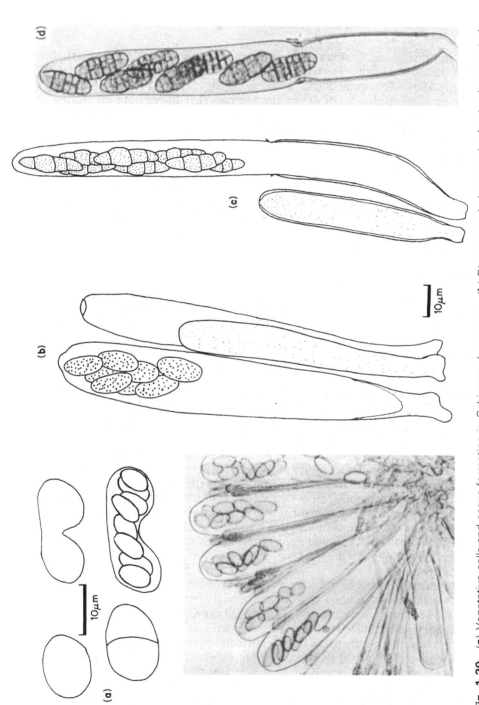

Fig. 1.20 (a) Vegetative cells and ascus formation in *Schizosaccharomyces*. (b) Diagram and photograph of unitunicate asci of *Coprobia*. (c) Bitunicate asci of *Leptosphaeria*. (d) Bitunicate ascus of *Pleospora infectoria* showing ruptured outer wall and inner wall extended.

10 μm

these asci there is no gradual enlargement and elongation to full turgor as there is in unitunicate types but a sudden extension. However, the end result is the same, the ascospores being violently discharged up into the turbulent air. The significance of these two modes of maturation can best be seen by considering the form of two types of ascocarps in which they develop. The Pyrenomycetes produce perithecia and the Loculoascomycetes pseudothecia. Both develop quite differently but when mature closely resemble each other in being flask- or pear-shaped, with a swollen base and a neck of varying lengths. For successful discharge the asci have to extend up a channel in the neck in order to release their spores through the pore at the apex. Both modes of ascus maturation achieve this.

Basidiospores

The Basidiomycotina are so-called because they produce basidia. There are a number of different types of basidia. The most common type is the holobasidium which is characteristic of the Hymenomycetes. This is a simple, single-celled cylindrical or club-shaped basidium with usually four basidiospores borne asymmetrically on fine stalks outside the basidium (Fig. 1.21a). The basidiospores are again violently discharged but, compared with the ascus, the basidium is not a very powerful spore gun. Thus they and the basidiospores tend to be held horizontally rather than vertically. If they were held vertically, after discharge the spores would fall back onto the basidial layer, as they have insufficient impetus to get up into the turbulent air. Immediately before discharge, a drop of liquid appears on a beak-like projection, the hilum, near to the spore's attachment to its stalk. It grows to a definite size and then the spore is launched horizontally 0.1–0.2 mm. As the basidia are usually borne aloft on structures such as gills, the basidiospores then fall vertically downwards into the turbulent air (Fig. 1.21b).

A number of mechanisms for basidiospore discharge have been suggested. These are discussed by Ingold (1971) and Webster (1980). One of the latest suggestions is that the spores are released by a 'springboard' mechanism, with the slender, curved, tapering stalks acting as the springboards. As the enlarging drop emerges from the adaxial side of the hilum, it adheres to the surface of the spore and spreads over the adaxial surface (Fig. 1.21c). The adhesion of the drop drags the spore towards the axis of the basidium, putting the stalk under tension and causing it to bend inwards. At this time a thin septum develops across the neck of the stalk. When it is complete, the spore is no longer attached to the stalk and breaks away, releasing the tension on the stalk. The stalk abruptly returns to its original position in a springboard manner projecting the spore away from the basidium. The springboard concept removes the need to postulate the involvement of an explosive discharge mechanism which would allow successive discharge of the four spores from any one basidium. It would also account for the fact that spores are consistently projected along identical trajectories.

The term 'ballistospore' has been applied to such violently projected spores. But not all basidiospores are ballistospores. In the Gasteromycetes,

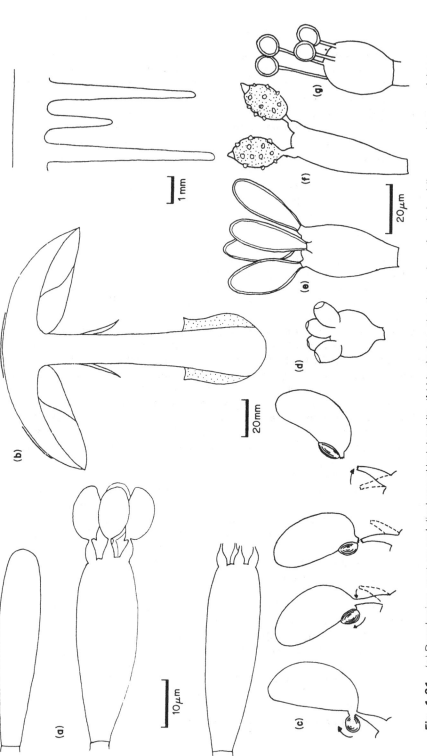

Fig. 1.21 (a) Developing, mature and discharged holobasidia. (b) Vertical section through an Agaric basidiocarp and tangential section of cap to show wedge-shaped gills. (c) Diagrammatic representation of the 'springboard' mechanism of basidiospore release. (d)–(g) Basidia of Gasteromycetes: (d) *Podaxis*; (e) *Hysterangium*; (f) *Hymenogaster*; and (g) *Lycoperdon*.

1 mm

20mm

10μm

20μm

(a)

(b)

(c)

(d)

(e)

(f)

(g)

the basidiospores are usually symmetrically poised on their stalks or are without stalks (Fig. 1.21d). The spores are released into cavities within the basidiocarp when the basidia break down. They are dispersed later by a variety of agencies (Ingold, 1971). Other Basidiomycotina produce hetero-basidia. These·may be transversely septate as in the Uredinales and Auricula-riales or longitudinally divided into four cells as in the Tremellales or unicellular but deeply divided and of the 'tuning-fork' type as in the Dacry-mycetales (Fig. 1.22). In most, the basidiospores are violently discharged.

Basidiocarp form and structure

In most Ascomycotina and Basidiomycotina, hyphae become aggregated to form microscopic or macroscopic reproductive structures, ascocarps and

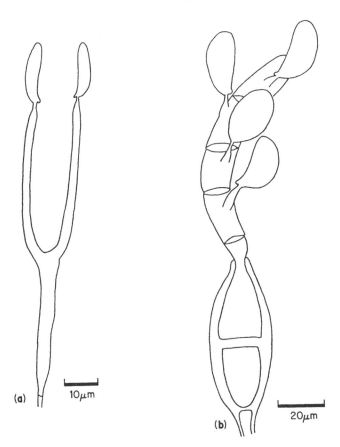

Fig. 1.22 Heterobasidia. (a) Tuning-fork type of *Dacrymyces*. (b) Transversely septate basidium of *Puccinia*.

basidiocarps, often of a more or less solid nature. Basidiocarps exceed all other reproductive stuctures in the fungi in their range of size, complexity and degree of hyphal differentiation. This is all achieved by hyphal aggregation, branching, fusion, swelling and in some cases wall thickening. Two examples from the Hymenomycetes, the fleshy, ephemeral agaric and the more durable, longer-lived bracket or polypore will serve to illustrated some of this range.

Agaric type basidiocarps

The Agaricales or agarics are the mushrooms and toadstools. The basidiocarp is some 50–150 mm tall consisting of a disc- or dome-shaped cap held aloft, as an umbrella, by a stalk and having its basidia usually lining radiating wedge-shaped gills (e.g. *Amanita*) (Fig. 1.23), or sometimes tubes (e.g. *Boletus*), or less frequently tapering spines (e.g. *Hydnum*), underneath. The development of such basidiocarps is extraordinarily variable but the final form seems to be governed by a number of factors, most of which relate to the basidium. Its range is such that the spores are discharged to such a distance that they clear the other spores but do not impact on the opposite gill, tube or spine face and they fall vertically. The stalk not only supports the cap but it is essential in order to provide a space between the cap and the ground so that the falling spores are caught in the turbulent air and carried away. The basidia are also prevented from functioning if covered by a water film. They

Fig. 1.23   Expanding and expanded basidiocarp of *Amanita phalloides*.

are not sufficiently powerful to project their spores through water. They are thus sheltered underneath the cap which protects them from the rain. This is also true of the polypores where they are sheltered underneath the bracket. Exceptions do occur in the Hymenomycetes, particularly in the Clavariaceae in such genera as *Ramaria*, where the basidia line the erect, uncovered branches of the basidiocarp. In addition, the basidia function only when the ambient air is close to saturation. Basidia must be turgid and the release of the spores from the turgid cells is very sensitive to atmospheric humidity. If the relative humidity of the atmosphere falls much below saturation spore discharge falls and almost ceases. Obviously closely packed, wedge-shaped gills or tapering spines or narrow tubes help to maintain constant high humidities around the basidia and make the spore-bearing layer less susceptible to fluctuations in the relative humidity of the outside air. Finally, unlike tubes which are fused to each other, gills and spines are free from each other and they can reorientate to some extent by geotropic movements if the whole basidiocarp tilts. Thus absolute rigidity is not essential.

Such basidiocarps show very little differentiation of their dikaryophase hyphae in the stalk and cap. Some hyphae may be wider, more elongated and less branched than others but throughout they are prosenchymatous in that hyphae are not recognizable as such. The basidiocarps lack any organized vascular system yet the stalk must possess some sort of supporting framework and must have some form of conducting tissue up to the basidia but these are not morphologically distinct. Such fleshy agarics are relatively short-lived, usually a matter of days, and rely mainly on the turgidity of their hyphae for support rather than on distinct mechanical tissues. Under dry atmospheric conditions, most lose water rapidly, shrivel and spore discharge ceases. Furthermore, they do not recover on return to more humid conditions. This reliance on turgor for spore development and discharge and for support may be why they are mainly restricted in their appearance to the damper days of autumn. Even so most are frost sensitive and so disappear with the first frosts.

Polypore type basidiocarps

The basidiocarps of the Aphyllophorales, the bracket fungi or polypores, are often membranous, leathery, corky or even woody in texture. They normally function for a longer period than most agarics and some such as those of *Ganoderma* are perennial. In these, three different types of hyphae may be recognized – generative, skeletal and binding. Generative hyphae are always present and are thin-walled, branched and septate, often with clamp connections. In the basidiocarp they give rise to the other two kinds of hyphae. Skeletal hyphae are thick-walled, with a narrow lumen, non-septate and usually unbranched (Fig. 1.24). They are of unlimited growth and are strengthening structures giving rigidity and support to the basidiocarp. In very hard basidiocarps, such as those of *Phellinus pomaceus* and *Ganoderma adspersum*, skeletal hyphae develop in considerable numbers to become the

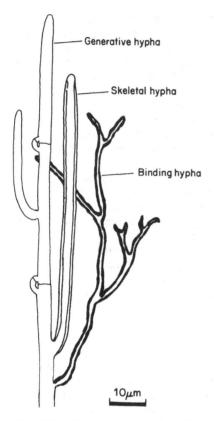

Fig. 1.24   Hyphae from the basidiocarp of a trimitic polypore.

dominant hyphal type. Binding hyphae may be produced as an alternative to skeletal hyphae but more often accompany them as in *Coriolus versicolor*. They are of limited growth, very thick-walled, relatively narrow, rarely septate and are highly and irregularly branched. Binding hyphae develop some way behind the growing margin. Thus in such basidiocarps, generative hyphae provide the ground plan, skeletal ones the firm constructional framework and the binding hyphae firmly lock these together. The relatively enormous strength which these hyphae impart to basidiocarps can readily be discovered by attempting to mutilate a basidiocarp of *G. adspersum*. Where only generative hyphae are present, the hyphal system is called monomitic, where a second type is present, dimitic and where there are all three, trimitic. In accordance with these terms, agarics would be described as monomitic with thin-walled, inflated generative hyphae.

The bracket fungi or polypores are restricted to woody substrates. Very few are centrally stalked like an agaric. Most arise on a tree trunk or branch above the ground so that the spores fall down into the turbulent air. Some have a short lateral stalk for attachment but most are broadly attached and thus bracket-like in form (Fig. 1.25). The majority have their basidia lining tubes which are fused together and thus cannot readjust to the vertical if tilted. The more skeletal and binding hyphae that are present, the more rigid the basidiocarp and the longer and narrower the tubes tend to be. If they are tilted the tubes may grow on by a meristem at their free edge. The newly formed portion of the tube is aligned to the vertical as it develops. If the basidiocarps are markedly displaced, for example from being strictly horizontal to being virtually vertical as might happen when a host tree is wind-blown, the free surface of the pores may be sealed off and a new basidiocarp may arise diageotropically on the old pore surface at about 90° to the latter (Figs. 1.26 and 1.27). The new tubes formed on the underside will be exactly orientated to the vertical by positive geotropic growth. Gravity may also have a profound formative effect on the basidiocarps of polypores. In *Piptoporus betulinus*, the birch polypore, if on a vertical trunk, the primordium grows out and develops into an almost semi-circular hoof-like

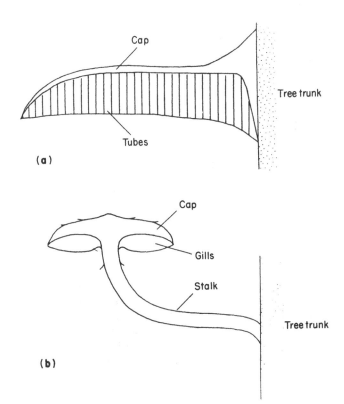

Fig. 1.25   (a) Vertical section through a broadly attached basidiocarp of a polypore. (b) Vertical section of an Agaric arising in a similar position.

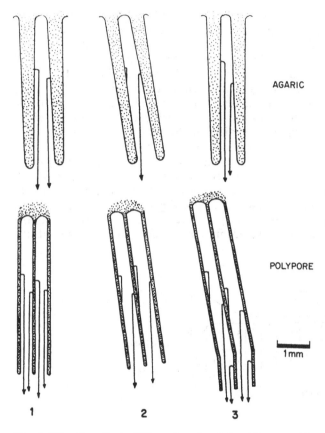

AGARIC

POLYPORE

1 mm

1   2   3

Fig. 1.26   The effect of tilting of an Agaric basidiocarp with wedge-shaped gills and spore release (**1** and **2**) and subsequent realignment (**3**); also (below) the corresponding response in a polypore showing realignment of newly formed parts of the tubes at their free edge (**3**).

bracket, laterally attached. If the primordium arises on the underside of a horizontal branch, it develops into an almost circular form, centrally attached; however, if a branch falls or is felled so that an already formed primordium becomes located on the upper side – it would not normally arise in this position – it grows out and upwards on one side and tubes arise from underneath (Fig. 1.28a). Gravity would have no such effect on wood-inhabiting agarics. They respond to these circumstances by stalk curvature (Fig. 1.28b).

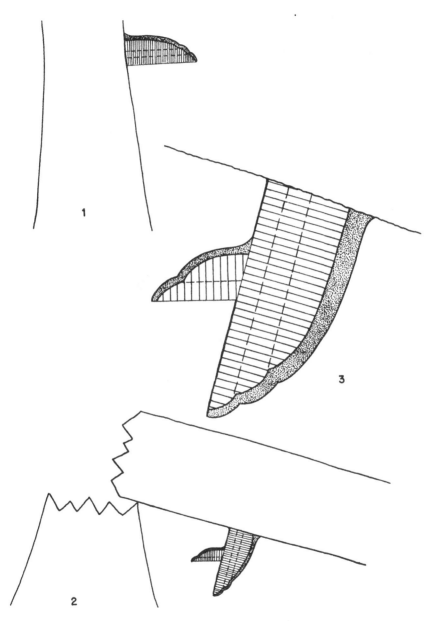

Fig. 1.27 Formation of a new basidiocarp diageotropically on the old pore surface of a displaced one. (**1**) Three-year-old basidiocarp of *Ganoderma* on an elm trunk. (**2**) The same two years after the tree had blown down. (**3**) As (**2**) in more detail showing new basidiocarp arising diageotropically on the sealed pore surface of the old one.

(a)

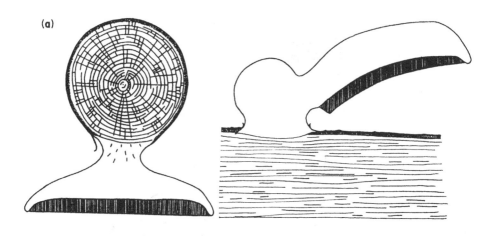

(b)

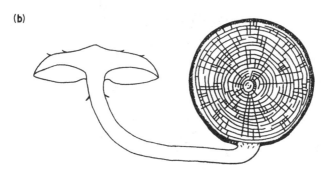

Fig. 1.28 (**a**) *Piptoporus betulinus*. Vertical section of a basidiocarp arising on the underside of a horizontal branch and on the upper side of a similar branch. (**b**) Vertical section of an Agaric basidiocarp arising from the underside of a horizontal branch.

Classification of the fungi. A modification of the scheme proposed by Ainsworth (1973) and adopted by Webster (1980).

Fungi   Free-living, parasitic or mutualistic symbionts. Saprotrophs, necrotrophs or biotrophs as devoid of chlorophyll. Cell wall composition very variable, majority contain chitin and glucans. Reserve materials, glycogen, oil and mannitol. Characteristic disaccharide trehalose. Some yeast-like but majority with thread-like filaments, hyphae, branching profusely to form the vegetative mycelium on which spores are produced, asexually or sexually, free on hyphae or enclosed in complex reproductive structures. Separated mainly on morphology of the latter. Two divisions.

Myxomycota   Wall-less and quite unusual organisms, only included in the fungi as mostly studied by mycologists. Possess either a plasmodium, a mass of naked, multinucleate protoplasm, which feeds by ingesting particulate matter and moves by amoeboid movement, or a pseudoplasmodium, an aggregation of separate amoeboid cells. Both of a slimy consistency, hence 'slime moulds'. Three classes.

I Acrasiomycetes   The cellular slime moulds. Vegetative phase free-living amoebae which aggregate into a pseudoplasmodium before forming spores within a sporocarp.

II Myxomycetes   The true slime moulds. Vegetative phase a free-living and often brightly coloured plasmodium which eventually becomes converted into one or more sporocarps.

III Plasmodiophoromycetes   The endoparasitic slime moulds. Biotrophic parasites producing plasmodia which develop into zoosporangia or resting spores. Zoospores with two, unequal, whiplash flagella.

Eumycota   True fungi, all with walls. Customary to recognize five subdivisions.

A. Mastigomycotina   Zoosporic fungi, many solely aquatic. Three classes each characterized by their distinctive type of zoospore.

I Chytridiomycetes   Zoospores with a single posterior whiplash flagellum. Walls chitin and glucan.

(i) CHYTRIDIALES   Vegetative phase either converted as a whole into reproductive zoosporangia or resting sporangia (holocarpic) or with a specialized rhizoidal vegetative system and one or more reproductive structures (eucarpic).

(ii) BLASTOCLADIALES   Vegetative phase usually more complex than Chytrids, often with an extensive rhizoidal system bearing a trunk-like body with numerous branches with reproductive structures at their apices. Alternation of haploid and diploid generations in some. Asexual reproduction by means of zoospores produced in thin-walled zoosporangia or from brown, pitted resting spores; sexual reproduction by means of isogamous or anisogamous motile gametes.

(iii) MONOBLEPHARIDALES   Characteristic delicate, highly vacuolated, much-branched hyphal system. Sexual reproduction oogamous with non-flagellated female cells (oospheres) and posteriorly uniflagellate antherozoids.

II Hyphochytridiomycetes   Zoospores with a single anterior tinsel flagellum. Walls cellulose and chitin. Otherwise closely resemble Chytridiales.

Classification of the fungi – *continued*

III Oomycetes   Zoospores with two sub-apically or laterally attached flagella; an anteriorly directed tinsel and a posteriorly directed whiplash. Sexual reproduction oogamous. Oogonia, containing one or more oospheres, and antheridia. Oospores, thick-walled resting spores. Vegetative phase diploid, usually mycelial with aseptate hyphae. Wall cellulose and glucans.

(i) SAPROLEGNIALES   The water moulds. Long cylindrical zoosporangia at tips of relatively wide aseptate hyhae. Usually with several oospheres per oogonium.

(ii) LEPTOMITALES   Hyphae usually constricted, with cellulin plugs. Oogonia with one oosphere.

(iii) LAGENIDIALES   Mostly parasitic on algae and water moulds. Endobiotic and holocarpic. Oogonia with one oosphere.

(iv) PERONOSPORALES   Primarily terrestrial, living in soil or parasitic on vascular plants, including the biotrophic downy mildews. Zoosporangia more or less globose, often detachable and may function as conidia. Oogonia with one oosphere.

B. Zygomycotina   Vegetative haplophase, hyphae aseptate. Asexual spores non-motile aplanospores. Sexual reproduction by the complete fusion of two multi-nucleate gametangia producing a zygospore. Wall chitin and chitosan. Two classes.

I Zygomycetes   As above.

(i) MUCORALES   The saprotrophic pin-moulds but also including the fungi of vesicular-arbuscular mycorrhiza. Aplanospores produced in globose multinucleate sporangia, in a row in narrow, cylindrical, sac-like merosporangia or in few-spored sporangiola or singly as conidia. Zygospores often thick-walled, black and warty resting spores. Large terminal chlamydospores common in mycorrhizal forms.

(ii) ENTOMOPHTHORALES   Some saprotrophic but mostly insect parasites. Vegetative phase tending to break up into segments (hyphal bodies). Asexual reproduction by means of forcibly discharged uni- or multi-nucleate conidia.

II Trichomycetes   Group of uncertain affinities. Mostly parasitic in guts of arthropods. Vegetative phase much reduced, attached by a basal cell to digestive tract or exterior of animal. Asexual reproduction by conidia. Resting spores produced by fusion of protoplasts.

C. Ascomycotina   Vegetative haplophase. Hyphae septate, with simple pore. Ascospores produced within an ascus. Nuclear fusion in the ascus followed by meiosis and usually a mitosis and the formation of eight ascospores. Asexual conidia often present. Wall chitin and glucan. Five classes.

I Hemiascomycetes   Asci one-walled (unitunicate), naked not borne on ascogenous hyphae, no ascocarps.

(i) ENDOMYCETALES   Filamentous and unicellular yeasts. Asexual reproduction by budding or fission. Sexual reproduction by fusion of two cells; product directly or after division forms the ascus.

(ii) TAPHRINALES   Mycelium of binucleate cells from which asci develop, Parasitic.

II Plectomycetes   Asci unitunicate, usually more or less globose, arising on ascogenous hyphae, within a closed, globose ascocarp (cleistothecium), and not accompanied by packing paraphyses.

(i) EUROTIALES   Mostly saprotrophs. The green and blue moulds. Asci very small, globose, non-explosive. Conidial states generally phialidic (*Aspergillus* state of *Eurotium* and *Penicillium* state of *Talaromyces*).

(ii) ERYSIPHALES   Powdery mildews, biotrophic parasites. Ascocarps with one to several oval- to club-shaped, explosive asci. Chains of conidia arising in basipetal succession from a mother cell on the superficial mycelium. Penetration of host by haustoria, confined to epidermal cells.

III Pyrenomycetes   Asci unitunicate, cylindrical, explosive, opening by an apical pore or slit, in a layer (hymenium) with paraphyses in a minute, flask-shaped ascocarp (perithecium). Perithecium opening by an apical pore, dark or bright coloured, soft and fleshy or leathery, produced singly, in clusters or within a stroma. Many with conidia.

(i) SPHAERIALES   As above.

IV Discomycetes   Asci unitunicate, cylindrical, explosive, arranged in a hymenium with paraphyses and exposed at maturity in a disc- or saucer-shaped ascocarp (apothecium).

(i) PEZIZALES   Asci operculate (opening by a distinct lid or operculum).

(ii) HELOTIALES   Asci inoperculate (opening by a slit).

(iii) TUBERALES   The Truffles. Asci indehiscent. Hymenium enclosed in a modified, subterranean apothecium.

(iv) LECANORALES   A group of mainly inoperculate discomycetes living in a mutualistic symbiosis with algae as lichens. Some 21% of all known species of fungi occur as lichens.

V Loculoascomycetes   Asci two-walled (bitunicate), outer wall thin and inextensible, inner thick and extensible, seen as outer wall ruptures and inner extends just prior to ascospore discharge. Ascocarp, an aggregation of vegetative hyphae, an ascostroma, resembling a perithecium and called a pseudothecium. Asci develop in spaces (locules) within.

(i) PLEOSPORALES   Pseudothecium uniloculate with interascal threads, pseudoparaphyses, attached to the ascocarp wall at top and bottom.

(ii) DOTHIDEALES   Pseudothecium without pseudoparaphyses.

D. Basidiomycotina   Vegetative dikaryophase, mycelium septate, with dolipore septa and often with clamp connections. Basidiospores produced on basidia. Nuclear fusion in the basidium followed by meiosis and the production exogenously of typically four basidiospores, each on a stalk, usually violently discharged by a drop excretion method (ballistospores). Wall chitin and glucan. Three classes.

I Hymenomycetes   Mushrooms, toadstools, bracket fungi or polypores, fairy clubs and jelly fungi. Basidia exposed at maturity on a palisade-like hymenium. Basidiospores typically ballistospores. Two sub-classes.

1 Holobasidiomycetidae   Basidium a holobasidium, a single cylindrical or club-shaped cell, undivided by septa.

(i) AGARICALES   Basidiocarps fleshy, being composed of thin-walled, inflated hyphae. Hymenium usually lining gills, sometimes tubes and less frequently spines.

(ii) APHYLLOPHORALES   Basidiocarps membraneous, leathery, corky or woody in texture being composed of thin-walled generative hyphae which give rise to either thick-walled, unbranched, aseptate skeletal hyphae or thick-walled, narrow, rarely septate, much-branched binding hyphae or both. Hymenium usually lining tubes opening to the exterior by means of pores, sometimes spines, anastomosing gills or smooth surfaces.

Classification of the fungi – *continued*

---

2 Phragmobasidiomycetidae   Basidium a phragmobasidium, divided by septa. Basidiospores often germinate by repetition i.e. by producing secondary spores.

(i) DACRYMYCETALES   Basidium forked or cleft, bearing two basidiospores. Basidiocarps gelatinous or waxy, often coloured yellow to orange.

(ii) TREMELLALES   Basidium longitudinally divided into four. Basidiocarps gelatinous, often brightly coloured, drying to a cartilaginous texture.

(iii) AURICULARIALES   Basidium ˙divided transversely by˙ three septa. Basidiocarps gelatinous to rubbery, drying to a cartilaginous texture.

II Gasteromycetes   Puffballs, earth stars, stinkhorn, bird's nest fungi and their allies. Basidia not exposed at maturity but enclosed within cavities in closed basidiocarps. Basidiospores not violently discharged but liberated by collapse of basidium and variously dispersed; as a dust, in a slime, or within packets (peridiola).

III Teliomycetes   Rusts and smuts. Mycelium septate with simple pore. Equivalent of basidia consist of thick-walled teliospores or chlamydospores within which nuclear fusion takes place, and promycelia, which usually give rise to four or more spores after a meiosis. Basidiocarps lacking.

(i) USTILAGINALES   Smuts, parasitic on angiosperms. Characteristic dusty, black spore masses (chlamydospores). Promycelium producing a large to indefinite number of spores.

(ii) UREDINALES   Rusts, biotrophic parasites of Angiosperms. Gymnosperms and Pteridophytes. Characteristic reddish-brown spore masses (urediospores). Promycelium from teliospore producing four spores. Many with complex life cycles involving five spore stages.

E. Deuteromycotina   So-called 'Fungi Imperfecti'. Fungi known only from their asexual (imperfect or anamorphic) or mycelial state. Their sexual (perfect or teleomorphic) states are either unknown or may possibly be lacking altogether. Solely conidial or mycelial fungi. Three classes.

I Blastomycetes   Yeast-like budding forms, true mycelium lacking or not well-developed.

II Hyphomycetes   Mycelial forms with conidia borne directly on the hyphae or special hyphal branches termed conidiophores, the latter borne singly or in clusters or cushions (sporodochia).

III Coelomycetes   Mycelial forms with conidia borne in flask-shaped pycnidia or saucer-shaped acervuli.

2

Fungi as decomposers of leaves

Leaves of all manner of types form suitable substrates for many fungi. Long before any leaf falls, the complex process of its decay is initiated. The greater part of this process takes place above ground in the litter covering the soil surface. In this chapter some facets of this are discussed but particularly the epiphytic leaf surface or phylloplane micro-flora, the leaf surface as a habitat niche for fungi, colonization by the common primary saprotrophs as leaves senesce and die, the attributes which make these primary saprotrophs such widespread and successful colonizers, and some of the subsequent events occurring in the litter.

The leaf as a spore trap

As any leaf unfolds it is a relatively clean sheet which immediately provides landing sites for air-borne particles such as bacteria, yeast cells and fungal spores but also pollen. Spore trapping by leaves is a natural phenomenon of nature. Spores may reach leaves in three main ways: wind-borne and deposited by impaction or by sedimentation under gravity; in falling rain drops; or in rain splash droplets. Air-borne spores are usually dry and often rough or spiny and readily detachable from their stalks, excellent examples being the urediospores of rust fungi. They are readily washed out of air by falling rain drops. Rain splashed spores tend to be wet or slimy and borne in a sticky liquid. Adaptations facilitating deposition are far less obvious than in spores of aquatic fungi (Chap. 4). Amongst dry spores, a larger size, as seen in the powdery and downy mildews, favours impaction and sedimentation. Rain splashed spores tend to be smaller and spherical. But these are only generalizations and there are many anomalies. The most ubiquitous and by far the most numerous of the phylloplane fungi are members of the Sporobolomycetaceae, the shadow yeasts. They produce air-borne spores which in relative terms are quite minute.

Leaf surfaces are differential spore traps. Their efficiency as traps depends upon whether they are horizontal or vertical, wet or dry, hairy or glabrous, glossy or mat, waxy or non-waxy and so on. Not all spores that land become securely attached. Some are washed off by rain, blown off by wind or redistributed by dew. Some have a two phase dispersal system. For example, the large sporangia of pathogenic species of *Phytophthora* and some other Oomycete Peronosporales are wind-borne and normally impacted onto leaf surfaces. Under moist conditions the impacted sporangia may germinate directly by a germ tube or indirectly to produce motile zoospores which may swim about in moisture or be redispersed further to other leaves in rain splash droplets.

Virtually any spore which may become air-borne can be found on leaves. If leaf surfaces are washed and the washings plated out onto nutrient agar, numerous yeasts and filamentous conidial Ascomycotina, some Zygomycete Mucorales and the occasional Mastigomycotina and Basidiomycotina develop on the plates. But microscopic examination of stained leaf surface impressions or peels reveals the presence of not only these but also spores of many other Ascomycotina and Basidiomycotina, including those of agarics, polypores and Gasteromycetes. They just do not grow or grow too slowly on the culture medium used. These impressions of peels can be made by spraying leaves with cellulose acetate in amyl acetate or painting with nail varnish or molten 1% agar, leaving to dry and then stripping off.

Many of these fungi and an even larger number of bacteria actively grow on the surface of the living leaf and have been called 'resident inhabitants' in contrast to 'casual inhabitants' which are unable to grow in such an environment because of the lack of essential nutrients, unfavourable physical factors, competition with or antagonism by others or some combination of these factors. This rather simple distinction can be extended by dividing the epiphytic fungi into three categories: non-pathogenic epiphytes; pathogens; and exochthonous or casual inhabitants. Exochthonous is a fitting, if somewhat clumsy, term as it is used for fungi found on or in a substrate which is not their habitual one.

Phylloplane inhabitants

Amongst the non-pathogenic epiphytes, two main groups, the phylloplane inhabitants and the common primary saprotrophs, can be recognized. The phylloplane inhabitants are able to complete their life cycle or a significant part of it on the living leaf without damaging it. *Sporobolomyces roseus* (Fig. 2.1a) not only is a very good example of such a fungus but is virtually omnipresent, being found on leaves of grasses, dicotyledonous herbs, trees and shrubs, wherever they grow. Its cells multiply very quickly by budding when conditions are favourable, forming distinct yeast-like colonies on the leaves. Budded cells can be redistributed on an individual leaf or from leaf to leaf by rain splash and are similarly locally dispersed to leaves of other plants. It also reproduces by ballistospore formation. The ballistospores are very effectively wind dispersed. They are produced under high humidities at night, as are the budded cells, and constitute the major component of the air-spora at that time. Other members of the Sporobolomycetaceae are also common. Species of *Bullera* behave similarly, whereas members of the genera *Tilletiopsis* and *Itersonilia* produce a sparse mycelium from which ballistospores

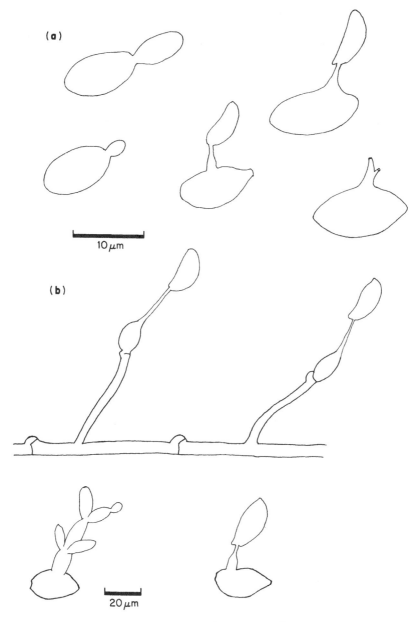

Fig. 2.1 (a) Budding and ballistospore formation in *Sporobolomyces roseus*. (b) Ballistospore formation in *Itersonilia perplexans*, ballistospores germinating by budding and ballistospore formation.

arise (Fig. 2.1b). Other yeasts, especially members of the Cryptococcaceae, the non-sporing yeasts, exist in the budding phase. They all complete their life cycle in the phylloplane. Such fungi are often called 'shadow yeasts'. Their presence can be demonstrated by suspending leaves from the underside of a Petri dish lid for about 12 h over 2% malt extract agar. In such a humid and still atmosphere, ballistospores are produced and discharged. They fall vertically onto the agar below and start budding. Tiny colonies, pink in *S. roseus*, become visible after 2–3 days. These form a mirror image of the distribution of the cells of the leaf.

Two conidial Ascomycotina also grow in the phylloplane. The conidia of *Aureobasidium pullulans* and several species of *Cladosporium* may germinate after impaction and develop into hyphae forming quite extensive colonies under favourable conditions. *Aureobasidium* more often grows by yeast-like budding with minimal hyphal growth (Fig. 2.2a). This is a modification of its normal cultural form. The budded cells may again be redistributed in moisture films or by rain splash droplets. The conidia of *Cladosporium* may germinate and produce secondary conidia from short germ-tubes rather than grow as hyphae (Fig. 2.2c). These conidia are dry and become air-borne. So both reproduce rapidly and complete a significant part of their life cycle in the phylloplane. Both eventually produce ascocarps to complete their life cycle. These are found only in the spring on overwintered

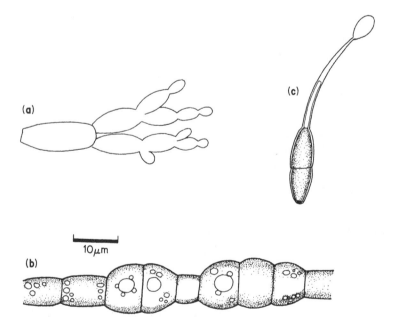

Fig. 2.2   (a) Yeast-like budding by *Aureobasidium*; and (b) chlamydospores of *Aureobasidium* on a leaf surface. (c) Conidium of *Cladosporium* germinating to produce a secondary conidium.

fallen leaves in temperate climates. Ascospores are discharged from these as new leaves unfold.

Aureobasidium and *Cladosporium* are also both very well-adapted to survive in this rigorous habitat. Their hyphal walls rapidly become thickened and melanized. This enables them not only to survive exposure to damaging ultra-violet light from the sun but may also help to prevent excessive desiccation and make them more resistant to bacterial lysis. *Aureobasidium* produces dark, thick-walled multicellular chlamydospores in chains or in clumps (Fig. 2.2b). *Cladosporium* produces more distinct microsclerotia, compact spheres of 10–100 cells with an outer layer of thick-walled cells with heavily melanized walls. Under favourable conditions, these produce clusters of conidiophores and abundant conidia which, like the ascospores, serve as a source of inoculum as new leaves unfold. In contrast *Sporobolomyces* does not appear to be able to withstand prolonged adverse conditions, such as low relative humidities. At relative humidities of 65% and below, it rapidly disappears but its population equally rapidly expands from reservoirs on more protected less exposed leaves, when favourable conditions return. In temperate climates with a combination of warm humid weather and aphid infestation producing honeydew on leaves, the so-called sooty moulds appear as black, soot-like coverings over leaves, especially of trees such as limes (*Tilia* spp.). These are the result of the profuse growth of *Aureobasidium* and *Cladosporium* using the trisaccharide melezitose in the honeydew as a carbon-source, together with aphid faeces, sloughed off parts and dead remains. In wet tropical climates, such as in Amazonia, parts of Africa, Australasia and the Caribbean, true sooty moulds occur. These, like the perfect states of *Aureobasidium* and *Cladosporium*, are also Loculo-ascomycetes and again grow as saprotrophs associated with honeydew from aphids. A wide range of species from several fungal families, especially the Capnodiaceae and Chaetothyriaceae are involved. They form distinct, dense, dark hyphal networks on leaves often in the form of a thick felt and each fungus produces abundant conidia of often two or even three types as well as ascocarps.

Nutrient sources

All the phylloplane inhabitants, the yeasts, the filamentous fungi and the bacteria, are chemo-organotrophs requiring organic nutrients for growth. Some of their nutritional requirements may be met by organic substances absorbed or deposited onto leaves such as detritus trapped in their superficial wefts of hyphae, as also happens with fungi living on paint films or glass. Most of their nutrients, however, must be derived directly or perhaps indirectly from the host. A great variety of substances exude or leak out of leaves. These include free sugars, amino acids and inorganic ions which are all essential for fungal and bacterial growth. For example, water droplets placed on leaves exhibit an increase in conductivity indicating exudation from the leaf; increased growth of some fungi in these drops shows that certain of these exudates are of nutritional value. Two sources of added nutrients are from pollen and other spores. Nutrients also leak out of these. Pollen added to leaf surfaces stimulates the development of *Sporobolomyces* and *Cladosporium* and probably accounts for the sudden increase in their population shortly after flowering on leaves of plants such as rye. Conidia of *Botrytis cinerea* placed in droplets on leaves leak out amino acids and sugars in sufficient quantity for phylloplane bacteria to develop in such numbers as to inhibit germination of the conidia themselves. This all occurs on the intact surfaces of healthy leaves. On aphid infested leaves nutrients may be derived indirectly from the host. Host sucrose is converted to melezitose in honeydew and this is used by the phylloplane inhabitants.

The number of phylloplane inhabitants increases with the age of the leaf. This association between population density and age of the leaf is usually explained by increase in leaf exudates with ageing. It is also assumed that the restricted availability of the nutrients is one of the main causes of the relatively poor development of the phylloplane inhabitants on immature leaves. There is also evidence that some of these fungi can slowly degrade the surface waxes and cuticle and so gradually increase the permeability of the epidermis. Their numbers are also far greater on leaves infected by pathogenic fungi, such as rusts and mildews. Four to five times as many colonies of *Sporobolomyces* can be isolated from mint leaves infected with the rust fungus, *Puccinia menthae*, as from healthy mint leaves. Here the injurious effect of the pathogen, especially perhaps the changes in cell permeability, cause an outflow of additional nutrients.

The phylloplane inhabitants are also not uniformly distributed over the leaf surface. Most are more prevalent on the upper surface and are usually more predominant along the veins, frequently with their cells orientated to lie parallel with the vein axis. They also tend to align themselves along the anticlinal walls, as with veins there is a slight depression there. They could be washed into these positions but there may also be more exudates released along the veins; also vein sheath cells may bring nutrients nearer the surface and thus facilitate exudation.

Common primary saprotrophs

The common primary saprotrophs are unable to grow to their full extent in the phylloplane until the onset of senescence. Their pattern of development is restricted until senescence and several rarely or never grow on the green leaf. Their spores accumulate on the leaf prior to senescence and remain dormant until the death of the tissues. If they do germinate they do so only to a limited extent. On senescence they very quickly take advantage of the changing conditions. Sporing colonies of these fungi are ubiquitous on newly dead leaves of the majority of plants. The phylloplane inhabitants and the common primary saprotrophs by no means form distinct groups. *Aureobasidium* and *Cladosporium* have to be included in both groups because, although they grow and reproduce by conidia in the phylloplane, they develop to a much greater extent in the dead leaf. Other fungi, all conidial Ascomycotina, in this group include *Alternaria alternata*, *Botrytis cinerea*, *Epicoccum purpurascens* and *Stemphylium botryosum*. In the Tropics the list can be extended to include species of *Curvularia* and *Nigrospora*. Spores of a great variety of other saprotrophs may also be present on the leaves. They germinate only on the death of the leaves or sometime thereafter.

Pathogens

Two distinct categories can be recognized amongst the pathogens found on leaves. There are those from the Plectomycete Erysiphales, the powdery mildews, which are wholly restricted to the phylloplane except for haustoria in the epidermal cells of the host leaf. All their very extensive mycelium, conidia and ascocarps are borne on the leaf surface. The second category, covering virtually all other pathogens, infect leaves and grow almost entirely within them with only their reproductive structures having access to or being produced on the outside. These latter exhibit all gradations, from those which produce an appressorium from their spore and penetrate immediately, to those which have a prolonged and relatively extensive phase of epiphytic non-parasitic hyphal growth on the leaf surface before they penetrate

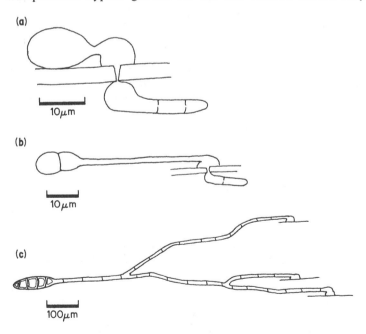

Fig. 2.3 Three different types of growth shown by pathogenic leaf-inhabiting fungi on leaf surfaces. (**a**) *Botrytis fabae*. (**b**) *Mycosphaerella ligulicola*. (**c**) *Cochliobolus sativus*.

(Fig. 2.3). The spores of many of these pathogens remain dormant for considerable periods only germinating as host resistance begins to fall prior to senescence or following a suitable change in the weather. Again no clear distinction can be drawn between pathogens of this latter type and some of the common primary saprotrophs. *Botrytis cinerea* is one of the latter but is also a necrotrophic parasite of some hosts under particular environmental conditions, such as prolonged very high humidities which favour it but not its host. Its conidia may then germinate on the leaf surface and after a phase of epiphytic growth penetrate and bring about a soft watery rot by means of its pectolytic enzymes.

Exochthonous fungi

Spores of pathogens which are unable to infect the leaves on which they have landed may also be present. They may remain dormant or they may germinate before they recognize that they are on the wrong host. They may contribute, like pollen, to the nutrients available on the leaf surface. They could be included with the exochthonous or casual fungi as they are found on leaves but do not grow there. The latter are unable to gain any nutritional advantage from the habitat which is clearly a dead end for many but by no means all. Any soil fungi with air-borne spores may be trapped on leaves and later washed off by rain onto the soil beneath and so they are successfully dispersed. Spores of many coprophilous fungi on herbivore dung are discharged onto grass leaves surrounding the dung and remain there until the grass is eaten by herbivores (Chap. 5). Passage through the gut of a herbivore may be necessary to trigger-off their germination. Direct dung to dung dispersal is abortive. Thus impaction onto leaf surfaces is important if they are to complete their dispersal and life cycle.

The leaf surface as a habitat for fungi

The leaf surface is a most inhospitable niche in both physical and chemical terms for fungi. Although transpiration may mitigate against extreme low levels of relative humidity, the fungi are repeatedly dried by the sun and wind and re-wetted by rain and dew. They are not insulated against temperature fluctuations and as such are subjected to marked and very rapid variations in temperature. Even in temperate climates in relatively still air, leaf surfaces may be 10–12°C above ambient in the sun at one moment and in the next 2° below ambient as a cloud passes over the sun. They are exposed to the harmful ultraviolet component in daylight. Nutrient sources must always be fluctuating and low and competition for them must be severe.

Microbial interactions in the phylloplane

Interest in the leaf surface as a habitat for fungi has centred mainly on the fact that it is here that any pathogen must spend a critical period of time until it can establish and infect. During this time it is not only subjected to such environmental stresses but it may also be subjected to antagonism from the phylloplane inhabitants as well as from the host itself. A great variety of microbial interactions occur in the phylloplane and in the applied field thoughts are turning to consider the possibility of achieving biological control of some leaf pathogens by building up sufficiently large populations of phylloplane inhabitants. This may be an exceedingly difficult objective to achieve but a consideration of some of the research which led to the development of such ideas gives further insight into the biology of phylloplane fungi.

Pollen as a nutrient source and competition for nutrients

The fact that phylloplane inhabitants such as *Cladosporium* and *Sporobolomyces* could benefit from nutrients leaked from pollen grains was convincingly demonstrated by Fokkema (1971). Rye leaves from two separate plots, in one of which the plants had their inflorescences removed or covered, so that no pollen fell onto the leaves below, were taken from plants and washed, twice a week from early June to September, in 1968 and 1969. The washings were plated out onto nutrient agar and the colonies of *Cladosporium* spp. which developed were counted, the assumption being made that each colony arose from a single spore (Fig. 2.4). In 1969, the number of colonies from leaves with pollen rose from 15 to 13 000 cm^{-2} two weeks after flowering. On leaves without pollen the numbers were 10 and 550 respectively. On leaf senescence the colonies recorded from all leaves reached the same levels. The leaves at this stage leaked more nutrients and the stimulating effect of the pollen wore off. The larger number of colonies of *Cladosporium* recorded in 1969 was shown to be due to more pollen on the leaves, a mean of 3450 cm^{-2} as against 300 cm^{-2}. There was frequent rain after flowering in 1968 so that the pollen was washed off. Stimulation in the presence of pollen was not restricted to *Cladosporium*. *Aureobasidium pullulans* and *Sporobolomyces roseus* were also stimulated. For the latter, two weeks after flowering, 33 600 colonies cm^{-2} were recorded from rye leaves with pollen and only 3800 colonies cm^{-2} from rye leaves without pollen.

The Loculoascomycete *Cochliobolus sativus* is a leaf pathogen of rye which makes a variable amount of epiphytic growth before penetration into the leaf (Fig. 2.3c). Fokkema inoculated rye leaves with conidia of *Cochliobolus*, together with pollen and without pollen. The effect of the pollen on successive stages of the infection process and the resultant necrosis is given in Table 2.1. Leaves inoculated with *Cochliobolus* and pollen had a significantly larger percentage of necrotic areas. He attributed this to the pollen leaking out nutrients which greatly stimulated the superficial growth of the mycelium. This increased the inoculum and so more young lesions per unit

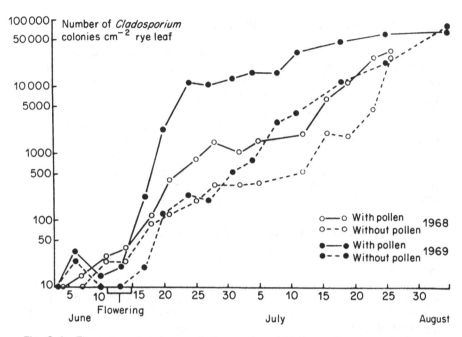

Fig. 2.4 The successive changes in the number of *Cladosporium* spp. colonies per cm² rye leaf during the season. The data are the means of the numbers of colonies from washings of eight penultimate leaves. (After Fokkema, 1971.)

Table 2.1 Effect of pollen on successive stages of the infection process of rye leaves by *Cochliobolus sativus*. (After Fokkema, 1971.)

Experiment	Pollen addition	Time after inoculation			
		2–3 days			7 days
		Mean number of germ tubes/ 100 spores	Mean mycelium length in μm mm^{-2}	Mean number young lesions 10 cm^{-2}	Mean % necrotic area
1	−	23	0	17	2
	+	137	3600	113	58
2	−	67	250	22	2
	+	102	3250	56	35

area were obtained and hence eventual necrosis. In these experiments both the phylloplane inhabitants and the pathogen were relying, in part at least, on the same nutrient source and in nature they might well compete for such a source. Some degree of biological control could be achieved if the phylloplane inhabitants could markedly neutralize the stimulating effect of the pollen when inoculated with both it and the pathogen. This is exactly the effect that Fokkema later observed. The relative inhibitory effect on surface mycelial development of *Cochliobolus* on leaves was depressed by 72% and the extent of necrosis by 75% on rye leaves inoculated with *Cochliobolus*, pollen and *Aureobasidium*, as compared with leaves inoculated with *Cochliobolus* and pollen and *Cochliobolus* and *Aureobasidium* only. Effective competition for nutrients by *Aureobasidium* appears to be an adequate explanation for these reductions.

Antagonistic reactions

A number of such interactions have been reported but competition for nutrients is not the sole explanation for some of these. In some cases inhibitory substances produced by the phylloplane inhabitants may also be involved and normal leaf exudates rather than pollen may be the nutrient source. Pace and Campbell (1974) found that *Aureobasidium pullulans* and *Epicoccum purpurascens* were common in the phylloplane of *Brassica* spp. and that they were antagonistic to the wound parasite *Alternaria brassicicola* in culture. Their growing colonies inhibited mycelial growth of *A. brassicicola* and their germinating conidia inhibited the germination of its conidia. They inoculated leaves of cabbage and Brussels sprout after wounding with *Alternaria brassicicola*, *Aureobasidium pullulans* and *Epicoccum purpurascens* separately and with *A. brassicicola* plus *A. pullulans* and *A. brassicicola* plus *E. purpurascens*. There was an 80–100% infection with *A. brassicicola* alone but none with the two saprotrophs alone. The percentage of successful infections by *A. brassicicola* was reduced when it was inoculated with either of the of the two saprotrophs (Fig. 2.5). The reduction was greater when the saprotrophs were

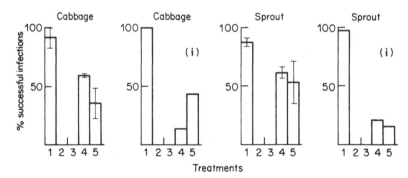

Fig. 2.5 Percentage (mean of three values) of successful infections obtained on cabbage and Brussels sprout leaves. Treatments: **1** = *Alternaria brassicicola* alone; **2** = *Epicoccum purpurascens* alone; **3** = *Aureobasidium pullulans* alone; **4** = *A. brassicicola* + *E. purpurascens*; **5** = *A. brassicicola* + *Au. pullulans*. The vertical bars are the standard deviations from the mean. (i) = after pre-incubating the antagonist for 14 h. (After Pace and Campbell (1974). Reproduced by permission of the British Mycological Society.)

inoculated 14 h prior to *A. brassicicola*. Both *A. pullulans* and *E. purpurascens* were capable of active growth on the leaf surface so might compete with the pathogen for nutrients in the form of leaf exudates. This could explain why they were more effective when inoculated 14 h prior to the pathogen. But they may also produce inhibitory substances. Some evidence for this is that a 50% reduction in successful infections was obtained when conidia of *A. brassicicola* were suspended in a culture filtrate of *A. pullulans* and used instead of water for inoculation. The culture medium itself enhanced infection.

Towards biological control

These sorts of antagonism must obviously be having some effect in the field but the question to be answered in terms of achieving biological control is how can their effects be maximized? One approach is to manipulate the system to stimulate the phylloplane inhabitants. Biological control using the indigenous population, rather than introducing others, would be possible if a sufficiently large population of antagonistic phylloplane inhabitants could be built up. Bashi and Fokkema (1977) have shown that continuous high humidities and nutrients in excess of those exuded by leaves are necessary to maintain a phylloplane population of *Sporobolomyces* dense enough to have sufficient antagonistic potential to control *Cochliobolus*. *Sporobolomyces* is particularly sensitive to low relative humidities. Populations on leaves decreased markedly when maintained at 65% RH. Any added nutrients would have a stimulatory effect on the pathogen as well. To be effective in stimulating only the saprotroph nutrients would have to be added just prior to any significant build up of spores of the pathogen on the leaf surface. This would require accurate disease forecasting. But even so it would be almost impossible to maintain in the field the necessary continuous high humidities. However if such control was possible, it would be applicable only to those pathogens which rely on the absorption of exogenous nutrients to enable them to make superficial mycelial growth on the leaf before penetration. Pathogens which normally penetrate the leaf immediately after germination or which have a very restricted superficial mycelial growth are probably less susceptible to competition for nutrients.

The alternative approach is to allow the natural control measures to proceed and to avoid the indiscriminate use of fungicides which may affect the phylloplane inhabitants more than the pathogen. Pace and Campbell (1974) noted that the systemic fungicide Benomyl gave a good control of many diseases but not the leaf spot of brassicas caused by *Alternaria brassicicola*. The pathogen is resistant to it. The two antagonistic saprotrophs which they used, *Aureobasidium pullulans* and *Epicoccum purpurascens* are inhibited by Benomyl. Therefore use of this fungicide could make the disease worse. Fokkema found that *Cochliobolus* is also relatively resistant to Benomyl, and that inoculation of rye leaves with *Cochliobolus* just after flowering (i.e. with pollen), resulted in 60% less necrosis on water-sprayed leaves than on Benomyl-sprayed leaves. At the time water-sprayed leaves had a natural phylloplane population of 10 000 spores cm^{-2} and Benomyl-sprayed leaves only 1200 spores cm^{-2}. This implies that the Benomyl had reduced the antagonistic capacity of the phylloplane inhabitants but it also provides direct field evidence for naturally occurring biological control.

Antagonism via lysis, antibiotic production or pH changes

Other forms of antagonism are exhibited in the phylloplane. Bacteria may lyse fungal spores. Chitinolytic enzymes are usually involved. Lenne and Parberry (1976) noted clusters of bacteria surrounding lysed conidia and germ tubes of the pathogen *Colletotrichum gloeosporioides* on leaf surfaces. Appressoria are necessary for penetration to occur (Chap. 10). The bacteria failed to lyse these. They have melanized walls and there are numerous reports of melanized structures resisting the lytic action of bacteria. The production of appressoria was enhanced in the presence of bacteria but was reduced by added nutrients, such as 1% glucose peptone solution. The stimulated production of appressoria in the presence of bacteria is a normal response of the fungus to a hostile environment. Desiccation and starvation also cause appressorial formation. This response serves as an important short term survival role during the infection phase. It should be noted that in this particular case added nutrients increased germ tube growth but fewer appressoria were formed. Since the latter are necessary for penetration, added nutrients may in this case enhance disease control.

Several phylloplane inhabitants, such as *Aureobasidium* and *Sporobolomyces* have been shown to produce antibiotics in culture although there is no direct evidence that they play a role *in vivo*. Although antibiotic production by bacteria on leaf surfaces does not appear to be very widespread, some bacteria have been shown to produce antifungal peptides which, under experimental conditions at least, reduce incidence of disease caused by a number of species of *Colletotrichum*.

Some fungal leaf pathogens are very sensitive to pH changes. In *Septoria nodorum*, for instance, spore germination is inhibited below pH 6. Conidia of this fungus, placed around the edge of a growing colony of *Botrytis cinerea*, failed to germinate. The pH of the medium fell to below 6 in advance of the hyphal tips of *Botrytis*. Such a mechanism could operate in the phylloplane.

Fungistatic substances produced by leaves

In addition to nutrients, leaves of many plants may exude fungistatic substances which cause inhibition of spore germination or restriction of germ tube growth. Phenols are the most widely known fungistatic substances produced by leaves. They are responsible for the inhibition of spore germination of the apple scab fungus, *Venturia inaequalis*, on some apple cultivars. Gallic acid has been identified as an antifungal component in droplets of dew obtained from sycamore leaves. Apart from these substances formed within the leaf cells and exuded onto the surface, some constituents of the cuticular waxes may also be fungistatic. An acidic ether-soluble fraction from the wax of apple leaves inhibits the growth of the apple mildew fungus, *Podosphaera leucotricha*. The properties of waxes on leaves will also affect the exudation of both nutrients and anti-fungal substances. Waxes with high proportions of more hydrophobic constituents will tend to limit the movement of exudates to the surface.

Thus at the leaf surface a series of complex interactions occur between pathogen/host/phylloplane inhabitants/environment. Numerous aspects and the outcome of many of these interactions are still to be discovered, but it is evident that the phylloplane inhabitants act in some sort of buffering capacity against some pathogenic fungi at least.

Distribution of the common primary saprotrophs

Eventually the leaf senesces either naturally or prematurely after supporting, in some cases, one or more pathogens. Of the multitude of fungal spores of a vast array of a species which are impacted onto leaf surfaces only relatively few succeed in colonizing the leaves as they senesce and grow as active saprotrophs within the leaf tissues after death. These common primary saprotrophs are virtually ubiquitous colonizers. On most leaves such as those of deciduous trees, shrubs, herbs, grasses including cereals and even bracken, most are usually present and exceptions are difficult to find. Pine needles are a particularly selective substrate and of these fungi only *Aureobasidium pullulans* is ever at all common. In the tropics, *Alternaria alternata* is less common and is replaced by *Nigrospora* spp., especially *N. sphaerica*, and *Curvularia* spp., especially *C. lunata*, as is evident from examining senescent leaves of guinea grass (*Panicum maximum*) and banana (*Musa sapientum*). The differences are also reflected in the comparison of the dry air-spora of tropical and temperate climates. The association of these particular fungi has been noted on other substrates, such as cereal stubble and cotton fabrics exposed to the weather. The blackening of the ears of cereals in a damp season is caused mainly by *Alternaria*, *Cladosporium* and *Epicoccum*. Christensen and Kaufmann (1965), in their studies on the deterioration of grain, designated these and others, such as *Chaetomium*, *Fusarium* and *Rhizopus* spp., as 'field fungi'. This is an appropriate term as they are almost always and constantly associated with exposed freshly decaying green parts of plants. On leaves they are usually associated with one or more other saprotrophs which are more restricted in the range of leaves which they colonize. These restricted primary saprotrophs may be confined to a particular host genus or a related group of plants. *Readeriella mirabilis* and *Piggotia stellata* appear to be restricted to *Eucalyptus*. Several species of *Leptosphaeria*, such as *L. microscopica*, are restricted to the Gramineae and *Fusicoccum bacillare* and *Sclerophoma pithiophila* are both very common on pine needles but the latter, at least, is also found on other coniferous leaves. In many of these substrate specificity might be synonymous with, and explained by, host specificity. Many of these, although very active saprotrophs, may have an additional advantage in that they can gain access as parasites. *S. pithiophila*, for instance, has been associated with the defoliation of the current year's needles of *Pinus sylvestris*.

As a group these common primary saprotrophs may be well-established in leaves long before leaf-fall. For example *Cladosporium herbarum* often colonizes and produces conidia on damaged necrotic parts of beech leaves in June, within two months of their unfolding. The duration of their persistence on leaves once they are in the litter is dependent upon many variables, one of which is the texture and another the composition of the leaves. In general, tree leaves, such as those of ash and sycamore, which decompose and disappear rapidly from the litter, support a more substantial growth of these common primary saprotrophs for a shorter time than do leaves of beech and oak which persist much longer in the litter. On beech leaves, for example, *C. herbarum* persists in high frequency through the winter after leaf-fall until the following June and disappears after September.

Similar sequences can be found on other substrates. Primary saprotrophs are the first fungi to appear on flowering stems of cocksfoot, *Dactylis glomerata*. They are present on the basal leaves in early summer and progress up the stems as successive leaves senesce. They are well-established by July and August on the upper leaf sheaths and internodes of stems which flowered in late May and June and they persist there until the following summer. On nettles, *Urtica dioica*, primary saprotrophs colonize the upper leaves in August or September of the year of flowering, at the onset of basipetal senescence. They again persist throughout the winter until the following spring and summer.

Attributes of the common primary saprotrophs

The intriguing aspect of this particular facet of fungal ecology is to ponder why so few of all the fungi are equipped to assume this role of primary saprotrophic colonizers of such exuberantly plentiful substrates.

Nutrients

In the well-known schema for fungal successions proposed by Garrett (1963), the primary saprotrophs to invade are 'sugar fungi'. They are non-cellulolytic and rely upon readily available sugars, such as hexoses and pentoses, and other carbon sources simpler than cellulose, such as pectins and starch. These fungi also normally possess a high mycelial growth rate and a capacity for rapid spore germination. The classic example of such fungi is the Zygomycete Mucorales, common on herbivore dung (Chap. 5). Primary saprotrophic sugar fungi are usually very ephemeral because of the transient nature of their substrate. The persistence of the common primary saprotrophs for months on leaves would suggest that they are not confined to such ephemeral substrates. The ability to utilize cellulose is often regarded as essential for saprotrophic fungi and the majority, except most Zygomycotina

and Mastigomycotina, can do this. Of this particular group of leaf saprotrophs only *Aureobasidium pullulans* is non-cellulolytic. It probably relies on
pectic substances for its carbon sources and this ability is often used to
explain its role as a primary colonizer. None of the others is markedly cellulolytic when compared with some of the Basidiomycotina which later colonize
leaves in the litter layer. For instance, Hering (1967) inoculated oak leaves
sterilized by γ irradiation with *A. pullulans, Cladosporium herbarum* and
Mycena galopus, a Hymenomycete agaric from the leaf litter, and measured
loss in mass after six months at 9–15°C. The two former brought about a loss
of 2 and 4% respectively and the latter 15–20%. Not all the loss in mass was
of cellulose but in the latter case the loss corresponded with the utilization of
about one sixth of the total cellulose present. On filter paper cellulose,
various isolates of *Alternaria alternata* brought about losses of 4–8% in 14
days and *Epicoccum purpurascens* about 4%. For comparison, under the
same conditions the vigorously cellulolytic *Chaetomium globosum* brought
about a 10% loss. It must be remembered that mass loss methods measure
only the amount of substrate, in this case cellulose, respired and lost as carbon dioxide and water and not the amount incorporated into fungal material.
The cellulolytic ability of these fungi thus varies and they all probably use
simpler carbohydrates, such as sugars and starch, as long as they last and then
go on to utilize cellulose even if to a limited extent and slowly. Thus they
persist. This may be placing undue emphasis on their carbohydrate nutrition
to the neglect of their nitrogen requirements. The nitrogen supply might be
extremely critical in determining their distribution. In culture they can all use
nitrate, ammonia or amino acids as their sole nitrogen source but nothing is
precisely known as to what sources are available to them within the leaf.
Indications of the over-riding limitations of their nitrogen supply are seen
when leaves are amended with an available source. Foliar applications of 5%
urea solution, after harvest but before leaf fall, prevent ascocarp development in the apple scab fungus, *Venturia inaequalis*, on the overwintering
leaves and is used as a control measure to limit the ascospore inoculum available to infect the newly emerging leaves in the following spring. Birchill and
Cook (1971) in studying the mode of action of the urea demonstrated that
both chemical and microbial changes occurred in the leaves after treatment.
Marked alterations occurred in the composition and density of fungal and
bacterial populations present on treated leaves. The urea in particular enormously increased the relative abundance, as assessed by the number of
conidia produced, of both *Cladosporium* spp. (Fig. 2.6) and *Alternaria* sp.
So marked was the development of the conidia and conidiophores of *Cladosporium* that they could be seen with the naked eye as olive green lawns. Many
times more conidia were produced overall on the treated leaves suggesting
that the added nitrogen enabled them to utilize more carbon sources and thus
outcompete *Venturia* for substrate. Application of urea to fallen pine needles
also dramatically changes the fungal succession. *Cladosporium herbarum*,
rather than being an occasional inhabitant, is again stimulated to develop to
such an extent that its conidiophores may cover the needles as a dense felt and
other common primary saprotrophs, such as *Epicoccum purpurascens*,

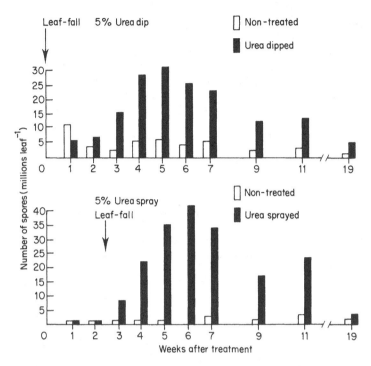

Fig. 2.6 Effect of urea dip and urea spray applied 17 October 1967 on numbers of *Cladosporium* spores washed from overwintering leaves. (After Birchill and Cook, 1971.)

which does not normally occur on pine needles, are stimulated to develop by the urea. The mode of action of the urea is not known in this case but is complex; its property of acting as an alkali may be one important aspect of its effect. For example, several agarics which have not been recorded from pine litter appear when plots are treated with urea or alkalis. An example is *Myxomphalia maura* which is characteristically found on the alkaline ash of bonfire sites on acid soils in coniferous woods. It is not found on woods on alkaline soils so is not a calcicole. *M. maura* is markedly encouraged by the addition of lime to pine litter. Pine needles treated with sodium carbonate are also colonized by *C. herbarum* and *E. purpurascens*. Urea and alkalis both produce similar effects on the litter. They both cause it to darken, become water-soaked and raise the pH from about 3.5–4.0 to 5.5–6.0. They both bring about the release of ammonia from the litter and its use as a nitrogen source may be another important factor inducing these changes.

Growth rates

These saprotrophs are not primary colonizers solely because they grow faster than any other would-be colonizers. Their only attribute with regard to growth rate is their great variability. *Aureobasidium pullulans* produces its slimy conidia very rapidly, but its yeast-like colonies are relatively slow growing. *C. herbarum* also sporulates rapidly but its growth rate is even slower. *Botrytis cinerea* grows very rapidly and *Alternaria alternata* and *E. purpurascens* not so rapidly but faster than *C. herbarum*.

Tolerance to desiccation

Senescing leaves on the tree and recently fallen leaves are very prone to drying out and also subject to strong sunlight. Webster and Dix (1960) compared the growth rates, latent period for germination, germ tube growth rate at 100% RH and the lowest RH at which spore germination occurred in three primary colonizers with two later secondary colonizers, *Torula herbarum* and *Tetraploa aristata*. They found that there was little difference between the capacity of the mycelium of the various colonizers to grow at low humidities and the primary colonizers did not make better growth at low humidities. But it can be seen from Table 2.2 that under favourable humidities (100% RH), *A. alternata* and *E. purpurascens* not only grew faster than the secondary colonizers but also had a shorter latent period before germination and their germ tubes grew faster. These features, coupled with the fact that their conidia can germinate at lower relative humidities, would give them an advantage over the others in that their conidia would germinate under less ideal conditions of humidity and they would quickly exploit, by virtue of their more rapid growth rate, any changes to more humid conditions.

Because of the rapidly fluctuating conditions on the leaf surface, germinating conidia may rapidly dry out before penetrating into the leaf. Diem (1971) has investigated the survival at low humidities of germinating conidia of *C. herbarum*, *A. alternata* and some casual inhabitants of the phylloplane. He

Table 2.2 Mycelial growth rate, latent period for germination, growth rate of germ tubes at 100% RH and lowest RH at which spores germinated. (After Webster and Dix, 1960.)

Primary colonizers	Mycelial growth rate (mm day^{-1})	Latent period (h)	Germ tube growth rate (μm h^{-1})	Lowest RH at which germination occurs
Cladosporium herbarum	2.96	6–12	4.2	89%
Alternaria alternata	6.41	3–6	29.1	89%
Epicoccum purpurascens	6.45	0–3	31.6	92%
Secondary colonizers				
Torula herbarum	1.41	12–18	0	Water
Tetraploa aristata	3.34	12–18	2.4	98%

found that germinating pigmented conidia, such as those of *Clàdosporium* and *Alternaria*, were more resistant than germinating colourless conidia of *Aspergillus* and *Penicillium*. The germ tubes of *Cladosporium* were remarkedly resistant. Some 90% grew on at 100% RH after 8 h in a desiccator over anhydrous calcium chloride and 99% did so after being kept at 40% RH for 8 h. This would indicate that if they germinated in the more humid conditions of the night and had not penetrated into the leaf by the morning they could survive the drier conditions of the day. Germ tubes of the conidia of *Alternaria* were equally resistant but some failed to grow on after periods at relative humidities below 65% but the conidia either germinated again from another cell or from a lateral branch from below the damaged part of the germ tube. In contrast, the germ tubes of the conidia of *Aspergillus* and *Penicillium* were no longer viable after periods at 85% RH. Species with coloured conidia are thus more likely to be successful in the phylloplane and as subsequent primary colonizers of the leaves. But it should be noted that the conidia of *Aureobasidium* and *Botrytis* are not pigmented. The biotrophic Erysiphales which produce their mycelium on leaf surfaces also have colourless hyphae and conidia (Chap. 10). Thus pigmentation is a useful but not an essential attribute to possess.

Primary saprotrophs also show an equally remarkable tolerance to desiccation in their hyphal tips as distinct from germ tubes. Hyphal tips are very delicate structures but in these fungi they survive periods of extreme desiccation, some as long as three weeks, above a saturated solution of potassium nitrate (a_w 0.45, Chap. 6). Thus they can rapidly exploit the return to favourable conditions of humidity with no apparent loss of previously synthesized biomass. Other fungi do not show this ability. This attribute may be critical in enabling such fungi to tolerate cycles of wetting and drying.

Survival structures

Once established on the leaf surface, most of the common primary saprotrophs produce some form of pigmented survival structure: *Cladosporium herbarum* minute micro-sclerotia; *Botrytis cinerea* and *Epicoccum purpurascens* sclerotia; and *Aureobasidium pullulans* aggregates of chlamydospores (see Fig. 2.2b). All have a pigmented mycelium. Such structures and pigmentation protect against desiccation, ultra-violet light and microbial lysis.

It is thus clear that the common primary saprotrophs possess a multiplicity of attributes by which they have become successfully adapted to this relatively inhospitable niche with each fungus possessing its own particular complex of attributes, not all necessarily the same.

As suggested, in temperate climates some of the common primary saprotrophs produce ascocarp initials in the late autumn in the year of leaf-fall as do a number of leaf pathogens, such as *Apiognomonia errabunda* on beech and *Venturia inaequalis* on apple. Ascospores are discharged from these over the period early April to early June. This is the time when the next crop of leaves is unfolding. The initially spore-free leaves become impaction sites for air-borne ascospores and under favourable conditions infection occurs. Such a life history is of particular significance in leaf pathogens with restricted periods of spore formation and release and where the host virtually frees itself of infection by shedding all its leaves prior to its dormant season. The requirement for an overwintering phase, a period of low temperature (5–8°C), before ascocarp initials mature is very common in these fungi. The maturation and release of the ascospores thus coincide with the breaking of bud dormancy of the host. Thus teleomorphic states of *Aureobasidium pullulans* (*Guignardia fagi*) and *Cladosporium herbarum* (*Mycosphaerella tassiana*), which are common on fallen leaves, may be regarded as additional survival structures adding an ascospore inoculum to the conidial inoculum available in the spring.

Subsequent colonizers and leaf decay

These initial colonizers gradually disappear, being replaced by other leaf-inhabiting saprotrophs which begin to reproduce in the late summer of the year after leaf-fall, reach a maximum in the autumn and persist over the winter, until the spring. These include a very wide variety of conidial fungi, such as *Polyscytalum fecundissimum* and *Chalara cylindrospora*, Ascomycotina such as *Microthyrium fagi*, and *Helotium caudatum*, and Basidiomycotina with minute basidiocarps, such as *Lachnella villosa* and *Pistillaria pusilla* on beech leaves. With fragmentation in the final stages of decomposition, the fungal flora becomes dominated by typical soil-inhabiting fungi, mainly Zygomycete Mucorales, especially species of *Mucor* and *Mortierella*, and conidial fungi, such as species of *Penicillium* and *Trichoderma*, together with litter-decomposing Hymenomycete Agaricales, such as species of *Collybia* and *Mycena*.

The soil-inhabiting fungi grow up from the soil via the continuum of organic debris. The role which they play in the decomposition process has not been fully elucidated. At this stage the Mucorales are certainly not using any simple carbohydrates initially present in the leaves as these would have been utilized already. They could be living in association with the cellulolytic Agaricales as commensals by taking a share of the hydrolytic products of cellulose and thus acting as secondary saprotrophic sugar fungi rather than primary ones. Alternatively, they could be primary colonizers of the wealth of faecal pellets produced by the micro-fauna, especially mites, as they are on pellets of *Glomeris* (Chap. 5). The H layer of the soil is particularly rich in chitin in the form of hyphal wall fragments and exoskeletons of insects or other chitinized remains of the micro-fauna. Species from several common genera of soil-inhabiting fungi, including *Mortierella*, *Penicillium* and *Trichoderma*, have the ability to break down this very resistant substrate and their activity may well represent one of the final stages of the mineralization of primary and secondary organic materials in the soil.

Decomposition of pine needles

The time period between leaf-fall and the final decomposition of a leaf varies enormously. In cool North temperate pine forests it may be 10 years or more, in ash and sycamore under 1 year and in tropical forests mere weeks. Pine needles are extremely durable and decay very slowly. Their decomposition most often results in the formation of a mor type of soil. The needles are shed mainly in August and September and there is an accumulation of considerable bulk of leaf litter, each successive leaf-fall burying the previous one so that a stratified litter layer is produced. The animals in the litter are sufficiently small not to disturb this stratification and such a litter layer well illustrates the diversity of organisms, fungi and animals, involved in the decomposition process. A very considerable amount of potential energy is available for micro-organisms in this litter. In *Pinus sylvestris*, production of needles accounts for about one third of the total productivity and accounts for about 60–80% of the total litter. Although the decay process is a continuous, if fluctuating, one, it is convenient to recognize a number of stages. The A horizon may thus be divided into L, F_1, F_2 and H layers. The L layer consists of freshly fallen, undecomposed needles, light brown to buff in colour and others somewhat darker in colour, which have fallen earlier. Needles remain in this layer for about six months. They all have a high tensile strength, a relatively low but fluctuating moisture content and form a loose, uncompacted layer on the litter surface. In this layer the needles are very susceptible to drying out and conditions are unfavourable for continuous fungal growth. In the upper parts of the F_1 layer, the needles are grey, becoming dark brown with depth but recognizable as needles. Their tissues become softened and they have a low tensile strength and a high moisture content. They remain in this layer for about two years. Below, in the F_2 layer, the character of the needles again changes. They are greyish, fragmented and compressed, but again still recognizable as needles. The mesophyll collapses and most bear dark amorphous faecal masses of the micro-fauna. Eventually the remains of the needles enter the H layer which consists of an amorphous mass of faeces and the remains of both the micro-fauna and fungi, the needles having undergone complete physical reduction. Below this layer is an intimate mixture of humus and mineral soil.

Two factors that may greatly influence the sequence of decomposer fungi on the needles are the time at which they fall and their previous history. Pine needles are far from being a homogeneous entity and at needle fall vary in age, physical structure, nutrient content and the presence or absence of fungal colonizers in or on the needles. The needles have a very thick, waxy cuticle and support a much sparser population of phylloplane inhabitants. *Sporobolomyces roseus*, although present on most attached needles, occurs in very low frequencies only. This contrasts markedly with its abundance on leaves of deciduous trees and herbaceous plants. It decreases rapidly on needle fall, whereas some other yeasts, such as *Bullera* spp., increase in frequency and persist. A number of other mycelial fungi such as the conidial *Sclerophoma pithiophila* may grow and sporulate on the leaf surface.

Vigorous ones, such as *Lophodermella sulcigena*, an apothecial Ascomycete, and *Coleosporium senecionis*, a rust, cause premature needle cast, either directly or by predisposing first year needles to infection by secondary pathogens. For example, *L. sulcigena* infects young first year needles and predisposes them to infection by *Hendersonia acicola* or *Lophodermium pinastri* and finally *Naemocyclus niveus*, which cause the needles to fall in their first summer. Such weak pathogens may colonize the needles directly but spread very little until senescence. They may also gain access via tissues damaged by insect pests. Living needles may also be colonized by *Fusicoccum bacillare* or *Sclerophoma pithiophila*. Needles infected by either of these two conidial fungi soon die and turn brown but remain attached to the tree. Such needles again fall in the summer. *S. pithiophila* is also a frequent colonizer of needles containing high nutrient levels, such as first year needles, shed while still green and of needles of felled pines. Clearly pine needles can fall at varying times of the year and may already be colonized by a variety of fungi which have already initiated the process of decomposition. *Lophodermella sulcigena* actively decomposes the mesophyll tissue and *Hendersonia acicola* may remove much of the cellulose, reducing the needle to a skeleton of epidermal waxes and lignified tissues. *Lophodermium pinastri* produces pigmented diaphragms across the needles delimiting the extent of its colonization (Fig. 2.7a). Such parts later escape extensive internal attack by saprotrophic needle-inhabiting fungi. This is often attributed to their inability to penetrate the melanized diaphragms but in culture at least *Lophodermium* produces powerful antifungal antibiotics and they may also play a part in restricting saprotrophic colonization. Such parts of the needles decay more slowly than uninfected parts and as a consequence accumulate in lower layers of the litter. The saprotrophic colonization of naturally fallen needles and needles shed after parasitic attack may thus be distinct.

Most needles, the bulk being second and third year ones, falling in August and September, are colonized by *L. pinastri* and somewhat fewer by *S. pithiophila*. Soon after needle-fall, a dark brown to black hyphal network develops on the surface of the needles (Fig. 2.7). A number of fungi may be involved, including the conidial *Sympodiella acicola* and *Helicoma monospora* (Fig. 2.7b and c) and, in drier situations, the ascocarpic *Kriegeriella mirabilis* (Fig. 2.7b and d). There is no apparent penetration of the needles by the surface hyphae although erosion of the needle surface does occur. The hyphal network shows marked linearity with the hyphae growing longitudinally along the cell boundaries. Internally the needles become colonized by *Desmazierella acicola*, which produces its conidial state from compacted, pigmented, hyphal cushions formed over the stomata. In spite of intensive grazing of the fungi by the micro-fauna, including mites, springtails and enchytraeid worms, all become more frequent as the needles become incorporated into the more moist regime of the F_1 layer. They persist for two years in the F_1 layer, that is for up to 2½ years after needle-fall. *D. acicola* produces crops of conidiophores in both the first and second summers after needle-fall. *L. pinastri* produces its ascocarps in the L layer over the period January to May, providing the inoculum to infect further needles on the trees. After about 10 months in the L and F_1 layers, it too disappears. In the F_2 layer, which the needles enter in the third year after needle-fall, the micro-fauna assume more importance. The external feeders continue to graze upon the fungal hyphae and reproductive structures whilst the internal feeders rapidly comminute needles attacked by *L. pinastri* and *D. acicola*. Any needle fragments which escape extensive internal colonization become colonized by more general litter inhabitants such as species of *Penicillium* and *Trichoderma* and by pine litter-inhabiting agarics. Needles remain in this layer for about 7 years, by which time the fungi and fauna reduce them to an amorphous mass, typical of the humus layer. The role of agarics in the decomposition of pine needle litter has not been extensively investigated. It is usually assumed that they colonize the litter when it is in a relatively late stage of decay. This is not always so. The tiny agaric, *Marasmius androsaceus*, is very common in pine needle litter. It is often called the 'Horse hair fungus' because its stalk is shiny black, like horse hair and is of about the same diameter. *M. androsaceus* colonizes the needles very shortly after needle-fall. Its delicate black, cotton-like rhizomorphs grow up from previously colonized needles below, binding them together in a loose tangle. Dense masses of basidiocarps may appear on the needles in the litter, any time from May to November (Fig. 2.7e, 2.8). It is both strongly cellulolytic and ligninolytic and causes very extensive internal decomposition. The role of such Basidiomycotina should not be underestimated. Their mycelium is often prolific in both the L and F layers. Long lists of agarics have been recorded from pine woods. Richardson (1970) has estimated the total productivity of these in a woodland of *Pinus sylvestris* in Scotland, to be between 0.25–0.5 million basidiocarps 10^4 m^{-2} y^{-1}. The majority are produced from August to September. However, because of our inability to distinguish species of Basidiomycotina from their mycelium and because we know insufficient about the biology of some of these, the problem is to assess the relative contributions of the litter decomposers and the mycorrhizal fungi. Since many agarics in the litter decompose both cellulose and lignin, it is probably delignification that reduces the needles to a greyish colour in the F_2 layer.

The role of the litter micro-fauna

As in other litter systems, the micro-fauna are important agents in the decomposition process. Mites and springtails cause considerable comminution of the needles and in so doing convert them to faecal pellets. It has been estimated that a pine needle with a surface area of 180 mm^2 would have a surface area of 1.80 m^2 after comminution to faecal pellets by micro-arthropods. Such comminution would present a much larger surface area to microbial enzymes and thus be expected to increase the decomposition rate. This may not always be so. Orobatid mite pellets persist longer than the source from which they are derived. This may be due to the nature of the substances cementing the particles of the pellets together and digestion by the animal of the more easily decomposable components of the litter. It is clear, however,

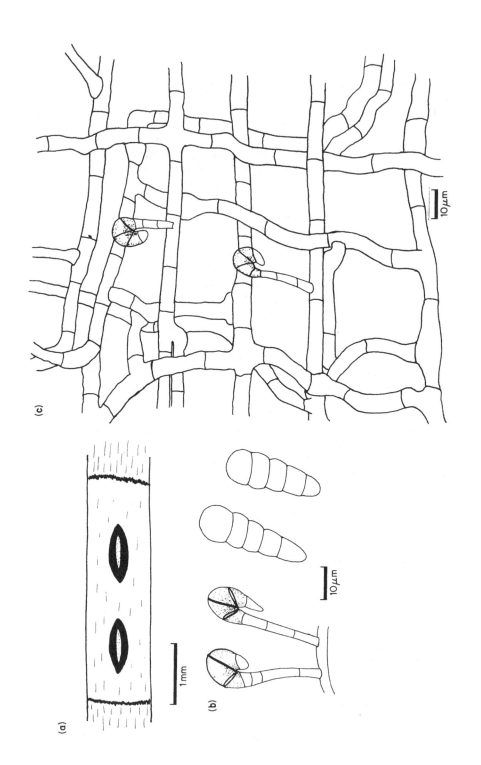

(a)

(b)

(c)

10 μm

10 μm

1 mm

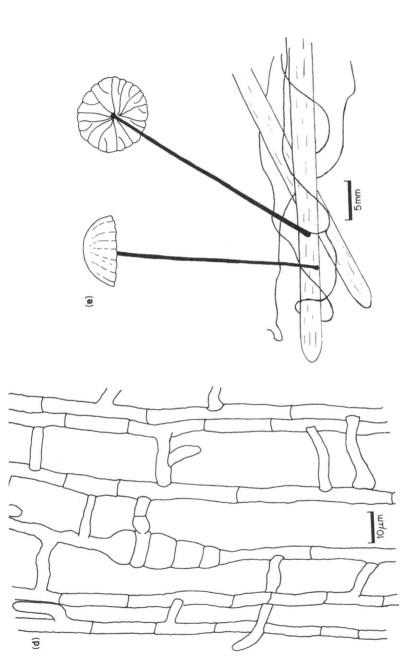

Fig. 2.7 (*Left*) (**a**) Two ascocarps of *Lophodermium pinastri* and diaphragms across a pine needle. (**b**) Conidia of *Helicoma monospora* and ascospores of *Kriegeriella mirabilis*. (**c**) Conidia and hyphal network of *Helicoma monospora* on the surface of a pine needle. (*Above*) (**d**) Ascospore and hyphal network of *Kriegeriella mirabilis* on the surface of a pine needle. (**e**) Thread-like rhizomorphs and basidiocarps of *Marasmius androsaceus* on pine needle litter.

Fig. 2.8   Basidiocarps of *Marasmius androsaceus* on pine needle litter.

that the water holding capacity of the pellets is higher and the rate of evaporation from the pellets decreased. This creates a higher and more stable moisture regime which again should favour microbial activity.

The enchytraeid worms, which are very abundant is podsols in northern coniferous forests, in addition to their grazing activities, play a vital role in the absence of earthworms in this mor type litter, in mixing the amorphous remains with the mineral soil. The major group of animals involved in the decomposition of pine needle litter are mites, many of which are strictly mycophagous, with some showing marked preferences for particular fungi. They are most abundant in the moister F layers where fungi are also more active. Protozoa and nematodes also occur in pine litter and feed on the contents of living fungal hyphae. There is no clear evidence to support or refute the hypothesis that mycophagy by the micro-fauna stimulate the growth of fungal mycelia.

Perhaps the most important effect of the micro-fauna is that they act as a reservoir of plant nutrients as they do in decaying wood, gradually making available the minerals which have become immobilized in the fungal hyphae. The large number of different species involved and their varying life spans mean that the nutrients contained in their tissues are only gradually mineralized (Chap. 3).

3

Fungi as decomposers of wood

Perennial woody plants are the predominant vegetation on earth. Forests form the climax vegetation of all parts of the world except where temperature and moisture extremes limit plant growth. Forests also contain the greatest biomass varying from 500 Mg 10^4 m^{-2} in tropical rain forests to 100–300 Mg 10^4 m^{-2} in northern temperate coniferous forests. Perennial woody parts above ground make up about three-quarters of this biomass. Woody tissues thus provide the bulk source of organic carbon for decomposer heterotrophs. Fungi are the major group of organisms responsible for wood decay and a number of groups of fungi are solely wood-inhabitants. They exist entirely on the components of wood. In this chapter, a detailed consideration of these fungi will further illustrate the versatility of fungi as saprotrophs.

The structure and components of wood

Technically, wood is the xylem cylinder inside the bark of trees. In many trees it consists of an outer, light coloured sapwood and an inner, darker heartwood. The bulk of the wood consists of dead and empty lignified vessels and/or tracheids and fibres but it also contains xylem parenchyma. Much of the parenchyma in the sapwood remains alive and unlignified and acts as a food store, mainly for starch but soluble sugars, proteins, peptides, and amino acids, lipids, nucleic acids and vitamins, such as thiamine, are also present. Once the tree is dead, these afford substrates for a wide variety of fungi but are all relatively minor and ephemeral components.

Wood consists of three major components – 40–60% cellulose, 10–30% hemicelluloses and 15–30% lignin. Although the biological decomposition of lignin is of critical importance in the continuous cycling of carbon, its degradation is incompletely understood. This can be attributed to many facts; not least of these are our lack of understanding of its precise chemical structure, the diversity of its structure in different woods, our inability to produce a pure form of lignin for cultural studies and a suitable assay for lignin degradation, the availability of a very potent lignin degrader, and the general cellular and chemical complexity of wood.

Lignin, in addition to making up about one-quarter of the dry mass of wood, is undoubtedly the structurally most complex of all the polymers and the most resistant of all to microbial decomposition. It is a three dimensionally branched aromatic polymer, formed by the oxidative polymerization of three different building blocks, not just one as in cellulose (Kirk, 1971). The building blocks are the phenyl propanes: coumaryl, coniferyl and sinapyl alcohol (Fig. 3.1a). The lignin of different plants may contain different proportions of the three building blocks. Conifer lignin consists of mostly coniferyl alcohol, with small amounts of coumaryl alcohol and minor amounts of sinapyl alcohol. In angiosperm lignin there are approximately equal amounts of coniferyl and sinapyl alcohol and minor amounts of coumaryl alcohol. These phenyl propane units are built up into a branched polymer by covalent bonding involving three major linkage types. By far the commonest, making up 40–60% of the total, and most important linkage type is the arylglycerol-β-aryl ether type. Phenylcoumaran structures form 10–20% of the linkage types and biphenyl structures another 10–25% (Fig. 3.1b). Thus in lignin there are three functional monomers, varying in proportions in the various lignins, and three major linkage types, but also other minor ones. There is no regular repeating unit as there is in starch or cellulose, nor are there bonds which are easily hydrolysed. Because of this structural complexity, decomposition must necessarily differ from that of most natural polymers where there is usually straight cleavage, often by hydrolysis, to produce the monomers. There is also the possibility of microbial enzymes bringing about a variety of limited changes to the intact molecule and only partially degrading it to substances which pass with little further change into humic materials. It appears that only the so-called white-rot fungi can completely decompose lignin to carbon dioxide and water. Lignin imparts rigidity and resistance to mechanical stress in woody plants and also resistance to microbial attack. Nevertheless it is degraded in natural environments but degradation is a very slow process.

Types of wood decay – white, brown and soft rots

As with other decomposing substrates, the form in which wood is presented to micro-organisms and the environment in which it occurs have a major influence on the path that degradation takes. The degradation of wood in the form of trunks and large branches of trees above the soil may be very different from that of the woody tissues of leaves and small roots in the soil. This may be quite different again from logs submerged in the sea.

The fungi which cause the decay of large masses of wood, such as tree trunks above ground, have been most thoroughly investigated because such wood is the natural material utilized in greatest quantity by man and any fungi which attack it are of potential economic importance. Three types of wood decay have been recognized – white, brown and soft rots. In white rots, the wall polysaccharides, such as cellulose and hemicelluloses, are attacked more or less simultaneously with the lignin and the wood becomes markedly paler and fibrous as the pigmented amorphous lignin is removed.

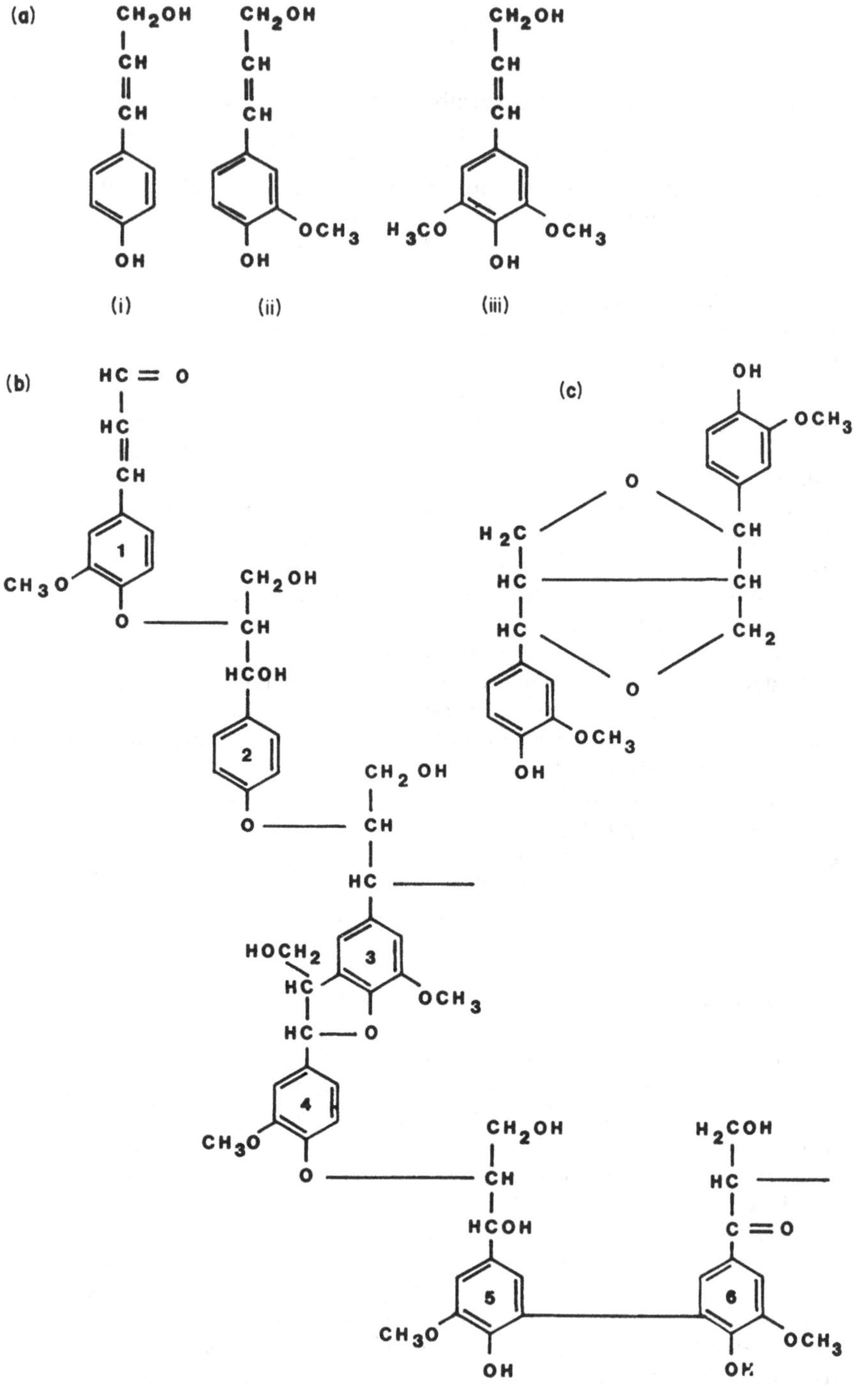

Fig. 3.1 (a) Building units of lignin: (i) coumaryl; (ii) coniferyl; and (iii) sinapyl alcohol. (b) Major linkage types of lignin. (c) Pinoresinol, a dilignol.

There is a general progressive thinning of the secondary cell walls of the xylem outwards from the cell cavity, the enzymes responsible acting in the near vicinity of the hyphae. Decomposition occurs uniformly in the region of attack. Fungi causing such rots preferentially attack hardwoods and simultaneously decompose all the components of the lignified cell walls. This type of rot has sometimes been called simultaneous rot and the term white rot used in a more restricted sense for rots in which the lignin is removed much more rapidly than the carbohydrates. The cellulose microfibrils in this latter case are unmasked and the cellulose utilized later. Brown-rot fungi preferentially attack softwoods. In these rots the wall polysaccharides are principally utilized. Very little, if any, of the lignin is used, although it may be altered structurally as, for instance, by the removal of methoxyl groups. With decay the wood becomes darker brown. There is no thinning of the walls. The enzymes responsible diffuse away from the hypha and act on the entire cell wall, often at some distance from the hyphae. The structural polymers are removed, leaving a framework of lignin to maintain the general cell shape so that there is little apparent damage until the cell walls collapse. Decomposition occurs in irregular patches in the attacked wood. This leads to the cubically cracked appearance of brown-rotted wood. It also crumbles readily to a powder when rubbed between the fingers. In both white and brown rots, the hyphae grow and branch in the cell cavities and penetrate the walls mechanically via pits or the surfaces in general, by coupling penetration with enzymic erosion, producing bore holes somewhat wider than the hyphae. In both, they penetrate deeply into the wood. Soft rots, on the other hand, are more conspicuous near the surface, advancing inwards after destroying the outer layers of the wood. They occur only in wood of unusually high moisture content, such as water-logged river and marine timbers. The soft-rot fungi again principally utilize the cellulose and the hemicelluloses of the walls, but their hyphae penetrate and grow within the secondary cell walls; here they enzymatically create chains of typically rhomboidal or elongated cylindrical cavities, with conically tapering ends (Fig. 3.2). Decomposition is restricted to the immediate neighbourhood of the hyphae.

Whereas soft rots are caused by Ascomycotina, such as species of *Chaetomium* and *Ceratocystis*, and anamorphic states such as *Alternaria* and *Phialophora*, white and brown rots are caused mainly by Basidiomycotina. Two good examples of these are *Coriolus versicolor* and *Piptoporus betulinus* respectively. The former is one of the commonest polypores and is found on a great variety of hardwoods whereas the latter is a facultative wound parasite restricted to birch (*Betula* spp.). *Coriolus* can degrade over 90% of the lignin in wood. Several hundred species in the Hymenomycete Agaricales, but more so in the Aphyllophorales, an order almost entirely confined to wood, cause white rots. But apart from these, only a very few Ascomycotina, including *Xylaria polymorpha* and *Ustulina deusta*, can cause a white rot. Somewhat fewer Basidiomycotina cause brown rots. Thus even in the fungi, the ability to degrade lignin completely is limited to the relative few.

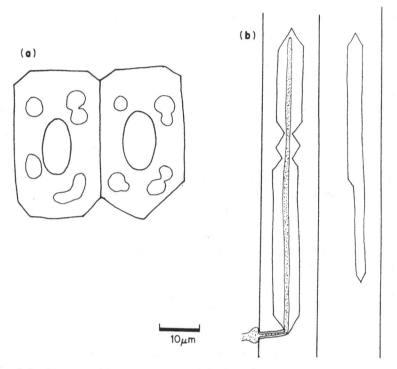

Fig. 3.2 Diagram of (**a**) transverse and (**b**) longitudinal sections of tracheids of *Pinus* with soft rot cavities in the secondary walls.

Lignin degradation

It is still not at all clear how these fungi act on lignin, in spite of the fact that mány studies have been made on the effects of white-rot fungi on lignin model compounds consisting of two phenyl propane units, such as the dilignol pinoresinol (Fig. 3.1c), or extracted lignin in liquid culture and the changes that occur in wood as it rots. Extracted lignins have been of little use in laboratory studies simply because of the physical and chemical changes brought about in its structure on extraction. Two extracted lignins, Kraft lignin and lignosulphate, have been widely studied because they are produced in such vast amounts as waste products of the paper industry. The compound which has been most widely used is a synthetic lignin designated DHP (dehydrogenative polymerizate). Like lignin, it is produced by condensation and it is chemically very similar but has a much lower molecular weight. ^{14}C labelled DHP is usually used. The amount of $^{14}CO_2$ evolved is the most sensitive measure of ligninolytic activity.

Role of extracellular phenolases

When cultured on agar containing phenolics, such as gallic or tannic acid, most – over 90% – of the white-rot fungi produce extracellular phenolases, such as laccase, peroxidase and tyrosinase, which oxidize these acids; a brown coloured diffusion ring appears around the colony margin. These catalyse the removal of electrons from phenols. They have long been considered as being involved in lignin degradation because lignin is a phenolic and so a substrate for these and lignin degradation is certainly an oxidation. Lignin is resistant to decomposition in anaerobic conditions. Also white-rot fungi produce these enzymes but the closely related brown-rot fungi, which do not decompose lignin, do not. Thus there is the apparent correlation between the ability to degrade lignin and the production of extracellular phenolases. The ability to degrade implies the formation of smaller compounds, yet it is usually considered that these oxidases act by coupling and polymerization to form compounds of higher molecular weight. There is, however, some scant evidence from experiments using white-rot fungi and model compounds that these enzymes can bring about limited depolymerization. It has also been shown that the continued action of these enzymes on wood itself leads to some degradation of the lignin. It is doubtful, however, whether they play any significant part in lignin degradation. In any case they can only be part of the enzyme complex which attacks lignin. It has been suggested that they have an indirect role in polymerizing and so detoxifying any toxic phenolics released during degradation – that is to suggest that they act after the monomers have been cleaved off. Simple phenolics are often toxic to fungal growth and they may have an important function in coupling these. This would be comparable to depside and depsidone formation from monocarboxylic acids in lichens (Chap. 7).

Cleavage of major linkage groups

One obvious step towards decomposition would be to cleave any of the major linkage groups between the phenyl propane units to release the C_6–C_3 monomers. *Coriolus versicolor* and a number of other white-rot fungi can cleave lignin models bonded by the arylglycerol-β-aryl ether bond. Although oxidative cleavage of the β-ether linkage occurs, there is again no convincing evidence that any great part of the lignin molecule is cleaved by white-rot fungi to produce the single monomers. There is also evidence that the monomers may be attacked while still bonded in the polymer, not by breaking the bonds between them but by directly attacking the aromatic rings, by either ring cleavage and/or demethylation of the methoxyl groups to hydroxyl ones. Demethylation may also be coupled with side chain oxidation. These are oxidized by the loss of two carbon atoms and the formation of new carboxyl groups. Both these, the formation of -OH and -COOH groups, would lead to increased solubility. Support for demethylation and side chain oxidation comes from two sources. Lignin degraded by white-rot fungi contains less carbon, slightly less hydrogen, fewer methoxyl groups but more oxygen and carboxyl and hydroxyl groups. Culture filtrates from white-rot fungi grown on extracted lignin contain small amounts of vanillin, vanillic acid and syringaldehyde. There is also slightly less vanillin in rotted wood compared with sound wood. This is taken to indicate that some phenyl propane units, either in the lignin or after cleavage, have had their side chains oxidized with the loss of two carbon atoms. Alternatively, it has been argued that vanillin and vanillic acid are attached as side groups along the main polymer and are released on hydrolysis.

A hypothetical scheme for lignin degradation

Although white-rot fungi unquestionably can use lignin as a sole carbon source and completely decompose it, we are by no means certain as to how lignin is degraded. A number of very hypothetical schemes have been put

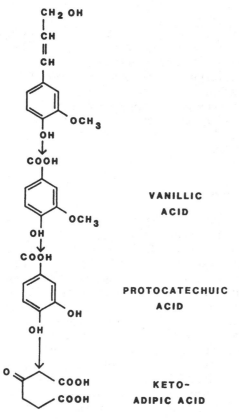

Fig. 3.3 Schema for lignin degradation.

forward. These usually assume initial cleavage of the arylglycerol-β-aryl ether bonds between the monomers (Fig. 3.3). This is followed by oxidative cleavage of the side chain with the loss of two carbon atoms and the formation of a carboxyl group to give vanillic acid. Vanillic acid is demethylated to protocatechuic acid. Ring cleavage then occurs to keto-adipic acid and this is used in the tricarboxylic acid cycle. It is most likely that these reactions occur simultaneously over the surface of the polymer with the oxidative cleavage of the side chains being centrally significant for fragmentation of the polymer.

Role of agents other than enzymes

Evidence is accumulating that agents other than enzymes, such as the hydroxyl radical (OH), may be involved in lignin degradation. In cultures of *Phanerochaete chrysosporium*, one of the most widely used white-rot fungi, hydroxyl dependent formation of ethylene coincides with ligninolytic activity. The radical is probably formed from hydrogen peroxide in the so-called Haber-Weiss reaction which is catalysed by iron and requires the superoxide radical (O_2^-)

$$O_2^- + Fe^{+++} \rightarrow Fe^{++} + O_2$$
$$Fe^{++} + H_2O_2 \rightarrow Fe^{+++} + OH^- + OH$$

The dependence of ligninolytic activity on the radical is verified by the fact that specific OH quenchers, such as mannitol, inhibit it. Wood rotting fungi produce sufficient hydrogen peroxide, by the action of a variety of oxidases, from the components of wood. Such oxidases are synthesized most rapidly when readily available carbon and nitrogen containing nutrients are low, when ligninolytic activity is at its peak. Sufficient amounts of Fe^{++} are also present in wood.

Cultural conditions are critical for ligninolytic activity. For example, to convert some 40% of DHP to carbon dioxide and water with *P. chrysosporium*, the culture must be maintained in the stationary phase at pH 4–5, with very low levels of metabolizable carbon compounds, low available nitrogen and high oxygen concentrations. The oxygen supply is a very critical variable as can be shown from the fact that incubation in pure oxygen increases lignin degradation 10-fold over incubation in air. The rate of degradation also increases if very thin mycelial mats are used rather than thick ones, diffusive supply being important. Cultures have also to be starved of carbon sources, such as glucose, sucrose and cellobiose, as well as nitrogen sources, such as ammonia, if high rates are to be maintained. Kirk and Fenn (1982) thus argue that lignin degradation is a strictly secondary metabolic function in that the products, as opposed to primary metabolites such as amino acids and simple sugars, are not essential for growth. But the process of degradation itself is of a selective value to the fungi. It gives such fungi a competitive ecological advantage in providing access to the cellulose and hemicelluloses masked by the lignin. White-rot fungi clearly produce a very elaborate and complex ligninolytic system, in part enzymatic and in part associated with the hydroxyl radical, to be able to degrade lignin completely to carbon dioxide and water. Brown-rot fungi lack the complete system. The most that many of these can do is to bring about a limited attack on the lignin molecule and cause such effects as demethylation.

Lignin as a physical barrier to cellulase

Both the brown-rot and soft-rot fungi decompose the carbohydrates, especially the cellulose of the wood. In the soft rots, the characteristic cavities are caused by decay being restricted to the immediate neighbourhood of the hyphae. The diffusion of their cellulase is definitely restricted. This situation contrasts markedly with the brown-rot fungi where the cellulase diffuses freely into the walls, hydrolysing the cellulose throughout and leaving a skeleton of predominantly lignin. It has often been suggested that the cellulo-lytic enzymes of the two groups differ in size and shape and that the lower diffusibility of the cellulase produced by the soft-rot fungi indicates larger molecular dimensions but this is not so. The cellulases produced by the two groups have similar dimensions and properties. Many actively cellulolytic fungi may be restricted in their ability to utilize cellulose in wood by virtue of the intimate nature of the association between the cellulose and the lignin. A particularly good example is *Chaetomium globosum* which rapidly degrades cotton and filter paper cellulose completely, but only attacks wood of high moisture content and merely produces soft rot cavities in the cell walls. Lignin appears to act as a physical barrier that prevents the cellulase from reaching sufficient glycosidic bonds in the cellulose to permit any large scale hydrolysis. Thus the accessibility of the cellulose to the degrading enzymes is a most important factor. The evidence for this comes from a number of sources. Increased accessibility can be achieved by breaking down the wood to much finer particles before adding cellulase. This exposes a larger surface area of the cellulose free of its association with lignin. For example, in experi-ments using sawdust and ball-milled sawdust, increased hydrolysis occurred in the latter when cellulase was added. It thus appears that brown-rot fungi possess some system – a pre-cellulolytic phase – which enables the cellulase to get at cellulose in wood. In the cell walls of wood, the cellulose microfibrils are encrusted with and surrounded by lignin and hemicelluloses. One suggestion that has been made is that brown-rot fungi produce enzymes which the soft-rot fungi lack. Some of these degrade the hemicelluloses and others disrupt the links between the cellulose and the lignin. There is also some evidence that one does not necessarily have to postulate enzymic disso-ciation of the lignin from the cellulose. Brown-rot fungi growing in wood, develop and maintain their own pH of between 2–4, whereas soft-rot fungi develop best in near neutral conditions. It may be just that acidic conditions are necessary to disrupt the association between the lignin and the cellulose. If wood is treated initially with acid and, after removing the acid, soft-rot fungi allowed to attack it, the fungi bring about a greater loss than in untreated wood.

Natural resistance of wood to fungal decay

Lignification

Lignification of the cell walls is obviously a very important factor that contributes to the natural resistance of wood to fungal decay. This is more important in soft rots. Softwoods are more resistant to these than hardwoods. This is usually attributed to the higher degree of lignification and the higher density of cell walls in conifers. It seems unlikely that mere abundance of lignin can account solely for the difference in resistance. The different proportions of the various phenyl propane units in the lignin, the degree of cross linkage with the cellulose and the nature of the hemicelluloses must also be important. Nevertheless lignification must act as some sort of physical barrier. Many very actively cellulolytic fungi and bacteria cannot attack wood because the lignin prevents their cellulase from reaching sufficient glycosidic bonds to permit hydrolysis on such a scale that they can grow on the proceeds.

Refractivity of cellulose

Many other factors contribute to decay resistance. The cellulose in wood tends to have a higher degree of refractivity or crystallinity than in the cell walls of herbaceous plants. The microfibrils are more highly ordered and there are correspondingly less amorphous or more randomly organized areas. The higher the refractivity, the smaller the surface immediately accessible to the components of cellulase.

Nitrogen content

In addition to being distinguished by its high lignin content, wood can also be distinguished from other plant materials by its very low nitrogen content. This also increases its resistance to decay. Woody tissues contain 0.03–1.0% nitrogen as compared to 1.0–5.0% in herbaceous tissues. The carbon: nitrogen ratio in most woody tissues is thus high, in the order of 350–500:1 and may exceed 1000:1. For most fungi a substrate with such a high carbon:nitrogen ratio would be nitrogen deficient and growth limiting. Wood-decay fungi are unusual in that they can grow in such substrates. They metabolize large amounts of carbohydrates (and lignin in white rots) in the presence of very small amounts of nitrogen. The mycelium of most fungi, grown on nutrient media, contains about 5.0% nitrogen and has a carbon:nitrogen ratio of about 10:1. The nitrogen content of the medium may fall, under starvation conditions, to around 1.0% before growth stops. The white-rot fungus *Coriolus versicolor* is unusual in that on high carbon:nitrogen containing media, the total nitrogen in the mycelium may fall as low as 0.2% before growth rapidly declines. The ability to grow under such conditions suggests a greater efficiency in its nitrogen metabolism. This may be achieved in a number of ways. Mycelial nitrogen may be re-used either by internal translocation from old to young hyphae or by autolysis and re-use. Extracellular lytic enzymes may be secreted which break down the old hyphal walls making the constituents, especially the nitrogen in the chitin, available for re-assimilation. Preferential allocation of available nitrogen to nucleic acids and enzymes may occur. For example, when growth of *C. versicolor* on media containing low and high carbon:nitrogen ratios was compared, the total nitrogen, expressed as percentage dry mass of the mycelium, fell from 4.4 to 0.2 but the percentage nitrogen in nucleic acids rose from 4 to 25 and the amount of cellulase produced per unit of mycelium was comparable in each. In fungi in general, cellulolysis diminishes with increase in the carbon:nitrogen ratio. White-rot fungi are unique in being able to produce cellulase at a carbon:nitrogen ratio of 2000:1 whereas in most other fungi this ability is negligible at a ratio of about 200:1 (Levi and Cowling, 1969).

Moisture content

Wood-decay fungi have higher moisture requirements for growth than fungi which attack most other plant materials. Their growth rate is very sensitive to changes in the water activity (a_w) of the medium. Whereas cotton is susceptible to fungal attack when it has a moisture content of more than 10% on a dry mass basis and cereal grains more than 13%, wood decay can be initiated only at moisture levels of about 26–32%. In standing trees and freshly felled timber, most of the cell cavities in the wood are water-filled. Such wood may have a moisture content of well over 100%. Such completely saturated wood is quite immune to fungal attack, presumably because the oxygen tension is too low to support active hyphal growth and the carbon dioxide content raised considerably above atmospheric levels. Many wood-decay fungi are very tolerant of high carbon dioxide concentrations. Whereas litter-inhabiting Basidiomycotina may be inhibited from growth by a partial pressure of 10 kPa, wood-decay species still grow at 30 kPa and some, including *Piptoporus betulinus*, still grow at 70 kPa. It appears that air equivalent to something more than 20% by volume of the wood is necessary for actual decay to take place. The existence of intact wooden galleys, submerged since Roman times, is adequate proof that completely waterlogged wood does not decay. The point at which all the free water has disappeared, but the cell walls are still fully saturated, is known as the fibre saturation point. For most woods this is around 26–32% (c.0.3 g g^{-1}, equivalent to an a_w of 0.97). Wood-decay fungi begin to grow at around this level and make optimum growth at about 40%. Tresner and Hayes (1971) tested just over 100 species of Basidiomycotina and found that 94 were unable to grow at an a_w of 0.97 and below and only one species grew down to 0.94. Other evidence suggests that the lower limit for growth of wood-decay fungi is about 0.97, with the linear growth rate reduced to about half normal even at 0.989.

Worked wood that has been thoroughly air-seasoned contains 15–18% moisture, which is far too low to support any fungal growth. Requirements for such high moisture contents, and thus water activities, obviously contribute to the resistance of wood to decay. Dry rot, caused by *Serpula lacrimans*, is an exception. Wood with a moisture content as low as 20–24% becomes liable to attack by *S. lacrimans*. Furthermore, as a brown-rot fungus, it produces metabolic water during cellulose degradation which considerably raises the moisture content of the wood on which it is growing. Once established on a small pocket of damp wood it can continue to colonize dry wood in this way. The exact relationship between water activity and wood decay is difficult to obtain because all, like *S. lacrimans*, degrade the cellulose in the cell walls. The complete degradation of 1.0 g cellulose liberates 0.56 g metabolic water. This is sufficient to alter the a_w of the wood significantly.

Toxic substances

All these factors contribute to decay resistance but the principal sources of such resistance in wood are toxic substances deposited during the formation of the heartwood. These are synthesized in the senescing parenchyma cells and diffuse out into the walls of the adjacent xylem elements. The distribution of decay resistance within a tree has been correlated with both the distribution and the nature of these toxic substances. They have been studied most in Gymnosperms. Most are phenolics. They fall into four main chemical groups, terpenoids, tropolones, flavonoids and stilbenes (Fig. 3.4). Of these the thujaplicins (tropolones) are the most inhibitory. They all provide protection from decay for many years but with time they may become lost by leaching or become inactivated. In spite of their toxicity, it is well-known that several fungi are able to destroy the heartwood, even in living trees, and also timber impregnated with similar phenols, such as pentachlorophenol and 2,4-dinitrophenol, which are used to protect less durable timbers and the

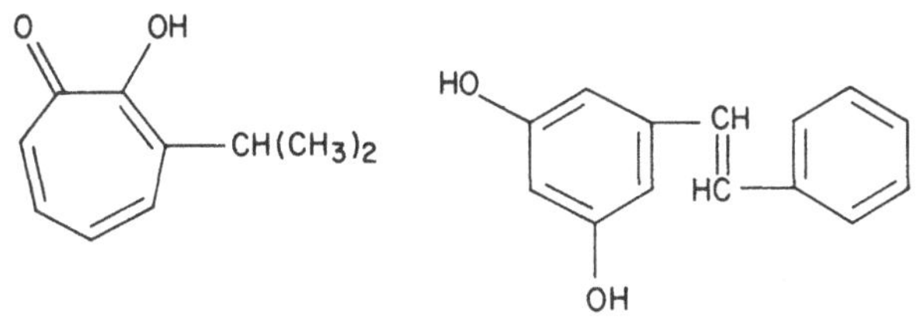

A TROPOLONE (THUJAPLICIN)   A STILBENE (PINOSYLVIN)

Fig. 3.4 Structure of two toxic chemicals extracted from gymnosperm wood.

sapwood of conifers. Such heartwood rotters are not insensitive to these toxins. They use their phenolases to oxidize them and polymerize the products to non-toxic melanins. Tannins are very common in the heartwood of Angiosperms and they play a similar role in decay resistance there. They inhibit fungal phenolases but decreased toxicity of the heartwood occurs with time by auto-oxidative polymerization of the tannins.

Sapwood is ordinarily very susceptible to decay but the resistance of different heartwoods is very variable. Trees with very resistant heartwoods include many oaks, cedars and the redwoods and those with non-resistant or only slightly resistant heartwoods include alders, beech, elms and poplars. A durable heartwood may be of survival value to the tree itself. Cedars live 2000 years or more whereas any of those in the slightly resistant category rarely live as long as 500 years.

Other wood-inhabiting fungi

Other fungi which inhabit wood occur chiefly in the sapwood where they obtain their food supply from the contents of the dead xylem parenchyma cells. These are the so-called moulds and stain fungi.

Mould fungi are mainly conidial Ascomycotina. They discolour the wood by producing pigmented conidia on the surface. Their hyphae accumulate within the ray parenchyma cells but may also be present in the cell cavities of most of the surface xylem elements, spreading from cell to cell via pits. This causes shallow discolouration and surface staining of the wood. For instance, surface blue-stain occurs most frequently on sawn timbers and on any wood surface exposed to the rain. It is caused by the surface growth of common air-borne fungi, but especially *Cladosporium* spp., with dark brown hyphae and coloured conidia. Such staining is easily removed during planing treatments.

Blue-stain fungi

Typical blue-staining is caused by pigmented hyphae that grow in the wood, whereas other stains, such as brown ones, are caused by chromogenic substances actually excreted by the hyphae into the wood. At least one fungal stain has been used commercially. The mycelium of the Ascomycete *Chlorosplenium aeruginascens* permeates the dead wood of oak and beech on the woodland floor and colours it a brilliant green. Such 'green oak' has been used for inlays and decoratively as Tunbridge Ware. The wood is unaltered in texture and resists decay.

Blue-stain fungi are common in coniferous sapwood, especially pines, but are also found in hardwoods. They are non-cellulolytic 'sugar' fungi, in that they utilize only the more readily assimilable carbon compounds, such as sugars and starches, which occur in the ray parenchyma cells of freshly killed wood. They do no structural damage to the wood as they move across it mainly through the pits. Blue-stain is thus not the first stage of a form of rot, but its occurrence does indicate that the wood has been kept moist and exposed to conditions favourable to the development of decay fungi. Although the structural properties of the wood are unaltered, blue staining of coniferous sapwood is responsible for large financial losses to the timber producer. The mere discolouration of the wood makes architects disinclined to use it and it is less acceptable to the manufacturers of packing cases and paper. In vigorously growing pine trees, the moisture content of the sapwood is too high to permit growth of blue-stain fungi. The low oxygen tension again appears to be the major limiting factor in the growth of these fungi in wood with a very high moisture content. In nature they may colonize standing pine trees which have been killed either by root-rot caused by *Heterobasidion annosum*, or some other disease, or by suppression. They are much more common on felled pine logs and will soon appear on these if they are left on the forest floor for any length of time. However they are rapidly replaced by wood-decay fungi. Both death and felling cause the wood to dry out progressively and such wood will support the growth of blue-stain fungi unless the moisture content falls below about 27%.

Blue-stain fungi are Ascomycotina, mainly of the genus *Ceratocystis*, most of which have both perithecial and conidial states. The majority present their spores for dispersal in the form of stalked spore drops, the spores, in this case, being insect dispersed. The perithecia of *Ceratocystis* have a swollen base and a very long, slender neck, some often 1 mm or more in length (Fig. 3.5a). The ascospores are not violently discharged but the asci break down within the ascocarp and the ascospores are forced up the neck; they are extruded in a mucilaginous drop at the apex where they are held in place by a fringe of hair-like hyphae lining a pore. A variety of conidial states are produced. The *Graphium* state has a thick sheath of dark hyphae forming the stalk (Fig. 3.5b). The component hyphae branch at their tips and produce masses of sticky conidia, whereas, in the stalked spore drop of the *Leptographium* state, the stalk is a deeply pigmented and very wide single hypha which branches profusely at the apex (Fig. 3.5c).

Blue-staining is usually associated with attack by bark beetles. Species of *Hylastes*, *Myelophilus* and others introduce spores in making their brood chambers at the interface of sapwood and bark. The spores germinate and grow radially and longitudinally in the sapwood forming wedges of blue-stained timber and then produce their conidia projecting into the brood chambers. These adhere to the young beetles as they emerge and are dispersed to other logs as they in turn make brood chambers.

Blue-staining becomes a problem where felled pine logs are left in piles on the forest floor for 2–3 months before being removed to timber depots. There are a number of ways of treating the problem, such as the use of insecticides and fungicides, but the most successful method of control, widely used in Europe, is to remove the bark immediately on felling. This not only prevents beetle attack but assists rapid drying out to moisture contents below those which will support fungal growth.

Depletion of the food reserves of the xylem parenchyma cells during ageing is one nutritional factor that tends to limit the susceptibility of sapwood to

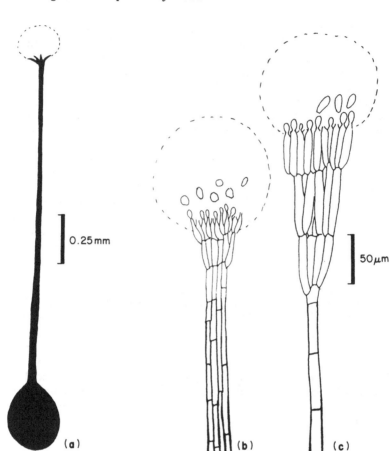

Fig. 3.5   Stalked spore drops of blue-stain fungi. (**a**) *Ceratocystis*. (**b**) *Graphium*. (**c**) *Leptographium*.

staining fungi. During ageing and the transition from sapwood to heartwood the parenchyma cells gradually die and become depleted of reserves, especially starch, and become less capable of supporting growth of blue-stain fungi. Similarly, during air-seasoning of wood, the parenchyma cells continue to respire reducing their food reserves and thus moulds and stain fungi are less common on seasoned than on unseasoned timber.

From the above, wood-inhabiting fungi can be conveniently divided into those which can live only on the cell contents, such as the moulds and the stain fungi, and those which in addition can degrade or partially degrade the cell walls, such as the white-, brown- and soft-rot fungi. Many are saprotrophs and can colonize only when the host tree has died or has been killed. Some such as *Piptoporus betulinus* and *Ceratocystis ulmi* are wound parasites, gaining entry at sites where the xylem is exposed. Still others, such as *Heterobasidion annosum* and *Armillaria mellea*, are necrotrophic parasites. They invade and kill the living root tissues and then degrade the cell walls of the xylem.

Dutch elm disease

Ceratocystis ulmi causes Dutch elm disease; it is spread in a similar manner to the blue-stain fungi by bark beetles, especially *Scolytus scolytus* and *S. multistriatus*. The beetles bore and breed within the bark of weakened, dying and dead elms, including those which are suffering from the disease. In infected trees, the fungus grows within the breeding galleries and produces either stalked spore drops of the *Graphium* state or rather smaller droplets of its *Cephalosporium* state. Perithecia are less easily found but in damp conditions develop on the surfaces of wood chips or partially immersed in fissures in the bark. The young adults emerge in May to October and the sticky conidia may adhere to their bodies. They fly immediately to feed on young, healthy, elm twigs and in doing so may introduce the conidia into the xylem in wounds made as they feed. Beetles thus spread the fungus from branch to branch and tree to tree. The fungus enters the xylem and grows in a yeast-like form. It can be carried up in the xylem in the transpiration stream as such or as conidia. Infected trees soon show signs of wilting and yellowing or drying out of the foliage. Fungitoxins may be involved but part of the symptoms may be explained by the occlusion of the xylem of the current year's growth by gums and tyloses.

Environmental factors and the decomposition of wood

The decomposition of wood under natural conditions is an exceedingly protracted process. Whereas leaves of the majority of northern temperate deciduous trees may decompose in one, two or three years, a tree trunk under the same conditions may take a decade or even two to do so. The low level of available nitrogen may be the overriding factor contributing to its slow rate of decay. The addition of organic nitrogen to wood blocks inoculated with various Basidiomycotina has been shown to increase their decay rate by over 60%. Other minerals, especially phosphorus and potassium, may also be limiting. The relatively high demand for such mineral nutrients combined with their relatively low availability places a limitation on the amount of fungal mycelium such a substrate can produce. Fluctuations, both diurnal and annual, in temperature and moisture content must also be important. However, in aseptic laboratory experiments decomposition of wood by a single species of decay fungus may be relatively rapid. The white-rot fungi *Lenzites betulina*, *Coriolus hirsutus* and *C. versicolor*, inoculated onto small blocks (20 mm³) of birch wood kept in sterile moist soil at 22°C, caused more than a 75% loss in dry mass in three months. *Piptoporus betulinus* and a number of other brown-rot fungi caused mass losses of 50–70% over the same period. These are substantial losses, the more so when it is borne in mind that birch wood contains some 20% lignin which is not available to brown-rot fungi. These facts may be contrasted with the observation that *P. betulinus*, which had killed birch trees in East Anglia, was still producing basidiocarps on these at least five years after they had fallen. The time period from infection to falling was not known but it may have been at least another five years. Even if the variable temperature and moisture regimes are taken into account, the decomposition of one of the least durable woods is very much slower in nature than in laboratory tests.

Most wood-decay fungi are mesophiles in terms of their temperature requirements, although some come into the category of cold-tolerant ones. The optimum temperature for the growth of most lies between 25 and 30°C. *P. betulinus* has an optimum at 25°C and its growth falls off very rapidly above and ceases at 30°C. Its minimum temperature for growth, which it must often experience in the field, lies between 7 and 9°C. For many others, the minimum lies below freezing point but decay at such temperatures would be very slow. The geographical distribution of a number of species is related to their temperature requirements. *Serpula lacrimans* has a low maximum of 25–26°C. It is absent from the tropics and other parts of the world with high summer temperatures.

In bulk wood, temperature is probably a more important variable than moisture content. With reference again to *P. betulinus*, it is able to decompose birch wood with a moisture content within the range 35–100% on a dry mass basis, although near maximum decomposition rates only occurred between 60 and 120% in laboratory experiments. Birch logs, stored outside in Central Sweden, had a moisture content of 85–91% on felling and after three years the moisture content was still 51–67%. Fluctuations did occur, with some drying in the summer and some water uptake in the winter, but over the whole period the moisture content was somewhere near the optimum for decay. However, if the bark peels off the position is quite different. In summer, rapid drying out may occur to moisture contents below those which will support growth and equally rapid soaking will occur in rain. A thick, highly suberized outer bark is not only a structural deterrent to fungi, but because of its high content of tannins, phenols and the like, also a chemical one. However, given this, if it remains intact after death of the tree, it helps to maintain a more equitable moisture regime within the wood and this will favour any decay fungi.

Habitat relations and specificity of wood-inhabiting fungi

The habitats of wood-inhabiting fungi vary from minute twigs, small and large branches to the most massive of tree trunks and stumps and from minute rootlets to major roots and include such man-made habitats as fencing posts, house timbers, sawdust and chip piles. Any one of these substrates is particularly complex in more ways than one. The trunk of any one tree will have varying proportions of bark, sapwood and heartwood along its length. These proportions will differ from those in the trunk of another species. The wood from different tree species differs structurally as can be seen by contrasting ring-porous with diffuse-porous types. Further marked differences occur between softwoods and hardwoods. Over and above these differences as already indicated the composition of the lignin varies in different wood. This complexity and heterogeneity make it difficult to generalize about the decomposition process. A number of successional studies have been made on woody substrates but these have to be interpreted with caution. A succession can be defined as the appearance of different fungi in sequence on the same part of the substrate. The fact that one fungus appears on one part of a log at one time and another fungus on another part, even an adjacent part, at another time does not necessarily prove a succession. Their habitat niches may be quite different, one growing on the sapwood and one on the heartwood or the latter may be colonizing a part of the heartwood not colonized by the former. For example, basidiocarps of *Daedaleopsis confragosa* or of *Hypholoma fasciculare* may appear on a birch trunk which has, for a number of years, supported basidiocarps of *Piptoporus betulinus*, but the mycelia of these would almost certainly be growing on parts of the wood not colonized by *P. betulinus*. The latter is specific to birches and is a wound parasite gaining entry where a branch has been fractured. Infected trees are usually killed by the fungus and the trunks of these often break off remarkably cleanly and transversely at a height of about 3 m in high winds. The structural polysaccharides in the walls are rapidly and completely removed leaving a cellular framework of amorphous lignin which has insufficient tensile strength to withstand the bending strains incurred. The fungus then continues to grow and to produce its characteristic kidney- or hoof-shaped basidiocarps on the fallen and standing parts of the tree. As with many woody substrates, *P. betulinus* is the primary and sole colonizer. It may completely permeate the wood of the whole trunk and persist there, virtually in pure culture, for several years, by which time the wood is in a very late stage of decay and extremely friable. In such a state the wood is unlikely to be capable of supporting fungi such as *D. confragosa* and *H. fasciculare*. They would not succeed *P. betulinus* but would be growing on parts not colonized by it.

Wood-decay fungi exhibit all degrees of specificity. Considering the white-rot and the brown-rot fungi as two groups, there are many more of the former than the latter. Those of the white-rot group primarily attack hardwoods and those of the brown-rot group softwoods. Some of these may be restricted to a single host genus. *P. betulinus* is a good example. *Fistulina hepatica* which causes a serious decay of the heartwood of oaks is another. The causes of such marked specificity are obscure. Other fungi may be restricted to the wood of a relatively small number of trees. *Polyporus squamosus* is, like *P. betulinus*, a wound parasite, in this case of elm in particular but it is often found on other trees, such as ash and sycamore. It causes a white-rot of the heartwood and may persist for a number of years on fallen trees which it has killed or which have been wind-blown, as a consequence of the rot. On elm trunks *P. squamosus* is replaced, but only in a temporal sense, by a number of other wood-decay fungi. Two in particular, *Auricularia mesenterica* and *Pleurotus cornucopiae* are rarely found on other wood. Basidiocarps of the former soon appear on any felled elms and production of these continues for up to eight years. Spatially it utilizes the bark and surface layers of the sapwood so it does not succeed *P. squamosus*. Basidiocarps of *P. cornucopiae* appear on elm trunks only some 3–10 years after they have fallen. The fungus then persists until the wood is well-decayed. Its mycelium appears to be confined to the sapwood not utilized by *P. squamosus* and again it does not actually succeed the latter. Whereas *P. cornucopiae* is most common on fallen elm trunks, *Flammulina velutipes* is most often found on standing dead elms, especially those killed by *Ceratocystis ulmi* and which have lost their bark.

Still other fungi, such as *Coriolus versicolor* and *Stereum hirsutum*, are much less discriminating and grow on a wide range of hardwoods. The former is entirely saprotrophic and is one of the commonest fungi found on fallen twigs, branches, trunks and dead stumps of hardwoods where it produces the most rapid of white rots but, like *Stereum*, is confined to the sapwood. It can actually replace, and therefore succeed, other established and less aggressive white-rot fungi. *Xylaria hypoxylon* and *Daldinia concentrica*, two Ascomycotina, produce black lines, zone lines, in the sapwood of ash delimiting areas which they have colonized. The hyphae of *C. versicolor* will penetrate these and grow on to replace them. Other fungi show a preference for coniferous wood. *Heterobasidion annosum*, *Paxillus atrotomentosus* and *Tricholomopsis rutilans* are characteristic of conifer stumps, *Hirschoporus abietinus* and *Stereum sanguinolentum* of coniferous twigs and branches and *Auriscalpium vulgare* of pine cones. These are all Basidiomycotina but similar examples can be found in the Ascomycotina. For example, *Daldinia concentrica* is very common on ash but is occasionally found on other hosts, especially beech and birch. *D. vernicosa* occurs on gorse, especially bushes which have been burnt and subsequently weathered. *Ustulina deusta* causes a white-rot of lime and beech, whereas *Xylaria polymorpha* and *X. hypoxylon* are very common on a wide variety of dead hardwoods.

Each tree species thus may have, within limits, its own particular wood-decay fungi. A number of these may enter as necrotrophic parasites at wounds above ground or along roots below ground and then persist as active saprotrophs after death. They would thus have a competitive advantage over purely saprotrophic fungi in being established first. As parasites they may only be able to overcome the host resistance of one or a few species of trees. This might account for some of the specificity noted. It may well be that the different naturally occurring tannins, terpenoids, etc. present in the different heartwoods further help to determine specificity. Only fungi which can tolerate or degrade these are able to become established.

Ecological studies on decaying wood

Numerous ecological studies have been made of fungi colonizing specific woody substrates including wounded living tree trunks and fallen dead ones, tree trunks after insect attack, fire-killed trees, branches and slash on the ground, tree stumps, fence posts, beech cupules, etc. (Hudson, 1968, Käärik, 1974). Changes with time in the fungal communities on these have been recorded and described, accurately or inaccurately, as successions. As might be expected, the sequences of fungi observed on these show considerable variation depending upon the species of wood, the type of substrate and, in addition, the environment in which decomposition is occurring. But fungi are not the only organisms found in decaying wood. A very wide variety of invertebrates and bacteria also occur and they, too, may play an important role in its decomposition.

Swift (1977) recognized three stages in the decay process – the pioneer colonization stage, the major decomposition stage and the incorporation stage, in which the products of decay are incorporated into the soil.

Pioneer colonization stage

Patterns of colonization may vary. In some cases, the Basidiomycotina which are going to dominate the decomposition stage are the primary and sole colonizers. In others, their colonization is preceded or accompanied by a variety of decay or non-decay fungi or bacteria. This may be illustrated with some specific examples. *Heterobasidion annosum* is a white-rot fungus causing butt- and root-rot of conifers. It may colonize via roots or the surfaces of freshly cut stumps. Infection of a healthy living root almost invariably occurs as a result of mycelial transfer from another infected root coming into contact with it. From the root the fungus grows up to the base of the stem and colonizes and kills the cambium, thus effectively girdling and so killing the tree. It then progressively rots the roots and the stem base. Rapid desiccation of the wood after death usually prevents extensive spread up the stem. In this case it is the sole colonizer. Alternatively it may colonize the surfaces of freshly cut stumps via its air-borne basidiospores. These stump surfaces are highly selective substrates and are initially colonized by a relatively small number but, nevertheless, a variety of fungi. These include, in addition to *H. annosum*, non-cellulolytic blue-stain fungi utilizing the contents of the parenchyma cells, cellulolytic fungi, such as *Phialophora* and *Trichoderma* spp., utilizing cell contents and any easily accessible cellulose, and other wood-decay fungi such as the white-rot fungus *Peniophora gigantea*. This is a much more competitive situation; whether or not it emerges as the major decomposer will depend upon a multiplicity of factors, including its ability to compete with these for the more readily available nutrients which are necessary if it is to become established. Similar patterns of colonization can be seen in the initiation of decay in trunks following wounding, such as by the branches breaking off in high winds. Again, in some cases the only fungi to colonize are the wood-decay Basidiomycotina which later become the dominant decomposers. This applies to most species of *Stereum*. They invade only freshly exposed tissues and are inhibited by the presence of other pioneer micro-fungi and bacteria. In other cases, such as with *Phellinus igniarius* invading wounds on poplars and other hardwoods, prior colonization by bacteria and micro-fungi such as the stain, mould and soft-rot fungi generally occurs and may even be a prerequisite if it is to attack and cause a progressive rot of the heartwood.

Insects, especially members of the Ipidae and Scolytidae, may attack living trees and introduce bacteria, yeasts, blue-stain or ambrosia fungi below the protective bark. The combined activities of the insects and the fungi may weaken or kill the tree. The wood-decay Basidiomycotina then follow. The attack of *Scolytus scolytus* on elms introducing *Ceratocystis ulmi*, followed by *Flammulina velutipes*, is a case in point.

Decomposition phase

The decomposition phase is dominated by the white- and brown-rot fungi but wood-boring beetles (Coleoptera) and wood-eating termites (Isoptera) may also contribute to decay. In many woody substrates only one fungus may be involved in the decomposition phase; examples of *Piptoporus betulinus* on birch, *Heterobasidion annosum* on pines and *Coriolus versicolor* on hardwoods in general, have already been given. In others, a number, but usually a very limited number, of fungi are involved. Each of these occupies discrete volumes of wood which are often clearly demarcated from each other by distinct dark zone lines. These colonies may intricately interlock but their mycelia do not intermix. They remain isolated by zone lines into virtually pure cultures. This balanced state may persist for a number of years, but, depending upon the relative competitive ability of adjacent mycelia, there may be eventually some replacement of one fungus by another, or aggressive saprotrophs, such as *Hypholoma fasciculare*, *Phallus impudicus* and *Phlebia merismoides*, may colonize from the surrounding litter and replace them (Rayner and Todd, 1979). For example, both *C. versicolor* and *Stereum hirsutum* are susceptible to replacement by any of these three in hardwood trunks and branches and *Heterobasidium annosum* by *Peniophora gigantea* in pine stumps. *H. annosum* is particularly sensitive to hyphal interference caused by the latter and this may be one factor involved in its replacement (p. 154).

The role of animals in degradation

Many wood-boring beetles and their larvae and termites are wood feeders depending upon microbial symbionts in their guts to semi-digest the wood. Some feed on sound wood, others on decaying and well-rotted wood. In the latter case, the fungi growing in the wood may be an important component of their food (Chap. 9). Many termites are polyphagous. When *Kalotermes flavicallis* is fed on wood, it decomposes 94–95% of the cellulose, 60–70% of the hemicelluloses and 3–4% of the lignin. Thus large populations of these wood-boring beetles would almost certainly contribute substantially to wood decay. Another important aspect of their degradative activity is the comminution of the wood as they attack it.

As the white- and brown-rot fungi exploit the wood, it softens and becomes friable and as such is more attractive to animals as a food source, as somewhere to live and as a breeding ground. Their access to it, if it is bulky, may be dependent upon the prior activity of wood-boring animals, such as the beetles. Their bore holes afford ports of entry for a very great variety of animals generally common in litter and soil. These include micro-arthropods, such as Acari and Collembola, and macro-arthropods, such as Diptera and Isopoda, as well as Oligochaetes, such as Enchytraeid and Lumbricid worms. Many of these, such as the Mycetophilid dipterous larvae feed mainly on the mycelia of the fungi. They all accelerate the process of comminution and carry spores from the surrounding litter and soil into the wood and thus inoculate it with common soil fungi. Many Zygomycete Mucorales appear for the first time on the decaying wood, along with a variety of conidial fungi including species of *Penicillium*, *Scytalidium* and *Trichoderma*. These now have a quite wide choice of substrates on which to grow. They may utilize the partly degraded wood, the dead hyphae of the wood-decay fungi, dead fauna or their faecal remains. Some may live as commensals sharing the hydrolytic products of the enzyme systems of the major decomposers. This is the incorporation stage. As the wood becomes more extensively decayed, the activity of the wood-decay fungi may decline and they are eventually replaced by such soil-inhabiting fungi. With time, the wood disintegrates and as it does so it is incorporated into the soil.

Wood decay and the cycling of mineral nutrients

Wood decay is important in regulating the cycling of mineral nutrients in the woodland ecosystem and contributes to the process of soil development there. Over the period of fungal decay, virtually all the important minerals, but in particular nitrogen and phosphorus, become immobilized in an organic form in the fungal hyphae and their reproductive structures, such as basidiocarps. Although the mineral content is low per unit volume, the sheer volume of decomposing wood means that it forms a very substantial part of the total minerals in the woodland ecosystem. Comminution of the decaying wood by the animal invaders leads to a release of some minerals. The small particulate form of the frass or faecal materials means that they are more effectively leached. As with the fungi, the animals themselves act as a further reservoir of plant nutrients in a considerably more concentrated form than in the wood itself. Their wanderings, after feeding, lead to some redistribution of minerals but by far the more major redistribution and export from the wood occurs when adult stages emerge from the broods reared in and on the decaying wood. But this is essentially only a redistribution. The adults eventually die and their tissues are mineralized elsewhere in the ecosystem, while any fungal remains are mineralized *in situ*.

Decomposition of lignin and humus in the soil

The white- and brown-rot fungi are, more often than not, associated with relatively large masses of wood, such as the dead tree trunk and decaying stump. This may be because only these contain enough energy resources for these fungi to amass sufficient to produce their relatively massive and conspicuous basidiocarps. Vast quantities of lignin are incorporated into the soil in the vascular network of the leaves, fine rootlets and so on. These are very different substrates for fungi and other micro-organisms and they are in a vastly different environment. The substrate is richer in terms of associated readily available carbon and nitrogen sources, as tissues other than the highly lignified xylem are present in relatively larger proportions than in bulk wood. It is also presented to a much more varied population of micro-organisms. As such it would support a more diverse micro-flora and any lignin decomposers would be competing for, not necessarily lignin, but other more generally assimilable components, which are necessary for establishment, and would also be exposed to antagonism by others. This situation is markedly different from the decaying tree trunk with its one or few decomposer fungi in isolation. Further large quantities of lignin may be introduced into the soil in the form of organic residues from wood decay, especially from brown rotted wood. The process of lignin degradation in the soil may be quite different from that occurring in a tree trunk. There is very little direct evidence that any Basidiomycotina degrade such lignin in the soil. This may be because we are ignorant of the facts. In studies on soil fungi, Basidiomycotina are only rarely recorded. They tend to be slower growing and so are easily overgrown on most widely used culture media. They are often very sensitive to antagonism by others and so suppressed. Most do not produce spores or possess any other readily recognizable feature so could easily be overlooked. Nevertheless, they may be equally important as lignin decomposers in the soil itself as they are in the litter and decaying wood. The number of soil-inhabiting micro-organisms which have been reported as being able to utilize lignin is very small. These include a few aerobic, Gram-positive, non-sporing, rod-shaped bacteria, in the genera *Bacillus* and *Flavobacterium*, and a few conidial fungi, in genera such as *Humicola* and *Phialophora*. The evidence for their ability to utilize lignin has again been obtained from the use of extracted lignin and lignin model compounds. These have been used incorporated in Kaolin pellets to enrich soil and fungi subsequently isolated from them and tested for their ability to utilize such compounds as vanillic acid and syringaldehyde. The ability to grow on and utilize these should not be taken as an ability to utilize lignin itself just as the ability to utilize carboxymethylcellulose is not taken as an ability to utilize native cellulose. They are partial degradation products and these fungi should be regarded as occupying a similar niche with regard to lignin as secondary sugar fungi do to cellulose. The latter do not possess the whole enzyme system necessary to hydrolyse cellulose (Fig. 3.6). They lack the C_1 component but possess the C_x component and β-glucosidase so that they can utilize the hydrolytic products. Some also lack the C_x component as well. Similarly with lignin, the fact that a fungus lacks one

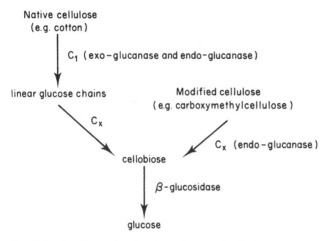

Fig. 3.6 A schema for the degradation of cellulose.

component of the multi-enzyme system does not necessarily debar it from participating in lignin degradation. Fungi may co-operate, sometimes synergistically, in the degradation of both cellulose and lignin. For example, it has been shown in experimental systems that a mixture of the C_1 component from one fungus and the C_x component of another is as efficient at cellulolysis as when both components are derived from the same fungus. It has also been shown using lignin preparations in culture tests that, in many cases, when two wood-decay fungi are grown together in mixed culture, degradation is more pronounced than when both fungi are grown apart.

Lignin in the soil decomposes very slowly and its degradation there is more of a joint effort. There may well be a pooling of enzymes from a variety of fungi and perhaps bacteria and actinomycetes – some enzymes capable of cleaving bonds between monomers, others of demethylation and still others of side chain oxidation and so on until the final products enter the respiratory pathways of one organism or another.

The nature of humus

With this breakdown there is a gradual accumulation of dark, amorphous, organic humus. The chemistry of humus has by no means been fully elucidated. It is very heterogeneous and can be separated into a number of molecular categories using extraction techniques. It forms a very dark solution in dilute NaOH and a black precipitate called humin. Acidification of the solution with HCl to between pH 1–2 precipitates out a fraction called humic acid, leaving fulvic acid. The humic acid fraction is the major molecular category and forms from between 50 to 80% of the soil humus. It usually contains about 5% nitrogen, mainly in the form of bound amino acids but also in amino sugars and heterocyclic purine or pyrimidine derivatives. The most favoured idea is that humic acid has a heterogeneous aromatic core with carbohydrates, peptides and proteins, phenolics and metals, attached peripherally. In some soils, especially under woodlands, the humic acid may originate from lignin residues, possibly the end products of the brown-rot fungi, which have been considerably modified by microbial action. Syringic and vanillic residues can often be detected in the degradation products of humic acid and the distribution of these residues is consistent with the composition of the lignin found in the vegetation above. Work with tracers has shown that as much as one third of the humic acid in the soil is derived from lignin and only about one twentieth from cellulose. Reductive cleavage of most humic acid fractions shows that they also contain units based on phloroglucinol. This suggests that seed plant flavonoids also contribute to humus. Flavonoids are phenolics with two aromatic rings and include pigments, such as anthocyanins. The phenolics are degraded to simple phenols which become polymerized into the humic acid fraction. But humus is not solely a product of degradation of the more resistant parts of seed plants. It is also in substantial part a product of microbial synthesis. When ^{14}C labelled glucose is added to the soil, 40–80% of the carbon is lost as carbon dioxide within a few days but, even after two years, about 5–10% is still present in the soil humus. Intracellular transformation of carbohydrates and other simple organic substances occurs to produce phenols, quinones and other aromatic substances. These are oxidatively polymerized and combined with peptides and other cell constituents to form humic-like pigments, melanins, inside or outside the cell. These serve several functions. They may be deposited in the walls of hyphae, spores or ascocarps, to protect against excessive ultra-violet light or as a water-proofing to prevent water loss. Eventually, on the death of these structures and with time, they become variously transformed and incorporated into the humus fraction. The existence of significant amounts of amino sugars and non-protein amino acids, such as diaminopimelic acid, also suggests that residues of bacterial cell walls may form part of the humus.

Turnover of humus in soil

Humus is extremely resistant to microbial degradation but nevertheless there is a very slow turnover with the rate depending upon the soil type. A sample from a chernozem soil from the USA was ^{14}C dated as 990 ± 60 y old.

In other soils, humus is less stable e.g. humus from a coniferous forest soil in Sweden was dated as 370 ± 100 y. A number of fungi have been found to decompose humic acid in laboratory tests. Humic acid was extracted from a Canadian soil. It contained 26% of the total soil carbon and was dated as 785 ± 50 y old. It was supplied as the sole carbon and nitrogen source as a 0.2% solution to a number of microorganisms, isolated by direct plating of the soil onto humic acid containing media. Four bacteria, in the genera *Bacillus* and *Pseudomonas*, and two conidial fungi, *Penicillium frequentans* and *Aspergillus versicolor*, could utilize the humic acid as a sole carbon and nitrogen source but no actinomycetes could. *P. frequentans* made the best growth and it appeared to utilize the humic acid by initially reducing carboxylic groups to aldehydes and then alcohols. Salicylaldehyde and salicyl alcohol appeared in culture filtrates. A number of Basidiomycotina, including *Coriolus versicolor*, *Hypholoma fasciculare* and *Trametes suaveolens*, all active white-rot fungi, can also utilize humic acid and this ability is always associated with the reduction of carboxylic acids to alcohols. This suggests that one of the first steps in the degradative process is an aerobic reductive one. Subsequent steps have as yet to be elucidated. Very little can be concluded from such studies about the process of degradation in soils. Resistance to degradation may not be so much that it is not susceptible to microbial enzymes but that its multi-dimensional complex structure physically restricts the access of such enzymes.

4
Fungi as inhabitants of aquatic environments

It is difficult to be precise in defining an aquatic fungus because almost any fungus, it it can be grown, will grow in liquid shake culture and, conversely, many fungi which are known to live in water will grow, but may not reproduce, in culture on solid media. Fungi, with the odd exceptions, are aerobic organisms. In aquatic environments, the rate of diffusion of oxygen is significantly slower than in aerial ones. In the former, unless there is turbulence or oxygen input from photosynthesizing plant communities, fungi do not really thrive because an oxygen deficit develops.

Although almost three-quarters of the earth's surface is covered by sea and a not insignificant part of the remainder by freshwater lakes, rivers and streams, only about 2% of all the fungi, so far described, are aquatic in that they normally complete their life cycle, or at least the known part of it, in water. These aquatic environments are often very rich in a wide variety of different types of organic matter, including decaying tree leaves and wood, yet, until quite recently, the study of the ecology of fungi involved in the decomposition of such substrates has been sadly neglected in comparison with similar studies on terrestrial fungi.

The diversity of marine fungi

Fungi were first described from marine habitats in the middle of the nineteenth century by Durieu and Montagne in France, but the study of marine fungi as a distinct branch of mycology did not begin until the publication of the now classic paper entitled *Marine Fungi: Their Taxonomy and Biology* by Barghoorn and Linder (1944). Less than 500 species of fungi, under 1% of the total number, have been described from oceans and estuaries. Just over one third of these have been described from woody substrates and a similar proportion has been recorded growing on algae.

Although true marine fungi occur in most classes of fungi, only four marine members of the Basidiomycotina have been described: three from submerged wood and one as a smut of leaf bases and stems of *Ruppia maritima*. In two of those from wood, the Hymenomycete *Digitatispora marina* and the Gasteromycete *Nia vibrissa*, the basidiospores are of the branched tetra-radiate form, similar to the conidia of many aquatic Hyphomycetes

(Fig. 4.1). The majority of the wood-inhabiting fungi recorded are Ascomycotina. In all, 149 species of Ascomycotina have been described. No truly
marine Discomycetes have as yet been encountered. The only one known,
Orbilia marina, is found on decaying algae but above the high water mark.

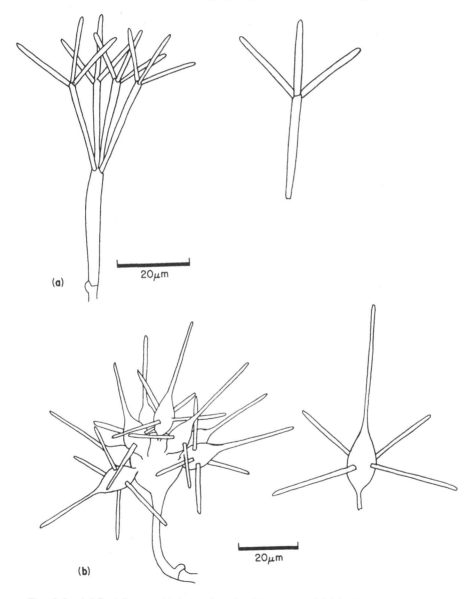

Fig. 4.1 (a) Basidium and tetra-radiate basidiospores of *Digitatispora marina*. (b)
Basidium and five-armed basidiospores of *Nia vibrissa*.

Only two Plectomycetes have been recorded, both from dead wood. The largest number of species are found in the Pyrenomycetes and Loculoasco-mycetes. Two orders, one from each of these two classes, contain the majority of marine species. The Sphaeriales contain 84 species and the Dothideales some 51 species. The majority of the Sphaeriales belong to the exclusively marine family Halosphaeriaceae. The ascospores of most species have appendages or gelatinous sheaths or both (Fig. 4.2a) and asci which swell and deliquesce at or before ascospore maturity so that the ascospores are not violently discharged. As with any aquatic propagule, the problem of anchorage after dispersal is a very real one and in this case the appendages and sheaths may aid impaction onto suitable substrates rather than help to

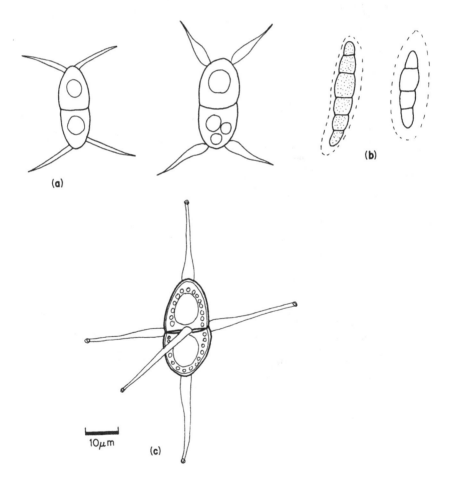

Fig. 4.2 Ascospores of marine wood-inhabiting fungi. (**a**) *Halosphaeria quadricornuta* and *H. salina*. (**b**) *Leptosphaeria neomaritima* and *L. contecta.* (**c**) *Ceriosporopsis calyptrata.*

maintain buoyancy. In contrast to the Sphaeriales, most species of the Dothideales belong to well-known terrestrial genera, such as *Leptosphaeria* (Fig. 4.2b), *Mycosphaerella* and *Pleospora*, but other genera, such as *Paraliomyces*, appear to be exclusively marine. In addition, 56 species of fungi have been described on purely conidial characters and placed in the Deuteromycotina. It is most likely that the majority of these are conidial Ascomycotina.

The remaining marine fungi consist of about 180 species of yeasts and 100 species of so-called lower marine fungi. The latter are all Mastigomycotina. Members of the Zygomycotina appear to be absent from the sea. Representatives from all classes of the Mastigomycotina occur but the most commonly encountered ones are simple chytrid-like Oomycetes which belong to the exclusively marine order Thraustochytriales (Fig. 4.3).

Kohlmeyer and Kohlmeyer (1979) attribute the relative paucity of different types of marine fungi to the fact that, compared to the land masses, the oceans provide a stable environment with small changes in temperatures and salinities. Organic substrates, such as algae and plant litter, which provide nutrients for fungi are concentrated mostly along the shores. They suggest that the open sea is a fungal desert where only yeasts or lower fungi may be found attached to plankton or pelagic animals and, in contrast to terrestrial habitats, the incubator-like stable environment of the oceans and the comparable small number of hosts and substrates found there have probably not exerted enough selective pressure during the course of evolution to induce the formation of a high number of different types of fungi. Be this as it may, a

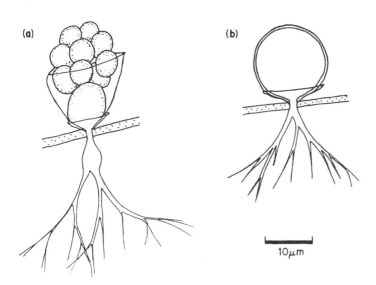

Fig. 4.3 *Thraustochytrium proliferum*. (**a**) Discharging zoosporangium. (**b**) Thick-walled resting sporangium.

wide diversity of types does occur in some groups such as the Halosphaeriaceae and Thraustochytriales.

Substrates and hosts of marine fungi

Wood

Many marine fungi are wood inhabiting. A large number of these have been
recorded from microbial free test blocks of wood after immersion in the sea
for varying periods but timber groynes and driftwood may also be colonized.
Barghoorn and Linder (1944) in their pioneer research on these fungi demonstrated that a number could produce typical soft rot cavities in the middle
layers (S_2 layer) of the secondary walls of the xylem elements in wood.
Eighteen of the 110 or so marine Ascomycotina so far tested caused a soft rot
of wood. Some of these also bring about substantial weight losses in the
wood. For example, after 18 weeks at 27°C, *Corollospora maritima*, a
strongly cellulolytic species, caused an average weight loss of 26% in beech
blocks incubated in 0.1% yeast extract seawater broth (Jones, 1971). In
contrast to the hyphae of these soft rot fungi, those of *Nia vibrissa* form bore
holes passing from cell to cell in submerged beech and pine blocks. Its extracellular enzymes are readily diffusible and cause a gradual thinning of the
wood cell walls giving a typical white rot decay pattern.

   Successional patterns of fungi on wood submerged in the sea have been
reported but it is not altogether clear from these whether these patterns are a
result of one species replacing another or whether they merely reflect the
variation in the times that different species take to produce their reproductive
structures. In Table 4.1, the time at which reproductive structures of
different fungi first appeared on beech test blocks submerged in the sea at
Brixham, Devon, is given. In culture under laboratory conditions, the
majority of these fungi reproduced in 2–6 weeks, so that the time taken to
reproduce is not the only factor responsible for their appearance on
submerged wood. Temperature and the type of wood are also important
factors in determining the colonization pattern. Certain fungi show

Table 4.1   Time of appearance of reproductive structures of marine wood-inhabiting fungi on beech blocks immersed in the sea. (After Jones, 1971).

	Weeks									
	6	12	18	24	30	36	42	48	54	60
Zalerion maritima										
Lulworthia purpurea										
Lulworthia floridana										
Lulworthia rufa										
Ceriosporopsis halima										
Ceriosporopsis cambrensis										
Halosphaeria appendiculata										

preference for a certain wood and in this respect are like many terrestrial Ascomycotina. For example, *Halosphaeria appendiculata* prefers beech and *Ceriosporopsis circumvestita*, pine.

Seaweeds

Some 50 Ascomycotina and 15 conidial fungi have been recorded as parasites or saprotrophs of seaweeds. Fucoids, kelps and red algae frequently support fungi. Two of the most interesting fungi occurring on seaweeds are *Spathulospora phycophila* and *Mycosphaerella ascophylli*. Five species of *Spathulospora* have been described. All are obligate parasites of red algae in the genus *Ballia* in the southern hemisphere. *S. phycophila* occurs on *B. callitricha* and *B. scoparia*. It does not possess typical hyphae but forms a crustlike thallus of irregular thick-walled cells surrounding branches of the host, with peg-like branches penetrating the algal cell walls, and almost polygonal thick-walled assimilative cells developing within the algal cells. From its peculiar reproductive structures its nearest relatives in the fungi are the Laboulbeniales, an odd order, restricted to insects.

M. *ascophylli* is universally associated with its hosts *Ascophyllum nodosum* and *Pelvetia canaliculata*. Every plant of these is infected and the association is initiated very early in the life of the alga. One month old sporelings are usually already infected. The fungus is systemic and perennial in the algae. A network of hyphae grows intercellularly in the cortex and medulla. The mycelium follows the growing tip of each branch of the algae remaining some ten cells behind the apical cell. The reproductive structures of the fungus are however restricted to the receptacles, the reproductive structures of the algae, in attached plants. Ascocarps and spermogonia develop, immersed in the walls of the receptacles (Fig. 4.4). There is no evidence that the fungus causes damage or destruction of the host tissues. This has led to the suggestion that these two algae are lichenized. The alga and the fungus appear to depend upon each other. They never occur separately in nature, indicating that the fungus is obligatorily dependent upon its host. The algae are also dependent upon the fungus as uninfected algal sporelings do not survive beyond a particular stage. Both algae occur in the upper intertidal zone and one suggestion is that, like lichens, the fungus and alga in association are more resistant to desiccation. A number of littoral true lichens, such as species of *Lichina* and *Verrucaria*, are very common, but several other more loose associations are also known. In these, such as the association between the sub-littoral *Ectocarpus fasciculatus*, epiphytic on the stipes of *Laminaria* and *Pharcidia laminariicola*, the hyphae of the latter loosely entwine the filaments of the alga. No new structural entity originates from such association and the algae, at least, are able to occur free-living. Such associations as these and the *Mycosphaerella/Ascophyllum* or *Pelvetia* associations are not lichens in the strict sense but it is likely that both partners benefit.

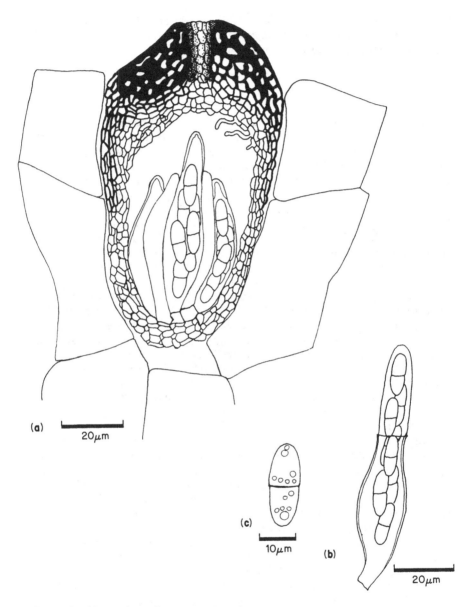

Fig. 4.4 *Mycosphaerella ascophylli*. (**a**) Vertical section of a mature ascocarp with asci in various stages of development. (**b**) Bitunicate ascus. (**c**) Ascospore.

Animals

Fungal parasites of sea fish appear to be quite uncommon. Those which are known are poorly understood, both in terms of their taxonomic status and the effects that they have on fish populations. *Ichthyophonus hoferi*, an aseptate fungus, is prevalent in herrings, *Clupea harengus*, in the Western Atlantic and probably plays a significant part in the control of the size of herring populations. Infection is systemic, affecting the viscera and musculature in particular. Internally, the fungus appears encapsulated as small white nodules, which in heavy infections is accompanied by extensive tissue invasion and necrosis, leading to the death of the fish.

Whereas the Zygomycotina appear to be absent from the seas, members of all classes of the Mastigomycotina occur. A relatively large number from the Lagenidiales, Saprolegniales and Peronosporales are known to infect the eggs and larvae of marine invertebrates, but one of the best known marine fungal diseases is the shell disease of oyster, *Ostrea edulis*, caused by the aseptate fungus *Ostracoblabe implexa*. The fungus is endemic in coastal waters of Western Europe and it bores into the shell, obtaining its nutrients from the organic matrix, especially the protein, of the shell. The growth of the fungus within the shell does not harm the living tissues until the mycelium reaches the inside. It then grows between the living mantle of the oyster and the shell. This irritates the tissues and they respond to this by the deposition of more shell material with a far higher protein content than normal around the site of irritation. This horny protein, conchiolin, is an ideal nutrient for *O. implexa* and it then grows faster. In severe infections such a response leads to extreme deformation of the shell.

A great deal remains to be learnt about such parasites. This is also true of another isolated group of simple chytrid-like Oomycete and exclusively marine fungi, the Thraustochytriales. They are the most commonly encountered fungi on marine sediments. Their major role seems to be as decomposers of moribund material derived in the main from terrestrial sources and the photic zone of the sea.

Most of the yeasts recorded from the sea have been obtained by isolation from seawater and not all of these may be truly marine. A number are parasitic on animals. For example, *Metschnikowia bicuspidata* var. *australis* parasitizes the brine shrimp, *Artemia salina*, and it appears to possess an active predatory mechanism. It has asci containing two sharp, needle-like spores which are expelled violently at its host (Fig. 4.5).

The diversity of freshwater fungi

In the majority of those fungi which are freshwater aquatics, their whole evolutionary history seems to have been in water. These are the Mastigomycotina. They all possess motile zoospores as the main units of dispersal and can obviously function only in a watery environment. The three distinct types of zoospore found in members of this sub-division suggest that the group is

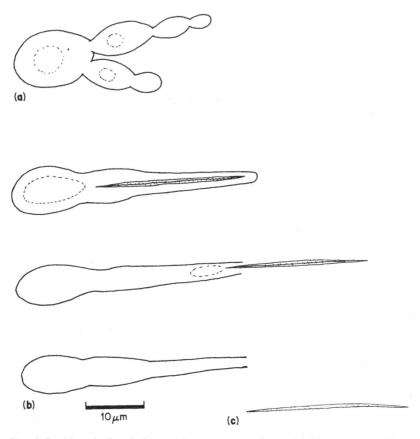

Fig. 4.5 *Metschnikowia bicuspidata* var. *australis*. (**a**) Budding cells. (**b**) Mature, discharging and discharged ascus. (**c**) Single ascospore.

not a natural one but is an assemblage, an extraordinarily diverse one, of organisms with particular features, such as asexual reproduction by zoospores, in common. Unlike in the sea, both Discomycete and Pyrenomycete, as well as Loculoascomycete, Ascomycotina occur. No fungi are known which produce basidiocarps in freshwater but some conidial Basidiomycotina do occur. As a spore-gun, the basidium is not powerful enough to project its spores away from the basidium in water, whereas the ascus is much more powerful and can project its spores into water. This feature of the basidium may account for the paucity of aquatic Basidiomycotina.

In freshwater systems, two major groups have attracted most attention: the zoosporic 'water moulds', mostly Saprolegniales, and the conglomerate of conidial fungi called aquatic Hyphomycetes, the majority of which are Ascomycotina and the remainder Basidiomycotina. Most of this attention has been centred on their diversity and modes of dispersal. These and the aquatic Ascomycotina are discussed below.

'Water moulds'

The water moulds, the Saprolegniales, are by far the best known of the freshwater fungi and can be isolated from almost any freshwater habitat – river, lake, stream, pond or persistent puddle. They also appear to be common in soils, especially those which are permanently wet. Water moulds occur on dead animal and plant remains, particularly dead insects, seeds and fruits. They rarely develop in sufficient abundance to be recognized directly in the field and baiting techniques, using such baits as the achenes of *Cannabis*, the seeds of Brassicas, the caryopses of grasses or ants' eggs, have provided data on species distribution in particular habitats. Relatively fewer numerical estimates of abundance have been made using plating techniques. Although we have a knowledge of the various water moulds present in a particular freshwater habitat and the patterns of their distribution, we know nothing of their activity in these habitats nor the basic causes of their distribution. As yet no attempts have been made to estimate, by biochemical techniques, the total productivity of any group of fungi in any aquatic ecosystem.

The nearest to a true assessment of their activity in any aquatic system that can be obtained is to use a quantitative method, such as that devised by Willoughby (1962), to assess the number of viable propagules present. The results give some indication of the relative activity of the different genera and species present, if only because they are all related fungi with similar modes of reproduction. But it should be borne in mind that if a viable spore count were made to estimate fungal activity in a habitat such as the soil, where a great diversity of different fungi occur, a very misleading picture would be obtained. Willoughby's method is an indirect one. Motile and encysted zoospores of the Saprolegniales are too small to count and lack any distinctive morphological features to enable generic and specific determinations to be made. He studied the occurrence and distribution of saprolegniaceous fungi in lakes in the English Lake District. Surface water samples were collected and five or ten 1–4 cm^3 aliquots were incorporated into molten but cool oatmeal agar. The agar was poured into Petri dishes and when set was divided into eight equal sectors, each of which was placed in a dish of sterile distilled water. Twenty mm diameter discs, cut with a cork borer, are easier to handle. Fringes of mycelium develop around the margins of the sectors on incubation. Any Saprolegniales can easily be recognized by their coarse, stiff, radiating and aseptate hyphae and identified to the genus by the manner of release of their zoospores (Fig. 4.6). By counting the colonies which develop, the number of viable propagules per unit volume of lake water can be estimated. Although it is a time-consuming technique, it is a considerable advance over baiting techniques usually used.

Using this technique, Willoughby found that the number of viable propagules in the water of Lake Windermere fluctuated widely. At the lake margin figures ranged from 25–5200 dm^{-3}. Figures were above 400 dm^{-3} in every month from August to January, whereas in the centre of the lake there was always less than 100 dm^{-3}. The counts were low in spring and higher in the

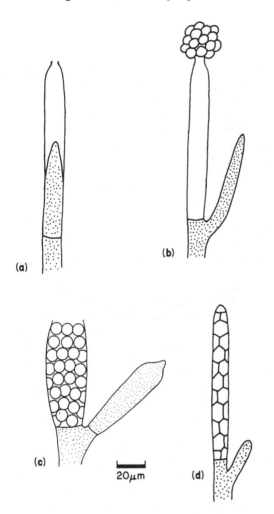

Fig. 4.6   Zoospore release in the Saprolegniales. (**a**) *Saprolegnia*. After release of primary zoospores, a second zoosporangium develops within the empty one. (**b**) *Achlya*. Primary zoospores form a clump of cysts at the exit papilla and a new zoosporangium develops laterally. (**c**) *Thraustotheca*. Primary zoospores encyst within the zoosporangium and are released by breakdown of its wall. A new zoosporangium proliferates laterally. (**d**) *Dictyuchus*. Primary zoospores encyst within the zoosporangium and release secondary zoospores through separate pores. Proliferation is again lateral.

late summer and autumn. The reasons for such differences are not at all clear. It could be that the open water towards the centre of the lake is a 'fungal desert' like the open sea. More suitable substrates may occur near the shore or they may drift there from the centre. There was some correlation between high counts and heavy rainfall and high lake levels. Speculations about the origin of such variability in counts is of little value until a survey of substrate availability, preference and distribution has been carried out. Apart from giving some concept of the magnitude of the numbers present, such figures tell us little else about the ecology of these fungi but do give us some insight into the preponderance of particular genera. In Lake Windermere and also Esthwaite Water, zoospores and cysts of *Saprolegnia* were easily the most common, followed by those of *Achlya* and *Aphanomyces*, whereas those of *Dictyuchus* and *Leptolegnia* were only occasionally encountered. Many of these are known to grow on small particulate substrates, such as seeds and the exuviae and dead bodies of insects, which may be freely suspended in the water.

Colonization of insect exuviae

Dick (1970) investigated the colonization of small insect exuviae in Marion Lake, B.C., Canada. *Saprolegnia diclina* and *Aphanomyces laevis* were by far the most important primary colonizers but there were some clear substrate preferences. Very much higher numbers of *Saprolegnia* were found on Trichoptera exuviae, whereas *Achlya* and, to a lesser extent, *Leptolegnia*, were more frequent on Anisoptera exuviae. *S. diclina* and *A. laevis* were also the most frequent colonizers of the exuviae of chironomid flies. Almost all the exuviae in shallow water were colonized within 24 h of being produced, whereas those released and maintained in traps at depth (4.5 m) were relatively little colonized. This is in keeping with the fact that the exuviae become available at the air/water interface and that the zoospores respond to high levels of oxygen in the water so that a distinct distribution of zoospores with respect to depth would be expected. Dick also investigated the distribution of the Saprolegniales in and around the same lake by baiting soil and mud samples. The maximum number of propagules and the greatest diversity of species were recorded from the submerged muds at the margin, the littoral zone, but there was an abrupt fall off in the submerged muds, one metre from the shore and beyond. Although the lentic muds contain viable spores, there is very little evidence for much, if any, metabolic activity of the Saprolegniales there, with the possible exception of one species of *Achlya*. There were also differences in the species composition at different sites. The sites above the water level, the terrestrial sites, had a different micro-flora characterized by *Saprolegnia litoralis*, *S. terrestris*, *Scoliolegnia asterophora* and *Achlya sparrowii*, but the littoral zone was characterized by the greatest abundance and diversity of especially species of *Achlya* and *Saprolegnia*. This is to be expected as this is the zone of high plant and animal activity, providing substrates for these. Other substrates formed in the lake continually drift towards the margin and these are added to by others introduced from terrestrial sites by drainage into the lake. Viable propagules of Saprolegniales in this drainage water may also add to the diversity of species and their abundance.

Other groups of essential aquatic fungi show similar distributions. For example, Willoughby (1961) compared the chytrid flora of similar zones at Esthwaite Water. He selected three arbitrary zones distinguished on their liability to inundation along a transect embracing soil, marginal sites and lentic environments. The field zone was above the highest recorded level of the lake, the marginal zone was subject to periodic inundation and the submerged zone was at about 2 m deep in the beds of the shore-weed, *Littorella uniflora*. Thirty six species of chytrids were recorded. Only one of these, *Phlyctorhiza variabilis*, occurred frequently in all three zones. All three zones had different chytrid micro-floras. As would be expected, the littoral zone contained species from both adjoining zones but there was very little overlap between terrestrial and lentic communities. A distinct, but limited, terrestrial chytrid micro-flora co-existed with a rather more species-rich aquatic one. For example, *Rhizophlyctis rosea* is essentially a terrestrial chytrid. It was never isolated from submerged muds. Thus the Chytridiales, like the Saprolegniales, are not all truly aquatic. Although they occur predominantly in freshwater, the soil has a regular and constant chytrid micro-flora. If about 3 g of soil is placed in a Petri dish, covered with sterile water and suitable baits added, chytrids will appear after 2–3 days at 20°C. *R. rosea* is strongly cellulolytic and appears on squares of cellophane which have been previously boiled to remove plasticizers and used as baits for alkaline soils from fields and hedgerows. Pine or other pollen dusted onto the surface of water covering soil samples from a pine wood produces crops of numerous species of *Rhizophydium* (Fig. 4.7).

Parasitic Saprolegniales

The few parasitic members of the Saprolegniales have been more intensely studied. *Aphanomyces astaci* causes a 100% mortality in attacked populations of the crayfish, *Astacus astacus*. It is a cause of considerable concern in Sweden where it is claimed that it is eradicating the crayfish. *Aphanomyces astaci* shows interesting adaptations to its host. It produces abundant chitinase which is necessary for infection and out of 25 common carbohydrates and organic acids, tested as sole carbon sources, it grew well only on glucose, the blood sugar of *Astacus*.

Species of *Saprolegnia* have been implicated in a number of fish diseases. But it is not absolutely clear whether these water moulds can infect uninjured fish or whether they are wound parasites. Handling or bruising freshwater fish certainly leads to fungal infection and high mortality rates. Also, inoculation of wounded platyfish, *Platypoecilus maculatus*, with zoospores of *Saprolegnia parasitica* under controlled laboratory conditions resulted in death within 24 h.

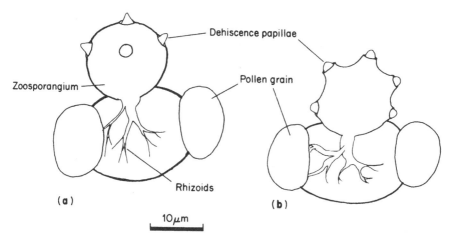

Fig. 4.7 Chytrids on pine pollen. (**a**) *Rhizophydium globosum*. (**b**) *R. elyensis*.

The role of *S. parasitica* in a serious skin disease of mature salmonid fish, such as salmon and brown trout, has always been controversial. The disease, ulcerative dermal necrosis, has been known since the early 1860s and a recent epidemic, which has now waned, began in 1964 in Co. Kerry, Ireland, and by 1968 had spread to almost all the British salmon rivers. In salmon, *Salmo salar*, it characteristically occurs as they enter freshwater on their spawning migration from the sea. The exact primary cause is not known but it may be a virus. Specific sites on the scaleless skin of the head are infected, causing a progressive necrosis. The lesions heal rapidly in the spring and summer. But at low temperatures, such as occur in winter, the disease spreads. The lesions become secondarily infected by *S. parasitica*. The musculature is exposed and the fungus penetrates deeply into this. It has been isolated from muscles 15 mm below the infected skin. Once this happens the fish soon dies from circulation failure as a result of the osmotic dilution of the blood. *S. parasitica* thus appears to be a secondary wound parasite especially at low temperatures. The condition is often complicated by bacterial infection.

Freshwater Ascomycotina

In contrast to the Basidiomycotina, many Ascomycotina are common in freshwater. The ascus as an explosive sporangium is capable of functioning in water. Many purely terrestrial apothecial forms can have their exposed hymenium submerged without ascus discharge being hindered, whereas the normal Hymenomycete hymenium, although exposed, is covered to protect it from rain as it is unable to function if wetted or submerged. The basidium has insufficient power to project its spores through water. Although there is a considerable micro-flora of submerged freshwater Ascomycotina, comparatively few studies have been made on it. *Leptosphaeria lemaneae* is a parallel to the marine *Mycosphaerella ascophylli*. It is only found growing in the thalli of species of the red alga *Lemanea* where it does not appear to cause any harm. Two very rich sources of freshwater Ascomycotina are decaying stalks of reed swamp plants, such as *Phragmites australis (communis)* and *Typha latifolia* and waterlogged twigs and branches. Perithecial, pseudothecial and, in contrast to marine habitats, apothecial forms abound on these. Many of these are from well-known terrestrial genera and show few, if any, modifications to the aquatic environment. Some, such as species of *Apostemidium* (Fig. 4.8a), essentially an aquatic genus, have very long thread-like ascospores reminiscent of the sigmoid ones of some aquatic Hyphomycetes. Others have mucilaginous appendages. In *Pleospora scirpicola* (Fig. 4.8b), the mucilaginous sheath of the ascospore swells enormously, following liberation from the ascus again to produce an elongate curved structure in which the spore is embedded. Such modifications are much commoner in marine forms and are probably of significance in underwater impaction.

Aquatic Hyphomycetes

The so-called aquatic Hyphomycetes occur in particular abundance on submerged decaying leaves of a very wide range of dicotyledonous trees and shrubs in well-aerated, shallow and fast-flowing streams. Their physical requirements for reproduction, at least, appear to be clear running freshwater where aeration is good and turbulence is pronounced. Although some of these fungi were first recognized by de Wildeman (1893), their existence as a very extensive and distinct micro-flora on such substrates was first reported in the early 1940s by Ingold (1942). Virtually all of the subsequent intensive research and interest in this remarkable group of fungi can be attributed to Ingold's infectious drive and enthusiasm. Excellent accounts of these are given in his publications (Ingold 1966, 1976) and in his illustrated guide to their identity (Ingold, 1975).

Conidial development

The septate mycelium of these fungi permeates the leaf tissues and conidiophores are produced, projecting into the water, especially from the veins, as the leaves are softened by aerobic decay and are beginning to become converted into leaf skeletons.

The two most striking features of these fungi are the form and the variety of modes of development of their conidia. In marked contrast to most terrestrial Hyphomycetes, very few of these have spherical, oval or slightly elongate conidia. The conidia of some are much elongated and curved, often sigmoid with the curvature in more than one plane. Others are multi-branched, but by far the majority have four arms usually diverging from a

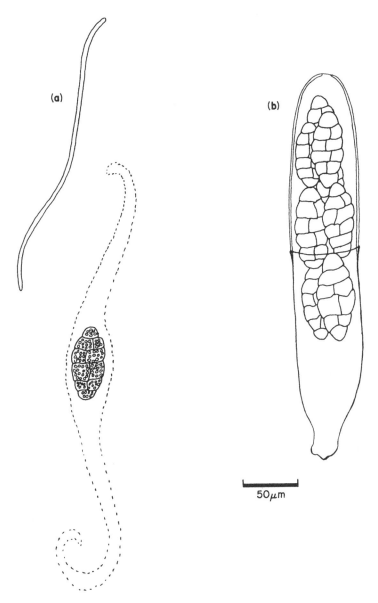

Fig. 4.8 Freshwater Ascomycotina. (**a**) Thread-like ascospore of *Apostemidium guernisaci*. (**b**) Ascospore with mucilaginous sheath and bitunicate ascus of *Pleospora scirpicola*.

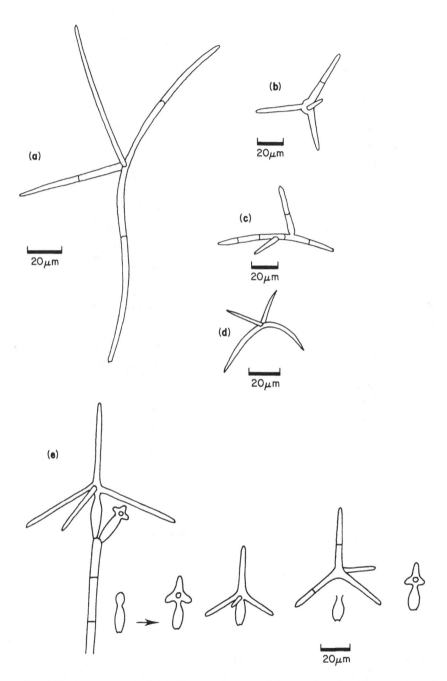

Fig. 4.9 (*Above and right*) Tetra-radiate conidia of Aquatic Hyphomycetes. (*Above*) (**a**) *Tetrachaetum*; (**b**) *Lemonniera*; (**c**) *Tricladium*; (**d**) *Alatospora*. Phialoconidial development in (**e**) *Lemonniera* and (*right*) (**f**) *Alatospora*. (**g**) Thalloconidial development in *Tricladium* After the release of one conidium another arises at a slightly different level.

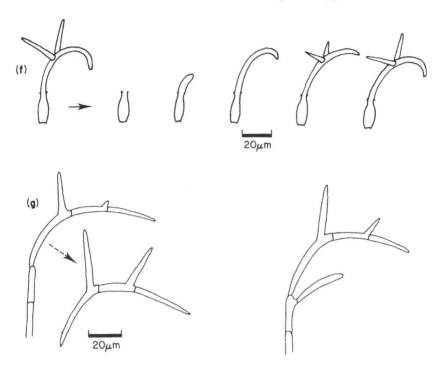

common point (Fig. 4.9a,b). These latter tetra-radiate conidia show a considerable diversity of developmental patterns. In the main there are two distinct types of conidia – phialoconidia and terminal thalloconidia or aleurioconidia. Blastoconidia form a minority type. In some the attachment to the conidiophore is at the tip of one of the four arms (Fig. 4.9f) and in others it is near the point of their divergence (Fig. 4.9e). In some the four arms arise synchronously and in others in strict succession (Fig. 4.9e, g). The phialoconidium is formed at the apex of a special bottle-shaped cell, the phialide. The conidia reach full size before they are cut off by a septum. As soon as this happens, another conidium begins to be formed from the same phialide (Fig. 4.9e, f). The thalloconidium is cut off by a septum from the conidiophore at an early stage in its development. After it is liberated, the conidiophore does not normally produce another directly at the same level (Fig. 4.9g). Although there are quite fundamental differences between phialoconidia and thalloconidia, the precise pattern of development in a phialoconidial genus may be exactly paralleled in a thalloconidial one. Blastoconidia are produced as buds from the conidiophore. They are produced singly from different sites and do not show any clear relationship to the apex of the conidiophore. Such variety suggests two things. First, they must have evolved repeatedly and independently and second, this type of conidium must have some especial biological significance in the aquatic habitat.

Teleomorphic states

Most conidial fungi are imperfect states of Ascomycotina but as yet only a few connections have been made between these imperfect states (anamorphs) and any perfect states (teleomorphs). Evidence linking anamorphs and teleomorphs in these aquatic fungi has been obtained in two ways. Some cultures derived from single conidia and incubated under particular conditions, such as low temperatures (10–15°C) and diffuse light supplemented with near ultra-violet wavelengths, have given rise to teleomorphs. Alternatively, single spore cultures have been started from natural collections of suspected teleomorphs and sub-cultures from these placed in sterile water and forcibly aerated to encourage the development of conidia. From such studies and other evidence, such as the type of septa in the conidia, it is clear that a wide variety of Ascomycotina and at least some Basidiomycotina have developed this type of conidia. For example, some isolates of *Flagellospora penicillioides* and *Heliscus lugdunensis* have been induced to produce different perithecial states, both referable to the genus *Nectria*. *Anguillospora crassa* has an apothecial perfect state referable to the genus *Mollisia*, with inoperculate asci, while *A. longissima* has a pseudothecial one referable to the genus *Massarina*. Ascospore derived cultures of another species of *Massarina* produced conidia of *Clavariopsis aquatica* (Fig. 4.10a), when placed in sterile water at 15–18°C, whereas the perfect state of *Actinospora megalospora* (Fig. 4.10b), which has the largest conidia of all the aquatic Hyphomycetes, some being 500 μm in width, has a perfect state referable to *Miladina*, an operculate Discomycete. Of all these anamorphs, *Flagellospora* and *Anguillospora* have conidia of the sigmoid type. The others have tetraradiate ones.

A number of others are believed to be Basidiomycotina on the evidence of the presence of dolipore septa or clamp connections at the septa. The conidia of *Dendrospora* (Fig. 4.11a) and *Dendrosporomyces* are both multibranched with a straight or curved main axis and laterals arising in succession from more than one level near the base. The former has septa with a single simple septal pore characteristic of Ascomycotina, whereas the latter has dolipore-type septa characteristic of the dikaryophase of Basidiomycotina. The tetraradiate conidia of *Ingoldiella hamata* (Fig. 4.11b) have clamp connections at each septum. The genus *Tricladium* produces thalloconidia, consisting of a long main axis and two laterals arising at different levels. One species has a single septum with a clamp connection in the main axis between the two laterals (Fig. 4.11c). Cultures derived from these conidia have produced crust-like basidiocarps identical with those of *Leptosporomyces galzinii*, which is terrestrial and occurs on wood and bark of numerous trees, such as *Acer* and *Quercus*, and conifers, as well as leaves in the litter. Conidia are produced only in submerged culture so this fungus appears to be amphibious.

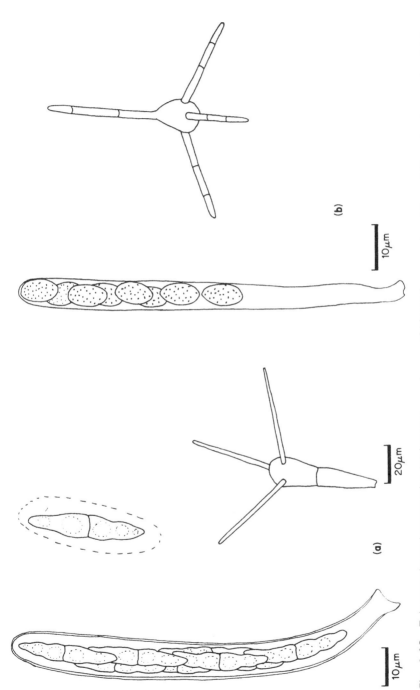

Fig. 4.10 Teleomorphs and anamorphs. (**a**) Ascospore and ascus of *Massarina* sp. and conidium of *Clavariopsis aquatica*. (**b**) Ascus of *Miladina lechithina* and conidium of *Actinospora megalospora*.

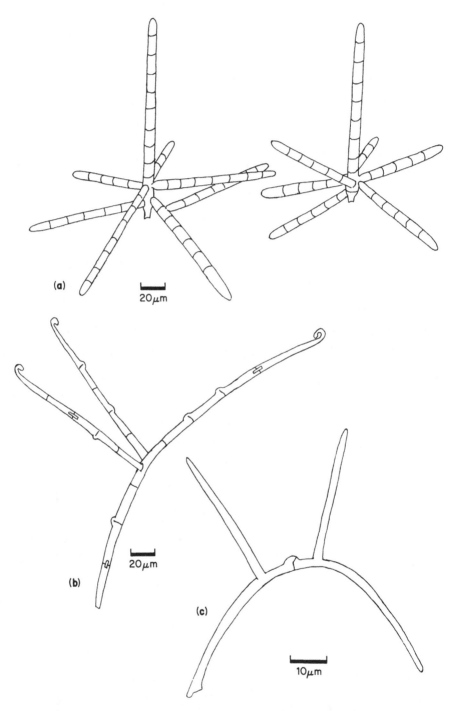

Fig. 4.11 Tetra-radiate conidia of Basidiomycotina. (**a**) *Dendrospora fusca*. (**b**) *Ingoldiella hamata*. (**c**) *Tricladium* conidial state of *Leptosporomyces galzinii*.

Other tetra-radiate spore forms

The tetra-radiate spore form is not limited to aquatic Hyphomycetes. Such shaped propagules have developed in other aquatic groups outside the fungi: in some *Sphacelaria* spp., brown algae and *Actinastrum hanzschii*, a coenobial green alga. Aquatic fungi from other taxonomic groups also produce tetra-radiate propagules; good examples are the conidia of some *Entomophthora* spp. from the Zygomycotina (Fig. 4.12a), and the cell groups of the yeast, *Candida aquatica* (Fig. 4.12b). Other examples, already mentioned, are the ascospores of the marine wood-inhabiting Halosphaeriaceae, such as those of *Halosphaeria quadricornuta* (see Fig. 4.2a) and *Ceriosporopsis calyptrata* (see Fig. 4.2c), but most striking of all are the basidiospores of the marine Hymenomycete *Digitatispora marina* (see Fig. 4.1) and the Gasteromycete *Nia vibrissa* (see Fig. 4.1). In *D. marina*, the basidiocarp is crust-like and the hymenium is exposed and bears basidia, each with four sessile tetra-radiate basidiospores. These are presumably released passively by wave

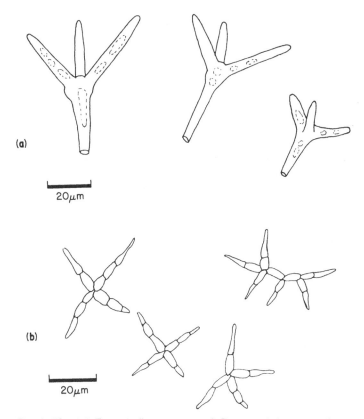

Fig. 4.12 (a) Tetra-radiate spores of *Entomophthora* sp. (b) Tetra-radiate and more branched cell groups of *Candida aquatica*.

action. In *N. vibrissa*, the basidia are enclosed in a small, up to 5 mm diameter, globose, orange-pink basidiocarp. The basidia are eight-spored and each spore has a spherical centre with five arms and looks like small replicas of the Hyphomycete *Actinospora megalospora* (Fig. 4.10b). These become immersed in a mass of mucilage, formed as the basidia break down. They are released when the wall of the basidiocarp eventually ruptures.

Biological significance of the tetra-radiate spore form

Since such spores are a special feature of aquatic environments, two quite plausible theories have been put forward as to their possible biological significance. The first of these is that the tetra-radiate spore settles more slowly than spores of a more normal form of the same mass and so are dispersed further. In planktonic organisms, a large surface to volume ratio has often been interpreted as of value in retarding sedimentation. Webster (1959) produced heavy concentrations of conidia of a variety of aquatic Hyphomycetes by agitating discs, cut from cultures on agar, in relatively large volumes of water. He found that the tetra-radiate types did not settle more slowly than other types. All settled at about 0.1 mm sec^{-1}, whereas the flow rate of the stream from which he isolated his particular fungi was in the order of 1000 mm sec^{-1}. Thus they sink in water at a rate which is insignificantly slow in relation to water flow. In any case general turbulence would mask any sedimentation.

The second theory is that they have evolved as impaction devices which are important in securing their anchorage on an appropriate substrate after dispersal. To test this, Webster (1959) used a water tunnel through which known concentrations of spores could be passed at controlled speeds and he determined the impaction efficiency of spores on rods inserted into the tunnel at right angles to the water flow. Impaction efficiencies for all types of spores were low but tetra-radiate conidia were impacted with greater efficiency than were those of a curved sigmoid shape and with even greater efficiency still than oval or rod-shaped spores. A possible reason for this is that tetra-radiate conidia make contact at three points, the tips of three arms making a tripod with the fourth arm projecting up into the water. This appears to make a very stable attachment. They also very quickly develop pads or appressoria at the tips of the three attached arms and are subsequently very difficult to remove.

It is a feature of these spores that they do not germinate while suspended in water but do so very rapidly if they come in contact with a solid object. Iqbal and Webster (1973a) have shown that the conidia of some aquatic Hyphomycetes produced in culture may remain viable for at least a month if continuously aerated. For example, conidia of *Tricladium splendens* retained 94% viability and those of *Articulospora tetracladia* 91%, after one month at 13°C. Another feature which helps to maintain these spores in suspension, but which also eventually brings them to the surface, is that as air bubbles move upwards through the water they temporarily lodge up between the arms and help to lift the spores upwards. The effectiveness of this uplift depends upon the shape of the spore. In general the branched type of spore is more readily removed from suspension than are spores of a more conventional shape, but again tetra-radiate spores are most effectively removed. Thus air bubbles carry spores to the surface. In streams, such bubbles collect as scum or foam at barriers formed by twigs and branches to stream flow, or as more persistent cakes of foam below rapids or miniature waterfalls. Such scum or foam is an exceedingly rich source of these spores. In such situations the spores themselves are in an ideal position to colonize leaves as they fall into the stream. The best way to collect a sample of spores is to scoop up the scum or foam with a spoon, gently agitate to burst the air bubbles and to transfer the resultant liquid to a vial or jar containing an equal amount of a fixative, such as formalin acetic alcohol. A fixative must be used, otherwise the spores will germinate when they touch the bottom on settling out. If some of the sample is then examined with a microscope, an idea of the variety of aquatic Hyphomycetes in the stream, above the site of collection, can quickly be obtained. Since most of the spores can be readily identified, a list of species occurring in the stream can be compiled, but, because the spores of different species are removed by air bubbles at different rates, a quantitative assessment may give an inaccurate impression of the comparative abundance of the species present. It can be shown that this is so by comparing the spore content of foam samples and quantitative assessments of the actual spore content of the water obtained by membrane filtration.

Aquatic Hyphomycetes are undoubtedly much more common and abundant in terms of spore numbers in rapidly flowing streams than they are in slower flowing ones. The faster flowing streams are better aerated and more turbulent. Webster and Towfik (1972) investigated spore production of a number of aquatic Hyphomycetes in relation to aeration. They grew these on agar and then cut out discs and aerated these in water. Webster and Towfik showed that in *Articulospora tetracladia* and *Lemonniera aquatica* there is an increase in spore production with increased aeration (Fig. 4.13), and they concluded that this enhanced spore production by forcible aeration is related to the mechanical effects of increased turbulence. They could find no evidence that it was due to either increased availability of oxygen or to the removal of a volatile inhibitor of sporulation. For instance, they found that if they aerated two sets of discs at the same rate but in one using a single orifice and in the other a number of smaller orifices, the discs were agitated more in the former, as there was greater turbulence, and such discs produced more conidia. The effect of forced aeration in stimulating spore production in these fungi has been interpreted by Webster (1975) to be due to two phenomena. The time taken for spore development and detachment is several hours shorter at high aeration rates ($1000 \ cm^3 \ min^{-1}$) than at low ones ($100 \ cm^3 \ min^{-1}$). The number of conidiophores per unit area of culture surface is significantly greater at higher aeration rates. This latter effect is probably due to increased hyphal branching induced by turbulence.

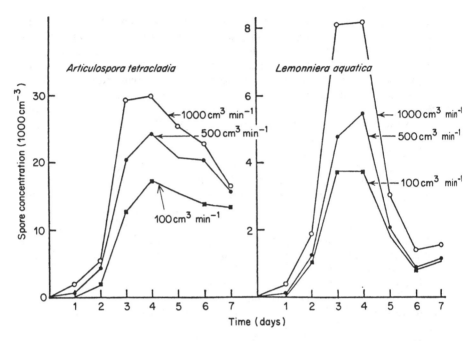

Fig. 4.13 Effect of aeration rate on daily spore output from culture discs of *Articulospora tetracladia* and *Lemonniera aquatica*. (After Webster and Towfik, 1972.) (Reproduced by permission of the British Mycological Society.)

Aquatic Hyphomycetes in other environments

It would thus appear that aquatic Hyphomycetes are adapted to turbulent flow by increased spore production. This leads to an intriguing question. Do they grow in other environments – aquatic or terrestrial – but produce spores less profusely or not at all because conditions are not conducive to spore development? Or, put another way, are they so common in fast flowing streams because increased turbulence induces increased spore production? Evidence is accumulating that some, perhaps many, of these fungi are not exclusively aquatic but that they might occur on leaves and other substrates, well away from water – flowing or standing. Bandoni (1972) collected fallen leaves of trees from well-drained sites on the campus of the University of British Columbia, Canada. When he submerged these in water, large numbers of tetra-radiate and sigmoid conidia were observed floating in the water. He listed nine species from these leaves. Two essentially tropical species, *Triscelophorus monosporus* and *Ingoldiella hamata*, are known to produce conidia on agar plates and also on leaves collected from the litter layer in well-drained sites without being submerged. Webster (1977) has reported on the seasonal distribution of fourteen aquatic Hyphomycetes on oak leaves, on the slopes of the valley through which the River Teign flows, in

Devon, England. He collected leaves at marked sites, at monthly intervals, throughout a year and then identified any aquatic Hyphomycetes which produced spores on the leaves, after they had been submerged in distilled water for 4 days at 10°C. Most records were made from leaves which had been collected from within 10 m of the river's edge, at sites which were occasionally inundated when the river was in flood. The numbers of conidia observed were often very low. From all the evidence available it would thus appear that some of these fungi have a terrestrial existence in the mycelial state.

One aspect which has yet to be satisfactorily explained is how these fungi manage to be dispersed upstream in spite of water flow. The most likely explanation would seem to be that the majority are amphibious with aquatic anamorphs and terrestrial teleomorphs. This is purely conjectural as so few teleomorphic states have been discovered so far; those which have, however, are either terrestrial or have developed only on substrates which have been kept damp rather than submerged, or allowed to dry out slowly. In terms of dispersal mechanisms, such a fungus as *Leptosporomyces galzinii* seems to be the ideal. Basidiocarps have been widely reported on varied substrates under terrestrial conditions. If they occurred on the banks of a stream, the conventional shaped basidiospores would be widely dispersed by wind, both towards and away from the stream's source. Some would become impacted onto leaves and when these fell they would grow within the leaves. If the leaves fell into a stream, tetra-radiate conidia would be produced and if they fell into the bank basidiocarps would eventually form. An amphibious habit would also explain long-distance dispersal from one well-isolated freshwater system to another.

Seasonal abundance

A very extensive literature has now been built up on the distribution and taxonomy of aquatic Hyphomycetes but relatively little quantitative work has been carried out on their ecology. As with other fungi, it is extremely difficult to obtain a meaningful estimate of their activity in nature. Iqbal and Webster (1973b) made quantitative estimates of their seasonal abundance by developing a technique which enabled them to count accurately the number of conidia present in a unit volume of water. The conidia, unlike the zoospores and cysts of members of the Saprolegniales, are of sufficient size and have such distinctive features that they can often be separated microscopically to the specific level. They selected three rivers, R. Exe, Creedy and Culm, around Exeter, Devon. Water samples, taken weekly from a number of sites, were filtered by suction through Millipore filters with 8 μm pore size. When dry the filters were treated with 0.1% cotton blue in lactic acid and heated at 50–60°C for 45–50 min. This treatment stained the spores and rendered the filters sufficiently transparent to be viewed with the low power of a microscope. The spores were then counted and identified and the number of spores per litre estimated. This provides a simple, yet elegant, technique for quantitative estimates of spore concentrations.

During the period August 1970 to October 1971, thirty-four species belonging to 24 genera were found in the three rivers. There were 20 species in the autumn spora but the winter spora was richest, with 32 species. The summer spora was very poor both in species and spore numbers. For example, spores of aquatic Hyphomycetes were not found in samples from the R. Exe during May and June. At other times the spore content was very high. For example, in the R. Creedy during October, the numbers increased from 1250 to 7500 dm^{-3} and during November remained stable at 7000 to 8300 dm^{-3}. In general, most conidia occurred during autumn to spring, with markedly fewer in the summer. These seasonal fluctuations can be partially explained in terms of substrate availability and preferences but physical factors affecting growth and spore production are obviously also important. The greatest output of conidia is closely related to the rapid addition of deciduous tree leaves as they fall in during the autumn. The spora take about two weeks to build up after leaf fall has started. The abundance of some species such as *Clavariopsis aquatica* (Fig. 4.14a) and *Flagellospora curvula* showed a clear correlation with the period of leaf-fall (August-December) and a continued spore production up to April. This reflected their preference for leaves of *Acer* and *Quercus*, which break down only slowly and so persist in the rivers over that period. *Lunulospora curvula* (Fig. 4.14b) showed a much shorter period of spore production from August to November. It is favoured by leaves of *Alnus* which are relatively soft and decompose quickly. If these leaves had fallen on the ground, it would have been in this autumn period that peak numbers of common primary saprotrophs would have occurred on them (Chap. 2). Thus a change in physical environment from terrestrial to aerated freshwater is mainly responsible for this change in colonization pattern. The above three fungi are clearly primary colonizers but *Tricladium chaetocladium* (Fig. 4.14c) had a different period of maximum spore production. It occurred in highest spore numbers from December to April. This may be because it is a secondary colonizer, i.e. it colonizes later, or it takes longer to produce its conidia or its growth and reproduction are stimulated by lower temperatures. It certainly has a lower temperature optimum for both growth and reproduction than does *L. curvula*.

At certain times of the year, especially in May, June and July, the spore concentration in these rivers, particularly the R. Exe, may be so low as to be undetectable by the filtration method. Yet a very rapid rise in numbers occurs again in August. It is assumed that very low, but effective, inoculum levels survive the summer when the substrate, in the form of leaves, is very scarce. Alternatively, they could persist elsewhere on other substrates, such as twigs and branches, perhaps in the teleomorphic state, on land. Too little is known about the ecology of the few teleomorphs so far identified to be categorical about this but *Nectria lugdunensis*, the teleomorph of *Heliscus lugdunensis*, has been observed on twigs from June to September, as they slowly dried out after submergence. Interpretation of these observations on seasonal abundance based on spore concentrations is difficult. In this case it is

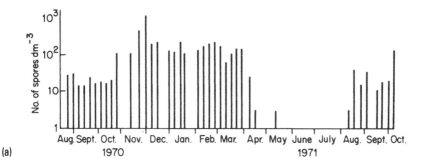

(a)

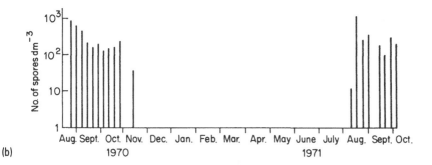

(b)

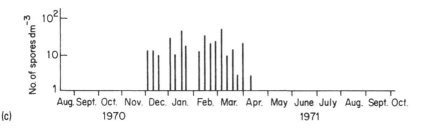

(c)

Fig. 4,14 Changes in concentration of spores in the River Creedy: (**a**) *Clavariopsis aquatica*; (**b**) *Lunulospora curvula*; (**c**) *Tricladium chaetocladium*. (After Iqbal and Webster (1973b). Reproduced by permission of the British Mycological Society.)

reasonably fair to suggest that they are indicative of the metabolic activity of the fungi concerned. Some assessment of mycelial activity would be preferable. As with other fungi growing in their natural substrates, this is impractical, not only because the mycelium is so intimately associated with the substrate as to be inseparable from it but also because several species may be growing in any one leaf at the same time.

Nutrition as decomposers in streams

Very little is known of what particular components of the leaves are metabolized by the aquatic Hyphomycetes. Thornton (1963) investigated eight species. All could use simple sugars, cellobiose and starch but none cellulose. Nitrate and ammonium ions both served as adequate sources of nitrogen for the aquatic Hyphomycetes which he used. It might be expected that leaves in streams would become colonized by the water moulds, the Saprolegniales. Most of the latter cannot use NO_3-N but can use NH_4-N. Thornton suggested that the aquatic Hyphomycetes benefit more from the NO_3 and NH_4-N leached from the soil into streams and so flourish.

Soluble carbohydrates are rapidly lost by leaching when leaves are immersed in streams and it would be surprising if polymers such as cellulose were not available to some at least. Jones (1982) has shown that several can utilize native cellulose in the form of solka floc. Species such as *Heliscus lugdunensis* and *Articulospora tetracladia* caused mass losses of between 20-25% - a quite appreciable cellulolytic activity. The former has also been recorded as causing a soft rot when grown on submerged wood. Suberkropp and Klug (1981) also showed that a number, including *Tetracladium marchalianum* and *Lemonniera aquatica*, could clear cellulose containing agar. They also found that when these two, and others, were grown on sterilized leaf litter in pure culture, they produced enzymes which liberated reducing sugars from carboxymethylcellulose, xylan and polygalacturonic acid. The leaves softened and the epidermal and parenchymatous cells separated, leaving only leaf skeletons. This maceration was associated with high polygalacturonase and polygalacturonate transeliminase activity. The production of such enzymes may be one of the reasons why fungal activity is most significant in the early stages of decomposition. One would certainly like some more positive data about their nutrition.

Almost 100 species of these so-called aquatic Hyphomycetes have now been described and it is becoming abundantly clear that they constitute the most substantial fungal element in well-aerated freshwater systems. It is equally clear that their major substrate, leaves of deciduous trees, form the most substantial element in the annual addition of organic matter to such systems. These fungi are invaluable members of these freshwater systems as they play a key role, not only in the decomposition process but also as intermediates in the food chain.

Role as intermediaries in the food chain in streams

Many streams have a low primary productivity. This may be especially so where they run through woodlands which are partially shaded from the direct rays of the sun by the trees. However, they are often well-supplied with substantial amounts of organic matter in the form of dead leaves and twigs from the adjacent terrestrial vegetation. It has been estimated that a stream in a wooded valley receives at least one kilogram of leaves (fresh mass) per metre of its length each year. Thriving stream animal communities develop, largely dependent upon this added organic material for their food supply. The contribution made by this material to the total energy budget of the stream communities has been estimated to range between 50 and 99%. Many observations have been confirmed that such imported organic material is of paramount significance in the food chain of streams.

Only a small fraction of the energy locked up in the leaf material in any ecosystem can be directly exploited by the micro-fauna. Usually more than 60%, and often as much as 80–90%, of leaf litter eaten by detritus feeders in woodlands, is returned in the form of faecal pellets (Chap. 5). A similar situation prevails in aquatic environments. For example, the common North American freshwater amphipod, *Hyalella azteca*, assimilates only 5% of the material which it ingests when it is fed on leaves of elm, *Ulmus*. To enable them to gain access to the remaining energy, the micro-fauna depend upon the activities of micro-organisms. These degrade the cellulose and other biopolymers, such as lignin, and convert them to their own biomass. Microbial carbohydrates, lipids and proteins are then digested by the micro-fauna.

It has long been recognized in terrestrial ecosystems that fungi as well as bacteria are important intermediaries in the food chain, but until recently it was assumed that in aquatic environments, especially streams, bacteria were the exclusive, or at least the predominant, intermediaries. This is in spite of the fact that fungi as a group are superior to bacteria in the degradation of solid plant materials, in particular cellulose and lignin, which account for up to 60% of the dry mass of leaves and which cannot be digested by most invertebrates.

The potential significance of decomposer fungi in streams was demonstrated by Kaushik and Hynes (1968, 1971). They found that the protein content of autumn shed leaves in streams significantly increased. In one sample of elm leaves the protein content doubled 14 days after submergence. They compared decomposition rates and changes in content of leaves in artificial stream systems when antibacterial or antifungal antibiotics or both were used. Increase in protein content was not significantly influenced by the presence or absence of antibacterial antibiotics, but was significantly depressed in the presence of antifungal ones. The loss of dry mass was usually high when only fungi were allowed to grow and low when fungal growth was suppressed. From such they concluded that fungi bring about substantial protein increment in decaying leaves and that they are more successful at degrading leaves than are bacteria.

Influence on food selection by detritus feeding micro-fauna

They also showed that several of the detritus feeding micro-fauna in streams favoured partly decomposed leaves rather than sterile fresh ones. Since many of the fallen leaves would have been colonized by the aquatic Hyphomycetes, the food of such detritus feeders would have consisted of two main components: leaf tissue and fungal mycelium. Bärlocher and Kendrick (1973a,b) have compared the relative merits of leaf and fungus in the diet of the amphipod, *Gammarus pseudolimnaeus*. They compared the efficiency with which it converted different types of food into its own biomass. Different groups of animals were given as their sole food supply, maple (*Acer*) or elm (*Ulmus*) leaves or the mycelium of one of ten fungi – five terrestrial ones and five aquatic Hyphomycetes. The actual amount of food consumed by those on the leaf diets was in the order of 10 times that of those on the fungus diets, but the greatest body mass increases were found on those feeding on four of the fungi, three of which were aquatic Hyphomycetes. Fungal mycelium thus represents a form of nourishment 10 times as concentrated as leaf material. Thus good fungal growth on leaves will very substantially improve their food value to the stream micro-fauna. In food selection experiments, all fungi were eaten in preference to maple leaves. Given the choice between maple leaf discs and the mycelium of *Tricladium angulatum*, all the gammarids used in the experiment fed on the fungus through the two hour test period. Preference for other fungi was less pronounced. Since such a strict separation of leaf and fungus is unlikely in nature, they investigated how individual fungi influenced the palatability of leaves. Sets of sterilized leaf discs were inoculated with pure cultures of particular fungi and incubated for 14 days. *Gammarus* was then given the choice of several combinations of each particular leaf and various fungi. If only leaf discs without fungi were provided, *Gammarus* preferred ash (*Fraxinus*), followed by maple and then oak (*Quercus*). It also preferred the mycelium of *Anguillospora* to that of *Tetracladium*. The preference of *Gammarus* for particular leaves could be changed, even reversed, by an appropriate choice of fungal inoculum. For example, oak, the least palatable, was chosen over maple and ash, provided that the oak discs had been inoculated with *Anguillospora* and the discs of the other two with *Tetracladium*. The fungal population on the leaves may thus have a decisive influence on food selection by *Gammarus*.

For such detritus feeding micro-fauna, fungi, in addition to being a source of major nutrients, may also fulfil other nutritional requirements. Fungal mycelium is also rich in choline, B vitamins and ergosterol, the commonest fungal sterol, which can easily be converted to cholesterol. The fungi growing in such substrates, as they do in wood, may concentrate biologically important mineral elements. *Gammarus* ensures a supply of these substances by preferential feeding on detritus which is already colonized by fungi, in this case aquatic Hyphomycetes. This is rather a less sophisticated strategy than that employed by other invertebrates such as Attine ants (Chap. 9), but in both cases the fungi are instrumental in transforming some of the vast amounts of energy stored in leaves into forms more acceptable to, and digestible by, the animal.

Traditionally, the main role of saprotrophic fungi in nature has been considered to be their decomposition of plant and animal remains into simple organic and inorganic substances which are then recycled by green plants. But it is less widely appreciated that, in both terrestrial and aquatic ecosystems, many animals depend on the accompanying increase in fungal biomass as nutrient sources as has been outlined above. However, this again is only a short part of a food chain in a complex food web. As a detritus feeder, *Gammarus* shreds its food and a large part of its intake is returned as faecal pellets. Much of the energy still remains locked up in the leaf and can be exploited only by further processing by subsequent decomposer colonizers, thus making them attractive as a food for coprophagous animals.

Aero-aquatic Hyphomycetes of stagnant waters

Another group of widespread aquatic Hyphomycetes which occur on the same substrate but under a quite different set of physical conditions are the aero-aquatic ones. They occur on deciduous tree leaves which have fallen into stagnant water. If such leaves are dredged from the bottom of a static pond, they are often characteristically black in colour. If they are rinsed and kept damp, but not submerged, for a few days in a Petri dish and exposed to the light, glistening clusters of conidia, mostly of a helicoid form, will develop on their surfaces. These belong to such genera as *Helicodendron*, *Helicoon* and *Clathrosphaerina*. Conidia of the last are not helicoid but clathroid, each consisting of a hollow sphere with a net-like multicellular wall (Fig. 4.15a). The clathrate structure is formed by repeated dichotomy of the arms of the developing conidium, which then curve inwards and join where the tips

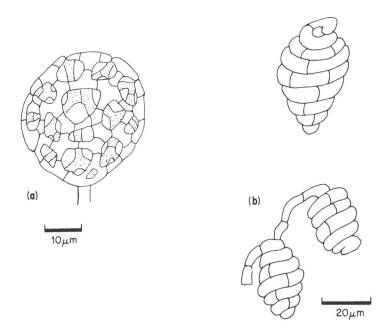

(a)

10 μm

(b)

20 μm

Fig. 4.15 Conidia of aero-aquatic Hyphomycetes. (a) *Clathrosphaerina zalewskii* (b) *Helicodendron conglomeratum*.

touch. The conidia of *Helicodendron* (Fig. 4.15b) consist of a many septate, narrow filament, coiled tightly to form a three-dimensional helix, often of six to twelve gyres. Although the mycelium of these fungi grows in the leaves under anaerobic or near-anaerobic conditions, conidia develop only on exposure to moist air. Where they grow in water only 10–20 mm deep, hyphae grow up to the surface and produce conidia there. They are also produced when the leaves become exposed as they would around the edge of a drying pond. In both cases, when the water level again rises, they float to the surface and, like the tetra-radiate conidia in the scum, are again in an ideal position to colonize leaves as they fall in. The conidia have a special flotation device. They are unsinkable. A bubble of air is trapped in the coil or the sphere of the conidium, so if it is temporarily forced under the water it bobs up again.

Very little is known as yet about the nutrition and physiology of these fungi so that no comparison can be made with the anaerobic members of the Oomycete Leptomitales (p. 145). Laboratory experiments have shown that the ability of these fungi to colonize leaves is in fact enhanced under well-aerated conditions. They also produce conidia in aerated water. It may well be that they have a poor competitive ability which restricts them to such habitats rather than that they require low oxygen and high carbon dioxide levels to grow. Aero-aquatic Hyphomycetes afford another good illustration of how the physical factors of the environment can have a decisive effect on determining which fungi colonize any particular substrate. Again they are not completely confined to leaves in such habitats. They also occur on well-weathered bark and on wet, very rotten wood in marshy places. Like several of the aquatic Hyphomycetes, some have been found to possess a perfect state. For example, the teleomorphic state of *Clathrosphaerina zalewskii* is *Hyaloscypha zalewskii*, an apothecial member of the Ascomycotina. Four species referable to the conidial or anamorph genus *Helicodendron* have been connected to perfect or teleomorphic states. Three of these belong to widely different genera of Discomycetes and one to the pseudothecial Loculoasco-mycetes. Thus the anamorph genus is not a natural one. Other aero-aquatic Hyphomycetes have teleomorphic states in the Basidiomycotina. The varied taxonomic position of these teleomorphs is indicative that the aero-aquatic Hyphomycetes, like the aquatic ones, have evolved independently among unrelated groups. Again if these teleomorphs were terrestrial distance dispersal would be no problem.

The aquatic Hyphomycetes, water moulds and other fungi may compete for substrates. When tree leaves fall into rivers and streams in the autumn, they are already colonized by a number of essentially terrestrial fungi, especially the common primary saprotrophs (p. 62). Newton (1971) has shown that on leaves in the R. Lune these quickly disappear and are replaced by other fungi. *Aureobasidium pullulans* disappeared within a month from leaves of alder, elm, oak and willow. *Epicoccum purpurascens* also rapidly declined in the two months following immersion and the leaves quickly attracted a number of other fungi, not all of which were aquatic Hyphomy-cetes. Species of *Fusarium* rapidly colonized all types of leaves but the other major colonizers, in addition to the aquatic Hyphomycetes, were the water moulds, the Saprolegniales, and related Oomycetes. Alder leaves submerged in August were colonized by species of *Phytophthora* and *Pythium*, in the first week, *Achyla*, within two, *Saprolegnia*, after four weeks and *Dictyuchus*, after eight months. The same alder leaves also produced *Anguillospora crassa*, *Articulospora tetracladia*, *Lemonniera aquatica* and other aquatic Hyphomycetes, after being submerged for only one week.

The fungi of stagnant waters

Fungi are generally thought to be highly aerobic but, in both the Chytridiomycetes and Oomycetes, there are a considerable number which are able to ferment sugars to lactic acid and thus synthesize ATP in the absence of oxygen or when its supply is severely limited. Even so, only two aquatic fungi, *Aqualinderella fermentans* and *Blastocladia ramosa*, are known to grow well under anaerobic conditions. They are also unusual in that both require high levels, 5 to 20%, of carbon dioxide for good growth. *A. fermentans* grows best at 20% carbon dioxide. Such levels inhibit the growth of most fungi. For instance, 10% carbon dioxide is used in grain storage to prevent mould growth. *A. fermentans* is a tropical or sub-tropical fungus. It has been isolated from surface growths of a great variety of fleshy fruits submerged in warm, still waters, such as the stagnant water at 28–35°C of hot delta regions and of swamps, ditches and lagoons. It is often found below dense mats of floating vegetation where carbon dioxide concentrations must be high. It is a member of the Leptomitales, a relatively little-known order of the Oomycetes. Less than half of the 30 or so described species in the order have been cultured and we have only a few clues about their ecology and physiology. They are all fresh water saprotrophs and usually occur on plant debris such as twigs, but especially in stagnant to distinctly foul water. Their morphology is odd. Their vegetative parts are constricted at more or less regular intervals. The constrictions are plugged by a carbohydrate called cellulin – it gives a cellulose reaction and is highly refractive. In some genera, such as *Leptomitus*, there is a true, if slender, hyphal system. Others, such as *Rhipidium*, are non-hyphal and their overall form closely parallels that of many of the Blastocladiales, in being tree-like, with a broad trunk-like portion, with a well-developed attaching rhizoidal system at the base and slender branches at the apex, terminating in stalked – because of the constrictions – zoosporangia and/or oogonia (Fig. 4.16). *Rhipidium*, other related genera and members of the Blastocladiales are quite common but like the majority of the Mastigomycotina are rarely seen. Their presence can be revealed by providing them with an appropriate substrate in the form of a bait. Unripe apples submerged in any relatively still, or stagnant stream, will produce on their surface, within 4–6 weeks, a crop of *Rhipidium americanum*, together with numerous species of *Blastocladia*. The best known member of the Leptomitales is *Leptomitus lacteus*, the sewage fungus, so-called because it occurs predominantly in water polluted by sewage. It is

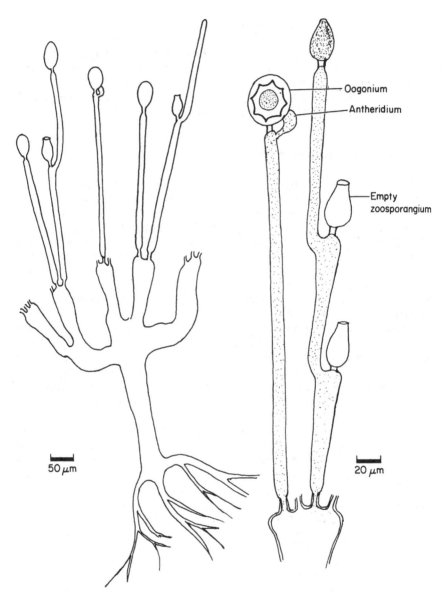

Fig. 4.16 *Rhipidium americanum* from an apple submerged in a stagnant pond

peculiar in that it cannot assimilate simple sugars. It is rare indeed to find a fungus which cannot utilize glucose. *L. lacteus* also requires organic sources of nitrogen and can utilize alanine and leucine as sole carbon and nitrogen sources. More suitable carbon sources include acetate, pyruvate, and fatty acids which are all readily available in sewage. *L. lacteus* requires efficient aeration. Other members of the Leptomitales, such as *Sapromyces elongatus*, can ferment but grow very slowly under anaerobic conditions. *Aqualinderella fermentans* can only ferment whatever the oxygen supply. It can use simple sugars only, such as glucose, fructose and mannose, as carbon sources which it degrades by homolactic fermentation, as do bacteria such as *Lactobacillus*. Oxygen levels have no effect on its growth or sugar consumption. It has lost its oxidative respiratory system. Only relics of mitochondria are present as double membrane bound vesicles with only slight invaginations of the inner membranes. It lacks the typical cytochromes found in aerobic fungi. For good growth it requires a supply of lipids. The biosynthesis of these requires molecular oxygen. It also has other rather elaborate requirements in the form of amino acids and nine or more vitamins. The obligately fermentative metabolism may be seen as an adaptation to environments poor in oxygen and rich in carbon dioxide and organic substances. Such environments exclude other fungi and are, apart from these few opportunistic intruders, the exclusive domain of the bacteria.

5

Fungi as inhabitants of animal faeces

Since representatives of most of the major groups of fungi, except the Masti-gomycotina, grow on dung, coprophilous or dung fungi are an excellent ecological group to use, not only to demonstrate the remarkably wide diversity and adaptability of fungal reproductive structures, but it is also one which readily lends itself to the experimental analysis of seral changes and of the various attributes which these fungi must possess, not only to survive but also to succeed in such substrates.

Succession of fungi on herbivore dung

Herbivore dung is a particularly rich substrate for fungi and produces in sequence, on incubation in a moist but adequately lit and aerated chamber, an exceedingly diverse array of strikingly elegant fungi, most of which are highly adapted to their habitat niche. It is also a particularly good substrate on which to observe a fungal succession. Virtually any kind of herbivore dung will suffice but in terms of mass, sheep or rabbit pellets are ideal for convenience of examination with a dissecting microscope. The Keys to Dung Fungi by Richardson and Watling (1968,1969) are useful for identifica-tions. On freshly deposited sheep or rabbit pellets, incubated for 1–2 days, sporangiophores of Zygomycete Mucorales appear. These belong to species of *Mucor*, *Pilaira* and *Pilobolus*. These may persist for 10–14 days but with time a number of fungal parasites, such as *Chaetocladium* and *Piptocephalis*, themselves Mucorales, grow on them. Some Ascomycotina may appear early in the succession. The minute apothecia of *Rhyparobius dubius* occur within 4 days but most begin to appear after 9–10 days. Characteristic copro-philous genera include apothecial forms such as *Coprobia*, *Cheilymenia*, *Ascobolus*, *Rhyparobius*, *Thelebolus*, *Lasiobolus* and *Saccobolus*, perithe-cial forms such as *Chaetomium*, *Coniochaeta*, *Delitschia* and *Podospora* and pseudothecial forms such as *Sporormia*. Apothecial forms tend to appear before perithecial ones and the latter may be produced over a period of weeks. On further incubation over 14–30 days, a number of basidiocarps of Agaricales, belonging to such genera as *Coprinus*, *Stropharia* and *Paneolus* appear. In the field, but far less commonly in the laboratory, basidiocarps of such genera as *Psilocybe*, *Psathyrella*, *Conocybe* and *Bolbitius* occur

(Webster, 1970). Many of these are not exclusively coprophilous. Even members of the very specialized genus *Pilobolus* have been recorded elsewhere, on river mud growing on decaying algae, but nevertheless certain groups have a very high proportion of coprophilous members. This is especially true of the Mucorales, Ascobolaceae, Sordariaceae and Coprinaceae.

Adaptations to habitat

Although these fungi are taxonomically very diverse, they show a number of common adaptations to their habitat. In many the spore-bearing structures are orientated by phototropic responses. This is usually coupled with a violent spore discharge mechanism such that the spores are discharged a considerable distance, 250 mm or more, towards the light and sufficient to carry them away from the staling substrate onto the surrounding vegetation. Such phototropic responses are shown by the sporangiophores of *Pilobolus*, the tips of the protruding asci of *Ascobolus*, the perithecial necks of *Podospora* and the stalks of the basidiocarps of *Coprinus*. The projectile discharged often consists of many spores. In *Pilobolus* the entire sporangium is discharged and in *Ascobolus* and *Podospora* the eight ascospores in each ascus are shot off as a unit. The larger the projectile the less wind resistance it encounters as it is propelled through the air. The projectile is often mucilaginous so that it adheres firmly to the surface on which it is impacted. The spores are often pigmented. Virtually all the Basidiomycete Agaricales which occur on dung have dark coloured spores. The pigment protects the protoplasm from the damaging effects of ultra-violet light. They thus can not only survive strong sunlight while exposed on vegetation after impaction but many are ingested with the herbage and also survive passage through the alimentary canal of the animal. Indeed, the spores of many coprophilous fungi, but not all, require the latter treatment before they will germinate.

Pilobolus

Pilobolus shows most of these adaptations. The asexual reproductive structure consists of a basal swelling, a cylindrical sporangiophore with a large pear-shaped sub-sporangial swelling surmounted by a conical columella, hidden by the shiny, black sporangial wall (Fig. 5.1a). The sporangium is discoid and filled with thousands of orange-yellow oval spores. A transparent ring of gelatinous material is found around the base of the columella between the sporangial wall and the spores. At the junction of the sub-sporangial swelling and the sporangiophore is a collar of carotene-rich cytoplasm having the form of a biconcave lens with a hole through the centre. When the whole structure is exposed to unilateral light, the upper part of the sporangiophore just below the sub-sporangial swelling makes a positive phototropic curvature so that the sporangium points accurately to the light. In unilateral light the sub-sporangial swelling acts as a lens. Rays of light are refracted through the wall into the interior of the swelling and converge to form a spot of light on the opposite wall near its base. The concentration of the rays at a particular point causes some stimulus to be transmitted to the cylindrical part of the sporangiophore below it which then grows faster at that point so that the sub-sporangial swelling is turned about its base through an angle. As it turns, the spot of light moves downwards on the wall of the swelling until it falls symmetrically on the carotene containing collar. At this point the sporangium is precisely orientated towards the light (Fig. 5.1b).

The sub-sporangial swelling contains a large volume of sap with an osmotic potential in the order of -0.5 to -0.6 MPa. It also possesses a highly elastic wall. When fully mature and turgid, its wall breaks transversely along a pre-determined line of weakness in a ring at its apex just beneath the sporangium and columella. Because of the elasticity of its wall and the pressure from within, the distended sub-sporangial swelling suddenly contracts squirting out its contents and carrying away the entire sporangium, with an attached drop of sap, some 1.5–2.5 m, depending upon the angle of inclination. The projection velocity varies widely with a mean of about 10 m sec^{-1}. The sporangial wall is unwettable and when the projectile strikes an object, such as a blade of grass, the sporangium flicks over in the drop of sap and faces outwards (Fig. 5.1c). As the drop dries the sporangium becomes firmly attached by the ring of gelatinous material and is virtually cemented to the leaf. The black melanized and reflective sporangial wall protects the spores within from excessive light. The release of the spores from the sporangium does not take place until the sporangium has been eaten by a herbivore, thoroughly moistened and compressed in the alimentary canal. They then survive passage through the gut and will germinate and grow only where the pH is 6.5 and above as in herbivore dung.

Herbivore dung as a substrate

Herbivore dung is a very specialized and highly complex substrate and is more favourable for rapid and continued fungal growth than are many others. It not only contains the comminuted and depleted remains of the ingested vegetation but also waste products of the animal, such as broken down red blood cells and bile pigments, in addition to the remains of a very large microbial population, especially cellulolytic bacteria from the rumen. Consequently it contains appreciable quantities of readily available carbohydrates and its nitrogen content may be as high as 4%, a three-to-four-fold increase over the ingested material, which is unusually high when the majority of substrates on which fungi grow are considered. These may be the major factors favouring initial colonization by the rapidly growing and relatively ephemeral Zygomycete Mucorales. Dung is also rich in water soluble vitamins, growth factors and mineral ions. Some of the gut micro-organisms are responsible for synthesizing such vitamins. Coprogen, an organo-iron compound, found in dung, is required for growth and reproduction in

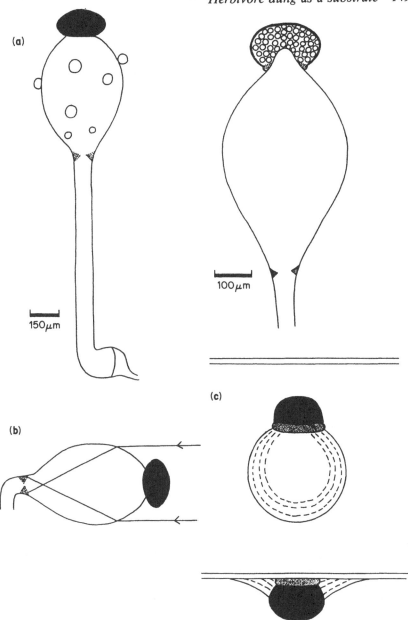

Fig. 5.1 *Pilobolus kleinii.* (**a**) Mature sporangium and sporangiophore and sporangium and sub-sporangial swelling in vertical section. (**b**) Sporangium precisely orientated to a source of lateral light. (**c**) Sporangium and sap from subsporangial swelling just prior to and after striking a blade of grass in its flight path.

Pilobolus. It is produced by a number of actinomycetes, bacteria and fungi in dung. Dung has a high moisture content, between 60 and 80% on a dry mass basis, and a capacity to retain moisture for a relatively long period as it dries out. Both these favour rapid and continuous hyphal growth. These features, together with the comminution of the material may lead to more rapid breakdown. The formation of semi-solid pellets and its peculiar physical structure also provide good aeration necessary for growth of fungi within. The pH is around 6.5 or above and this high value, as in pyrophilous fungi, is reflected in the pH optima for coprophilous fungi. These are usually near neutrality rather than on the acid side.

Analysis of the fungal succession

The nutritional hypothesis

The succession of fungal reproductive structures which occurs on herbivore dung is often quoted as the classic example of a fungal succession. It is well known yet it is difficult to establish who first observed and described it. In the early years of this century it was familiar to Massee and Salmon (1901,1902) but it was not until the mid-1960s that the succession was experimentally analysed (Harper and Webster, 1964). Up until that time the succession had been interpreted as a strict nutritional sequence fitting the very generalized schema for fungal successions on decomposing plant tissues proposed by Garrett (1963). In the decomposition of organic matter, such as dung, simple soluble sugars, starches and proteins disappear first, followed by hemicelluloses, then cellulose and finally lignin. The Zygomycete Mucorales, which have the capacity for rapid spore germination and hyphal growth, exploit the dung initially. They are one of the best examples of primary sugar fungi. None can utilize natural cellulose or lignin. When the simple carbon sources are exhausted, they disappear and it is believed that the Ascomycotina utilize the cellulose and are dominant on the substrate. These are in turn replaced by the Basidiomycotina. A number of these coprophilous Agaricales are known to utilize both cellulose and lignin. In addition the spores of these two latter groups are said to germinate less rapidly and grow more slowly than those of the Mucorales. This so-called nutritional hypothesis, being based on the known nutrition and physiology of the various groups of fungi involved and the nutrient content of the dung at the various stages of decay, appeared to afford a satisfactory explanation of the succession. Although the hypothesis is an attractive one, hypotheses should be tested. It also neglects some important ecological factors.

Explanation based on time taken to reproduce

The observed succession is one based solely on the time of appearance of reproductive structures, either sporangiophores, ascocarps of varying types or basidiocarps of the agaric type. The sequence of mycelial development may not be the same. The speed of spore germination and the ensuing rate of mycelial growth are of importance in determining the rate of exploitation of any substrate and have often been cited as being important factors in the colonization of fresh dung. Although there may be differences in the time taken for spores of the different fungi to germinate when placed directly onto nutrient agar, this overlooks the fact that the spores of many coprophilous fungi are stimulated to germinate by passage through the gut. The spores of some, but by no means all, fail to germinate without such pretreatment. They develop on the dung from spores which have survived passage through the gut and this process breaks their dormancy. For example, in *Pilaira anomala* the majority of the spores in any population germinate only after a digestive treatment. About 5% of the spores will germinate after being placed in water at 37°C for 3 hours, whereas 100% will if they are placed in alkaline pancreatin at 37°C for 3 hours, a treatment considered equivalent to passage through the gut of a rabbit. Harper and Webster showed that after pretreatment of spores of a range of coprophilous fungi with pancreatin for 5 hours at 37°C, most germinated within 6 hours and the various groups of fungi showed no appreciable differences in the latent period for germination (Table. 5.1). In addition, for the Mucorales and the Ascomycotina which they studied, there was no clear correlation between the growth rate of germ tubes and the time of their appearance in the succession but the

Table 5.1 Latent period, germ tube extension rate and mycelial growth rates of coprophilous fungi. (After Harper and Webster (1964). Reproduced by permission of the British Mycological Society.)

	Latent period of germination (h)	Germ tube growth rate (μm h^{-1})	Linear growth rate (mm day^{-1})
Mucor mucedo	5–8	18.7	9.1
Pilaira anomala	4–6	14.1	4.8
Pilobolus crystallinus	6–9	10.0	4.8
Ascobolus glaber	4–6	44.1	12.0
A. stictoideus	9–12	52.5	10.9
A. viridulus	6–10	21.6	10.6
Rhyparobius dubius	6–8	10.8	1.8
Chaetomium caprinum	5–10	7.5	2.2
Podospora minuta	4–6	27.5	1.8
Sordaria fimicola	4–6	63.7	19.0
Sporormia intermedia	6–10	14.2	2.3
Coprinus heptemerus	5–8	15.8	3.2
C. patouillardii	10–12	9.7	3.7
C. radiatus	6–12	7.5	4.4

Basidiomycete Coprini did show slower germ tube growth rates. Similarly, no good correlation was found between mycelial growth rate and the successional appearance. For example, the growth rate of *Sordaria fimicola* was 19.0 mm day^{-1} and of *Ascobolus glaber* 12.0 mm day^{-1} respectively. These appeared after 9 and 11–12 days respectively. Whereas, the growth rate of *Mucor mucedo* was 9.1 mm day^{-1} and *Pilaira anomala* 4.8 mm day^{-1}, yet both appeared after 2 days. The growth rates of the Coprini were rather low, 3.2–4.4 mm day^{-1}, but even so exceeded the growth rates of a number of Ascomycotina which preceded them in the succession. Thus it does not appear to be a sequence because the spores of the Mucorales germinate more rapidly and their germ tubes and hyphae grow faster than those of the Ascomycotina, which in turn germinate and grow faster than those of the Basidiomycotina.

The succession as described, as it has been stressed, is a succession of the appearance of reproductive structures. Harper and Webster produced some good evidence that each particular fungus does take a characteristic minimum time to produce its reproductive structures. They compared the time taken for reproductive structures to appear on inoculated rabbit pellets brought in from the field with that on incubated sterilized pellets inoculated with germinating spores and on incubated pellets from rabbits fed with sterilized food sprayed with a spore suspension of each fungus individually. It should be noted that such treatments not only altered the substrate but also eliminated or reduced competition. The time taken for particular fungi to reproduce in each treatment was very similar (Table 5.2) with the exceptions

Table 5.2   Number of days elapsing before reproduction by coprophilous fungi. (After Harper and Webster (1964). Reproduced by permission of the British Mycological Society.)

	On fresh dung from time of field collection	On sterile dung monoculture	On fresh dung from feeding experiments
Mucor hiemalis	2	2	2
M. mucedo	2	2	2
Pilaira anomala	2	2	2
Pilobolus crystallinus	4	4	4
Rhyparobius dubius	6	6	6
Ascobolus stictoideus	9–10	8–9	8–9
A. viridulus	10–11	7–8	8
A. glaber	11–12	12–13	13–14
Chaetomium caprinum	9	9–10	9
Sordaria fimicola	9	9	9
Podospora minuta	9–11	10	10
Sporormia intermedia	10–11	10–11	11
C. brasiliense	28–34	34	—
Coprinus heptemerus	9–13	7–8	7–8
C. patouillardii	37	14	11–12
C. radiatus	—	14	11–12

of *Coprinus patouillardii* and *C. radiatus*. There may have been a low spore inoculum of these in the fresh dung from the field and a high inoculum may be necessary for rapid reproduction. Thus the minimum time taken to reproduce in itself provides a simple and satisfactory explanation of the observed succession. Obviously, it is to be expected that it would take an active mycelium of *Coprinus* longer to amass sufficient reserves to produce its relatively massive basidiocarps than it would a similar mycelium of *Mucor* to produce its relatively simple sporangiophores.

In the coprophilous succession it would appear that the majority of fungi are present when the dung is deposited. They also begin growth virtually together soon after the dung is deposited and initially make preferential use of, and compete for, soluble and simple carbon sources while they last. On their disappearance the Ascomycotina and Basidiomycotina produce hemicellulases and cellulase. By this time, the Mucorales, at least, have reproduced. The Ascomycotina would then reproduce and persist as long as there was sufficient available cellulose. Ultimately all that would be left would be the lignin which only the Basidiomycotina could fully degrade but they begin to produce their basidiocarps well before all the cellulose is utilized. In its simplicity the idea of a progressive run down of nutrients may be basically correct but in actuality the situation may be exceedingly more complex and indeed substantially unconfirmed until accurate sequential biochemical analyses have been made on dung. There must also be differential depletion of other nutritionally important compounds, such as nitrogenous ones, and growth factors which would affect growth and reproduction of some fungi. For instance, ammonia stimulates sporangial production in *Pilobolus*. It often grows best in the presence of other micro-organisms. *P. kleinii* produces more sporangia on synthetic media in mixed culture with *Mucor plumbeus* than it does in pure culture. This is due to release of ammonia by *Mucor*. The luxuriant growth and reproduction of *Pilobolus* on dung may in part be due to release of ammonia by micro-organisms in the dung or to the specific growth factor coprogen. Simple sugars, such as pentoses and hexoses, by comparison with fatty acids are poorer carbon sources for *Pilobolus*. The latter are known to be present in dung but it is not known how they affect the growth of *Pilobolus* there.

The effect of competition for nutrients and antagonism

Dung is not just a substrate for fungi. It supports vast populations of bacteria, protozoa, nematodes, annelids, insects and the like. They must compete with fungi for nutrients, graze on them and in turn provide substrates for them. Competition between fungi and with other micro-organisms present in the dung apparently has very little effect on their actual time of appearance but it may well be important in limiting the duration and intensity of reproduction. Harper and Webster convincingly demonstrated this by inoculating sterilized rabbit pellets with pre-germinated spores of *Pilobolus crystallinus* only, of *P. crystallinus* and *Ascobolus viridulus*,

P. crystallinus and *Coprinus heptemerus* and of all three fungi. They then counted the numbers of sporangia of *Pilobolus* produced daily and scored their frequency on samples of 10 pellets in each case (Table 5.3 and Fig. 5.2). In all three treatments *Pilobolus* reproduced after four days. The presence of *A. viridulus* reduced both the yield of sporangia and their frequency but the effect of *C. heptemerus* was much more drastic. It brought about a marked diminution in the number of sporangia produced and a premature cessation of reproduction. The presence of *A. viridulus* tended to reduce the effect of the inhibition caused by *C. heptemerus*. Most of the coprophilous Agaricales, especially species of *Coprinus*, are antagonistic to fungi such as *Pilobolus* and *Ascobolus*. Ikediugwu and Webster (1970) investigated the nature of the antagonism. It can be observed only after contact between hyphae has been made and has been termed hyphal interference. Within minutes of being contacted, the contacted cell and the immediately adjacent ones undergo vacuolation and suffer a loss of turgor. There is a drastic alteration in the permeability of the cell membrane which leads to the death of the cell. This can be seen as a very effective form of antagonism and is probably widespread in fungi in general. It would certainly help to explain the dominance of the Agaricales in the later stages of the succession. The antagonism shown by species of *Bolbitius*, *Paneolus* and *Stropharia* is almost as strong as that of *C. heptemerus* and the majority of fungi which precede them in the succession are sensitive to them. Other forms of antagonism exist. The conidial *Stilbella erythrocephala* is insensitive to *C. heptemerus* and it produces a diffusible antibiotic which inhibits the growth of *Coprinus*. If *Stilbella* is very prevalent on a pellet, there is usually a relatively poorly developed associated fungal community.

Preferences for particular dung types

Webster (1970) pointed out that although large numbers of species had been reported from the dung of animals and birds, relatively few have been reported from the dung of amphibia and reptiles. The evolution of the coprophilous habit in fungi appears to be associated with the warm-blooded condition in animals. Although there is a great variation in the feeding preferences, habitats and digestive systems of herbivores, very few coprophilous

Table 5.3 Numbers of sporangia of *Pilobolus crystallinus* grown singly or in the presence of other fungi (daily totals). (After Harper and Webster (1964). Reproduced by permission of the British Mycological Society.)

| | Days | | | | | | | | | | |
	4	5	6	7	8	9	10	11	12	13	14
Pilobolus only	97	78	167	292	307	324	302	491	193	193	248
Pilobolus + *Ascobolus*	25	47	52	76	68	43	39	48	30	32	31
Pilobolus + *Coprinus*	51	20	12	7	5	1	0	1	2	0	1
All three	65	28	18	9	4	4	0	1	0	0	0

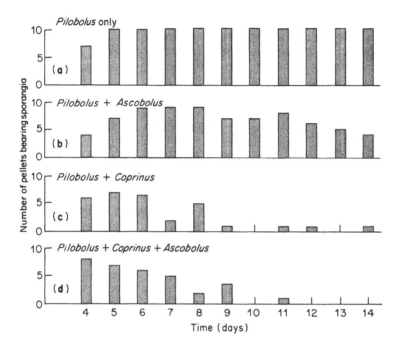

Fig. 5.2 Frequency of production of sporangia by *Pilobolus crystallinus* on samples of ten pellets inoculated with (**a**) this fungus; (**b**) with *Ascobolus viridulus*; (**c**) with *Coprinus heptemerus*; or (**d**) with *Coprinus* and *Ascobolus*. (After Harper and Webster (1964). Reproduced by permission of the British Mycological Society.)

fungi are restricted to grow on a particular or a few particular types of dung. The majority are thus cosmopolitan. However some do show preferences for the dung of particular animals. Richardson (1972), in a survey of the coprophilous Ascomycotina on different dung types, found that some species, for example *Podospora curvula* and *Ascobolus immersus* were associated predominantly with ruminant dung, while *Podospora appendiculata* and *Thelebolus stercoreus* occurred more frequently on lagomorph dung. *Coprobia granulata* is usually found only on cow dung. Others are even more specialized. The perithecial *Phaeotrichum hystricinum* has been found only on porcupine dung and all three species of *Dimargaris*, one of the Mucorales, occur on mouse dung. Nevertheless it is usually not possible to tell what the type of dung is by the fungi growing on it. In this coprophilous fungi show much less specificity than, for instance, do many wood-inhabiting fungi.

Basidiobolus and two-phased animal dispersal

Basidiobolus ranarum is one of the very few fungi which have been reported from the dung of amphibia. It is found only in the dung of frogs and has a most unusual life cycle. It is one of the Zygomycete Entomophthorales. It occurs in the gut of frogs as large spherical cells or spores, some 20 μm diameter. In the faeces these grow into a coarse septate mycelium from which conidiophores develop. These are markedly phototropic and resemble the sporangiophores of *Pilobolus* in gross form, except that the terminal black sporangium of the latter is replaced by a colourless single-spored, pip-shaped projectile, the conidium (Fig. 5.3). When mature, a line of weakness develops around the base of the vesicle. It ruptures at this point and the conidium and upper part of the vesicle are discharged to a distance of 10–12 mm as the upper elastic part of the vesicle contracts, squirting sap out backwards. The conidia may then be eaten by beetles but remain unchanged in their guts. If the beetles are in turn eaten by frogs, the conidia complete their development in their guts by dividing into a number of spherical cells which are released and develop into a mycelium in the faeces. By such a two-phased animal dispersal mechanism the fungus may achieve a more widespread distribution. Such two-phase dispersal systems may be relatively

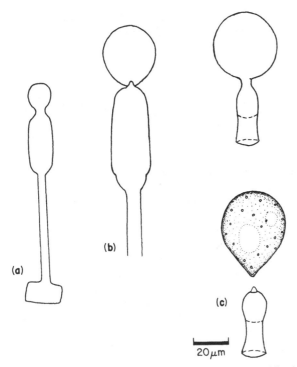

Fig. 5.3 *Basidiobolus ranarum* (**a**) Developing conidiophore with a single terminal conidium. (**b**) Mature conidiophore with line of weakness towards base of vesicle. (**c**) Discharged conidium with the upper part of vesicle still attached and separated below.

common. Seven out of 20 droppings of the blackbird, *Turdus merula*, contained viable spores of the chytrid *Rhizophlyctis rosea*, and 13 out of 20, viable spores of the hyphochytrid, *Hyphochytrium catenoides*; both fungi are widespread in soils. They have no air-borne dispersal units. In soils they are dispersed in space by zoospores and in time by very thick-walled resting spores. Earthworms browsing in the soil may ingest the latter and if they themselves are then eaten by birds, such as the blackbird, the spores will survive passage through the gut of both animals and be deposited in the bird droppings. The frequent finding of these two species, and many others, in droppings and their general distribution gives some credence to the suggestion that their long-range dispersal takes place by internal transport in the guts of birds.

Decomposition of the faecal pellets of arthropods

Faecal pellets of arthropods form another rich substrate for fungi. Much of the annual litter fall in woodlands is eventually eaten by the litter micro-fauna but usually more than 60% and often as much as 80–90% of that eaten is returned as faecal pellets. This includes all the lignin and varying proportions of the cellulose. The exact amount of the latter depends upon such factors as whether or not the animal, or its intestinal micro-flora, produces cellulase or not, the amount produced and the extent to which the material is fragmented, thus allowing enzymes access to it.

Nicholson, Bocock and Heal (1966) studied the decomposition of the faecal pellets of the millipede, *Glomeris marginata*, fed on leaves of hazel, *Corylus avellana*. *Glomeris* returned in this case over 90% of the ingested leaves as faecal pellets and these included over 70% of the cellulose. The pellets contained more ammonia nitrogen and total nitrogen g g^{-1} than did the leaves. Thus, like other members of the litter micro-fauna, *Glomeris* eats relatively large amounts of leaf litter of little nutritional value and excretes most of it unchanged chemically but nevertheless greatly fragmented and richer in nitrogen. As in herbivore dung, the overall rate of decomposition of the leaf litter is usually greatly increased when it is converted to faecal pellets. The fragmentation by the micro-fauna provides a larger surface area for extracellular enzymes to act upon. This particularly favours bacterial activity. The pellet form ensures that they have a better water holding capacity than the natural leaf litter which is prone to drying out until it is compacted. In the *Glomeris* pellets the moisture content never fell below 60%, whereas intact leaves would have been virtually air dry at times. This favours rapid and continuous mycelial growth. The higher nitrogen content, 1.7%, as against 1.4% in leaves, also favours synthesis of microbial protein and thus increases mycelial growth and as a consequence the decomposition rate. In all these respects the pellets show striking similarities with herbivore dung which are reflected in the fungal succession which occurs on them.

There is a peak in microbial activity, as indicated by the rate of consumption of oxygen, during the early stages of decomposition. Bacterial numbers rose to peak values during the first two weeks of decomposition but afterwards decreased to a relatively constant level. Zygomycete Mucorales, such as *Mucor hiemalis* and *M. ramannianus*, and their parasites such as *Piptocephalis* spp., were common up to the end of the second week after deposition but were then replaced by a wide range of members of the Deuteromycotina, especially *Endophragmia hyalosperma*, and a few members of the Ascomycotina, especially *Chaetomium* spp. Their appearance was coupled with both the slow utilization of cellulose and a marked increase in the amount of mycelium present. Fresh pellets contained about 1000 m of mycelium g^{-1} dry mass. This increased to about 2500 m g^{-1} at 28 days. In the later stages of decomposition, other members of the Deuteromycotina, such as *Polyscytalum* spp., became more prevalent and a number of sterile forms, including at least two of the Basidiomycotina, were present. These presumably remained sterile as the individual pellets did not contain sufficient energy sources for them to amass sufficient reserves to reproduce. Thus, in outline, with regard to the major groups of fungi concerned, the succession is very reminiscent of that on herbivore dung but in this case none of the component species was typically coprophilous. They could all be found on a number of substrates.

Although in this specific case there was no indication of a higher rate of decomposition in the pellets than in the intact leaf litter, the pellets are undoubtedly a much more favourable habitat for decomposer micro-organisms than are whole leaves. Greater microbial activity and higher numbers of decomposer micro-organisms have been recorded in freshly produced faeces of a number of detritus feeders than in comparable undigested food. Microbial activity, much of it due to fungi, has been estimated as about seven times as high in the faecal pellets of the terrestrial caddis fly, *Eniocyla pusilla*, as in the oak leaves on which it fed. Filamentous fungi and yeasts increase rapidly above normal soil levels in fresh earthworm casts. These casts contain more than twice as much total nitrogen as does soil from the same area and, moreover, more of it, about 96%, is in an available form as ammonia. In these, as in faecal pellets, herbivore dung and compost systems (Chap. 6), the increased level of available nitrogen is a major factor in stimulating fungal activity. The pattern of succession of coprophilous fungi on herbivore dung is both well-known and unique. Other plant remains decomposing above ground lack an initial Zygomycete phase. Whether or not the absence of Mucorales from such substrates is due to their lower nitrogen content is worth investigation.

6

Fungi as inhabitants of extreme environments

THERMOPHILY IN FUNGI

Temperature is one of the cardinal factors which determines the distribution of many fungi. Fungi as a group can live in a wide range of temperatures but by no means such a wide one as the prokaryotic bacteria and cyanobacteria. Although fungi can be classified as psychrophiles, mesophiles and thermophiles, there is no clear dividing line between these groups. If the temperature characteristics, in terms of minimum, optimum and maximum, of a very large number of fungi were considered, these groups would undoubtedly merge imperceptibly and become a continuous cline. By far the majority of fungi are mesophiles, growing between 5 and 37°C.

A thermophilous organism is defined as one that grows at temperatures above those considered to be the maximum limits for most forms of life. Although a number of prokaryotic organisms are able to live and thrive at temperatures of over 90°C and whereas for some of these 60–62°C is near the minimum temperature at which they will grow, the upper limit for eukaryotic organisms is near 60°C. The most widely used definition of a thermophilous fungus is one whose minimum for growth is at 20°C or above and maximum for growth at 50°C or above. Such fungi have optima around 40–50°C. There are a number of fungi which grow well in temperatures below 20°C and which may grow in temperatures of up to 50°C. They have optima at or around 40°C and grow slowly or not at all above 50°C. These are best regarded as thermo-tolerant mesophiles rather than as true thermophiles (Fig. 6.1).

Basis of thermophily

Cooney and Emerson (1964) comprehensively reviewed all facets of thermophily in fungi and stimulated a surge of interest in thermophiles as a group. Nevertheless we still do not really understand the basis of thermophily in fungi or indeed why many of these cannot grow at ordinary temperatures. Numerous hypotheses have been put forward. Crisan (1973) discusses four of these: lipid solubilization; rapid resynthesis of essential metabolites; molecular thermostability; and ultrastructural thermostability. He argues that the last is the only hypothesis that can explain satisfactorily the existence

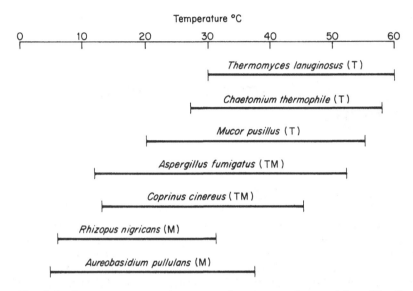

Fig. 6.1 Temperature-growth range of common thermophiles (**T**), thermo-tolerant mesophiles (**TM**) and mesophiles (**M**).

of thermophilous fungi. Solubilization of cellular lipids may occur at high temperatures to such an extent that the cell loses its integrity. It has been shown that in a number of different types of organisms the cellular lipids contain more saturated fatty acids as the growth temperature increases. By virtue of the higher melting point of these, thermophiles are more able to maintain their cellular integrity at higher temperatures than are mesophiles which contain less of these saturated lipids. However in fungi the major function of cellular lipids is as an energy store and as such would have very little effect on the thermal stability of the cell. A very much smaller but highly significant part of the lipids occurs as phospholipids in cellular membranes. It is here at the ultrastructural level that changes in the saturation of fatty acids could affect the metabolic functioning of the cell. At maximum temperatures excessive fluidity of the lipids would decrease the ability of the membranes to serve as both selective permeability barriers and sites of enzyme activity. At minimal temperatures, however the lipids would be at the point of minimum fluidity such that the passage of essential metabolites into and toxic byproducts out of the cells would be too slow. In this context, the ultrastructural thermostability hypothesis is a refinement which embodies the lipid solubilization one. Any system involving the rapid resynthesis of essential thermolabile metabolites at high temperatures would of necessity require thermostable components. This, coupled with the fact that thermophiles do not show any unusually active metabolic pathways, tends to weigh against the rapid resynthesis hypothesis. Numerous thermostable enzymes have been isolated, especially from the thermophilous bacteria, but most enzymes of such organisms are thermolabile. The enzymes of thermophilous fungi do not appear to be particularly thermostable. Indeed no single thermo-stable essential macro-molecule common to all thermophiles has been found whereas ultrastructural differences from mesophiles have been detected in the hyphae of some thermophiles. In some of these, but not in all species examined, spherical cytoplasmic membrane-bound lipid containing bodies have been found. These are often associated with profuse membrane-bound systems continuous with endoplasmic reticulum. These are more common in hyphae grown at higher temperatures. For example, they have been found in hyphae grown at 50°C but not in those grown at 25°C. Typical lipid storage bodies occur in hyphae grown at 37°C and below but not in hyphae grown at 50°C. Whether these lipid containing bodies produced at high temperatures are special lipid storage structures or whether they carry out some physiologi-cal function related to thermophily is as yet unknown.

Variety and distribution of thermophilous fungi

Some thirty species of thermophilous fungi have so far been described. A similar number of fungi are thermo-tolerant. All but two of the thermophiles are members of the Ascomycotina and Deuteromycotina (Fig. 6.2). The odd exceptions are the two Zygomycetes, *Mucor miehe* and *M. pusillus*. The latter was also the first thermophilous fungus to be described by Lindt in 1886. A number of other Zygomycotina are markedly thermo-tolerant as are a number of Basidiomycotina, especially some members of the genus *Coprinus*. There are no known truly aquatic thermophilous fungi. The temperature characteristics of these fungi provide a key to their isolation. Incubation temperatures of 40–50°C are most suitable. There is no evidence that they have any more special substrate requirements than mesophiles. Cooney and Emerson favoured yeast starch and yeast glucose agar for their isolation. Most grow well on these. The majority are autotrophic for vitamins and can use nitrate and sulphate as sole nitrogen and sulphur sources respectively.

They are generally world-wide in their distribution as a result of the world-wide occurrence of self-heating masses of organic debris. They are well repre-sented in aerobic and semi-aerobic self-heating organic matter whenever this occurs in sufficient volume to be relatively well-insulated. By far the majority of these habitats are man-made and include self-heating garden and mushroom compost, hay, municipal and cotton waste, piles of wood chips, moist stored grain and rapeseed, stacks of oil palm kernels, stored peat, curing tobacco leaves, retting guayule and coal spoil tips. They also occur in natural habitats where geothermal heat, sun-heating, microbial metabolism or body heat of warm-blooded animals provide the elevated temperatures necessary for their growth. Such habitats include hot spring effluent channels, such as occur in the Yellowstone National Park, Wyoming, USA; and similar habitats in Iceland, Japan and New Zealand; volcanic soils, such as those of the Ural mountains; sun-heated muds; self-heating piles, such as alligator nesting material in the Everglades National Park, Florida, USA, the

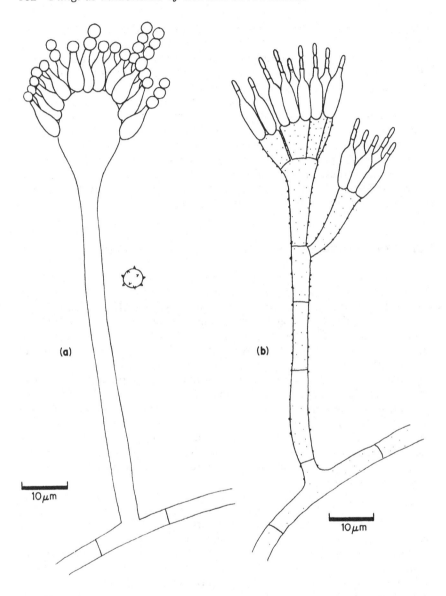

Fig. 6.2 (*Above and right*) Conidia of thermophilous fungi. (*Above*) (**a**) Conidiophore and phialoconidia of *Aspergillus fumigatus*. (**b**) Conidiophore and phialoconidia of *Penicillium emersonii*, the anamorphic state of *Talaromyces emersonii*. (*Right*) (**c**) Conidiophore and phialoconidia of *Paecilomyces crustaceus*, the anamorphic state of *Thermoascus crustaceus*. (**d**) Aleurioconidia of *Thermomyces lanuginosus*.

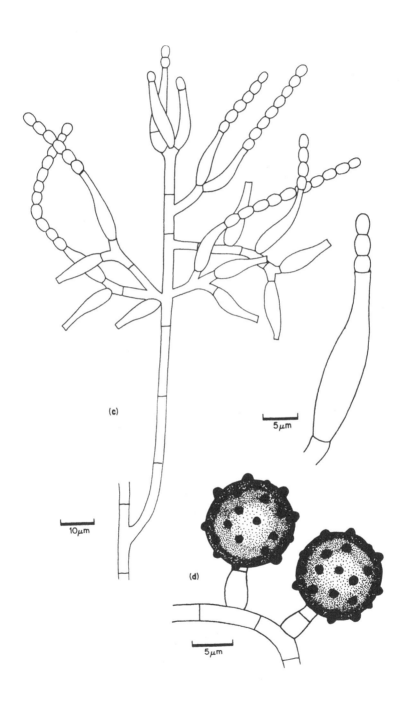

maintained mounds of the incubator bird or mallee fowl of Australia and the islands in the Southwestern Pacific Ocean and nests of passerine birds.

Succession of fungi in wheat straw compost

The composition of the active myco-flora of any self-heating system shifts from predominantly mesophiles in the early stages of thermogenesis to one of predominantly thermophiles at the peak of the heating cycle as long as the material remains aerobic. The thermophiles become active at about 35–40°C by which time the competition from mesophiles has markedly declined. On self-heating substrates as on any other a succession of fungi develops. For instance, in experimental wheat straw composts in which chopped wheat straw is induced to heat by moisturizing with a dilute solution of ammonium nitrate and stacking loosely in wooden bins (1 x 1 x 1 m) or within large compounds made of straw bales, the temperature rises quickly in the centre to reach 60–72°C by the fifth or sixth day. It is maintained there for two or three days before it begins to fall slowly. It may remain at 40–50°C for 3–4 weeks, the plateau period, before it finally drops to ambient (Fig. 6.3). In wheat straw there are many soluble carbohydrates and others, such as starch, which readily become available to micro-organisms as soon as it is moistened. This, and the added nitrogen source in particular, stimulate rapid microbial growth. At harvest, wheat straw would have already been colonized by the mesophilous common primary saprotrophs, *Cladosporium herbarum*, *Aureobasidium pullulans*, *Alternaria alternata* and *Epicoccum purpurascens*. These and others develop profusely for about two days and because of the packing raise the temperatures to levels where thermophiles, such as the non-cellulolytic *Mucor pusillus* and *Thermomyces lanuginosus*, and the thermo-tolerant *Aspergillus fumigatus*, are stimulated to grow. Their

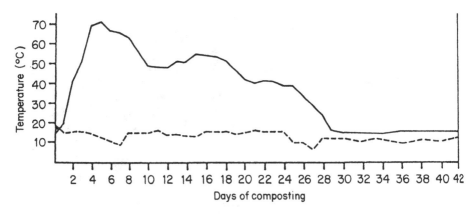

Fig. 6.3 Temperature curve for wheat straw compost (------ air temperature). (After Chang and Hudson (1967). Reproduced by permission of the British Mycological Society.)

metabolism, in turn, raises the temperature still further to above that which mesophiles can tolerate and these latter are either killed off or persist only around the edges, where the heating is less intense and may not reach 40°C. The thermophilous fungi are active up to 55–60°C but above this level thermophilous actinomycetes and bacteria take over and raise the temperature to its peak. At the peak, the central areas of the compost are devoid of fungi. After the peak, the number of actinomycetes and bacteria decline and thermophilous fungi recolonize from cooler, more peripheral areas and maintain the temperature above 40°C during the prolonged plateau period. This is the period of maximum activity of the markedly cellulolytic *Chaetomium thermophile* and *Humicola insolens* but also of the non-cellulolytic *T. lanuginosus*. As the temperature finally drops below 40°C several mesophiles colonize. The most notable of these are a number of small Coprini, especially the moderately thermo-tolerant *Coprinus cinereus*. Some of these are capable of utilizing both cellulose and lignin.

Similarities with the coprophilous fungal succession

In broadest outline the succession is similar to the coprophilous one with an initial, but shorter and less species-rich Zygomycete phase, followed by numerous Ascomycetes and Deuteromycetes and terminating with a Basidiomycete phase (Chap. 5). Although temperature is undoubtedly the major environmental factor determining the sequence, the resemblance to the coprophilous fungal succession would suggest that additional environmental factors interact with temperature. Herbivore dung has an initially higher nitrogen level than the ingested material and this has an influence on the succession. Similarly, the addition of ammonium nitrate to the straw might also help to determine the sequence in composts. The moisture content of the composting straw, like herbivore dung, is also higher and subject to less marked fluctuations than in naturally decomposing straw. This would not only favour more rapid decomposition but would also make it a more favourable substrate for fungi in general.

Nutritional capabilities of the fungi involved

The nutritional capabilities of the fungi concerned are also obviously important in determining the succession. *Mucor pusillus* occurred only in the early phases of composting. It can use only simple carbon compounds as its carbon source and is unable to utilize hemicelluloses and cellulose. Its failure to reinvade after peak heating was probably in the main because all the initially available simple carbon compounds had been utilized at this stage. *M. pusillus* is a primary saprotrophic sugar fungus relying solely upon simple carbon sources initially available. In contrast, *Thermomyces lanuginosus* did return after peak heating and was most active during the plateau period. It appears then to be dependent upon the cellulolytic fungi for its carbon supplies. In culture on cellulose, many cellulolytic fungi, especially *Chaetomium thermophile* and *Humicola insolens*, produce sufficient simple sugars as a result of cellulolysis, not only to support themselves but also to support active growth of such fungi as *T. lanuginosus*. In composts, *T. lanuginosus* probably grows commensally with cellulolytic fungi taking a share of· the secondarily derived sugars. It may also rely upon the cellulolytic fungi for its nitrogen sources as they release a number of amino acids into their environment which it can also utilize. In the absence of cellulolytic fungi, *T. lanuginosus* occupies the same niche as *M. pusillus*, utilizing any simple carbon sources initially present. The inability of *M. pusillus* to act as a similar secondary saprotrophic sugar fungus may well be due to the fact that it is much more sensitive to the antagonistic conditions which develop with time in the compost. If its spores or mycelium are added to the compost after peak heating they are rapidly lysed. The development of a commensalistic relationship is not therefore correlated entirely with the cellulolytic ability of the potential partner. In the later stages of composting *Coprinus* spp. become established and the compost may become almost a pure culture of one of these, *Coprinus cinereus*. It is markedly cellulolytic but much more efficient at absorbing the products of cellulolysis such that insufficient sugars accumulate around its hyphae to support the growth of fungi such as *T. lanuginosus*. In addition it is markedly antagonistic to it and many other fungi in the compost by causing hyphal interference. This affords a further parallel with the final stages of the coprophilous succession.

Beneficial and detrimental activities of thermophilous fungi

Although relatively few in number, thermophilous and thermo-tolerant fungi are of importance in contributing to the decomposition of dead plant tissues whenever a suitable temperature occurs. Their activities may be seen as both beneficial and detrimental to man. Among their beneficial activities are: the preparation of mushroom compost until it has reached a suitable stage of decomposition such that *Agaricus bisporus* can colonize; high temperature composting for the disposal of municipal waste; the conversion of cellulose wastes to microbial protein for animal feeds; and the retting of guayule, in which, during World War II in the USA, *Parthenium argentatum* was subjected to microbial decomposition at high temperatures, to produce a high quality rubber with great tensile strength. Amongst their detrimental activities may be listed the deterioration of stored cereal grains such that they become mouldy and valueless for malting or for cattle feed and the self-heating, leading to the possible spontaneous combustion, of hay, peat and wood chips. The spontaneous heating and ignition of hay led to the first detailed study of thermophilous fungi by Miehe in 1907. A number have been reported as causing diseases of warm-blooded animals, including man. *Mucor pusillus* has been associated with mastitis in cattle and with a variety of mycoses. The thermo-tolerant *Aspergillus fumigatus* is well known as causing aspergillosis in birds, especially poultry, but also in gulls and ostriches. Aspergillosis is a respiratory disease and when severe the lumina of the tracheae, the bronchi and the air sacs become completely lined and almost blocked with a bluish-green felt of conidial heads. The source of the inoculum for many outbreaks has been traced to mouldy straw or composts. The same fungus causes mycotic abortion in cattle and egg infection in poultry. In the latter, conidia are picked up on the egg at the time of laying. They germinate in the high humidity and temperature regime of the incubator, especially if the eggs are dirty, or partly covered by faeces or the contents of broken eggs. The shell is penetrated and the fungus grows in the membranes. Conidiophores often develop protruding into the air space. The embryo is killed after about six days and affected eggs, if broken, emit clouds of conidia which may lead to aspergillosis developing in the newly hatched chicks. *Dactylaria gallopava* causes encephalitis in young turkeys and chicks. It is widespread in many natural and man-made heated habitats. It has been isolated from acid thermal soils, effluents of acid hot springs and self-heating coal spoil tips. Temperatures in the range of mammalian and avian body temperatures exist in these and other sites which thus may be natural reservoirs for such pathogens.

Cultivation of *Agaricus bisporus*

The majority of all the known thermophilous fungi have been isolated from mushroom compost and they play an important part in the preparation of a suitable compost for the growth of the cultivated mushroom, *Agaricus bisporus*. At first and until well into the twentieth century, mushroom growing was a very hit-or-miss affair, use being made of underground caves and quarries, unused cellars and abandoned sheds or anywhere where it was cool and as humid as an autumnal night, when wild species of *Agaricus* normally appear in pastures and on well-decomposed manure heaps. Little or no control was exerted on the environment because so little was known of the ecology of the mushroom beyond the fact that it produced basidiocarps in temperatures from 10–14°C. Today with the world production in excess of 300 000 tonnes annum^{-1}, mushroom growing has become a very sophisticated technology almost independent of outside climatic conditions, with mushroom houses equipped to control temperature, humidity and aeration.

Preparation of the compost

The preparation of a suitable substrate, the compost, is probably the most difficult and certainly the most critical phase of mushroom growing. The compost has traditionally been prepared from wheat straw and horse manure. Pig manure, hay and straw and ground corn cobs may also be used. Other alternatives to wheat straw which may be used in the future include bagasse, wood waste, pulverized tree bark and municipal garbage. Any such mixtures must undergo a certain amount of decomposition before *Agaricus* will grow well on it and produce a worthwhile crop of basidiocarps. A mycelial inoculum or 'spawn' will grow on such mixtures before composting but crops are exceedingly poor and erratically produced. This can be attributed in the main to three features. *A. bisporus* is a mesophile and it would be eliminated by the heat inevitably generated as the mixture decomposes. Although it will permeate and utilize such substrates when they are sterile, it fares badly as a competitor with the indigenous population of micro-organisms. Composting is a form of pasteurization. Like many other Basidiomycotina, *A. bisporus* is especially demanding in its requirements for carbon and nitrogen in order to complete its growth and reproductive cycles. Composting is a fermentation specific to mushroom cultivation. It involves the sequential establishment of different groups of micro-organisms which modify the nutritional status of the substrate such that it particularly favours the growth and development of the mushroom. A number of additions are made to the mixture prior to composting. The nitrogen content of a suitable compost should be between 2–2.5%. Nitrogen-rich supplements are added initially to activate the fermentation. Potassium may also be deficient in wheat straw and is added as potassium chloride or sulphate. Various 'conditioners', such as gypsum, may also be added. Gypsum improves the texture of the final compost. It produces a short, fibrous, well-aerated, brown coloured compost of pH 6.5–7.5. It prevents over-compacting and improves the ability of the compost to retain water.

The composting process

The actual composting process can be divided into two phases. The first occurs outside in long, narrow piles, about 1.0 m tall and 1.8 m wide. During this phase, which lasts about 7–10 days, microbial thermogenesis takes place and the temperature of the compost may reach 70–80°C. Aerobic conditions are maintained by restricting the cross-sectional dimensions of the piles and by turning over the compost at regular intervals. The importance of this first phase is to produce a compost unsuitable for the growth of what are called 'weed' moulds which may, if allowed to develop, cause extensive loss of crop. The second phase occurs inside an insulated building. The compost is packed into movable trays, 150–200 mm deep, and these are stacked one above the other with adequate air spaces in between. During this phase peak heating occurs. This may be initiated by the introduction of live steam but is raised by thermogenesis of the thermophilous micro-organisms within. The temperature is prevented from rising above 50–60°C by controlled ventilation of the room. This phase lasts for 3–7 days. Often the compost becomes powdery white, caused by what has been called 'fire fang'. Much of this is due to thermophilous actinomycetes but many thermophilous fungi are also very active at this phase. The idea behind this phase is to ensure controlled microbial decomposition to produce nutrients, especially proteins, which are necessary if the mycelium of the mushroom is to grow. At the same time any residues of ammonia and amines left at the end of the first phase are decomposed or converted. They are toxic to the mycelium of *Agaricus* but not to 'weed' moulds which would subsequently colonize the compost. This phase also effectively pasteurizes the inside of the compost and at some stage the air temperature around the compost is raised to 60–62°C for a short period. This completes the process of pasteurization, killing any spores on the surface as well as pests such as mites, other insects and nematodes.

Production of inoculum and inoculation

At the end of this phase, the compost has reached a stage at which an inoculum of the mushroom will actively grow on it. It is then capable of sustaining the rapid development of the mycelium of such a strongly cellulolytic species as *A. bisporus* to the exclusion of all other fungi. In addition its mycelium produces an antibiotic which represses bacteria and other fungi so that once it has colonized it is virtually a pure culture. To achieve this pure culture the compost is 'spawned'. The production of a pure culture inoculum or 'spawn' is another key operation in the success or otherwise of a commercial venture. Spawn is produced by growing the fungus on sterilized rye or wheat grain with added chalk to maintain alkalinity. This produces a granular spawn with sufficient reserves of nutrients in the grain to support active mycelial growth after spawning to ensure rapid establishment. The granular nature also allows it to be thoroughly mixed throughout the compost. Once spawned, the compost is transferred to controlled temperature rooms with ducted ventilation to provide optimum temperature, aeration and humidity. The mycelium from each particle of spawn grows rapidly and hyphal fusions occur between the hyphae from adjacent particles so the mycelium quickly grows through and completely permeates the compost well before contaminants or 'weed' moulds can develop. The limited mycelium will again cause limited thermogenesis but temperatures are kept down to below 35°C by circulating air over the compost at 25–28°C. Otherwise the temperature would rise to above that tolerable to the mushroom.

Casing and the switch from vegetative to reproductive growth

After 10–25 days, the heating subsides as colonization is complete and the fungus has accumulated sufficient reserves to produce basidiocarps. These are produced only if the compost is covered by a casing layer. It is thus removed to production rooms and cased. Numerous materials are used for casing. In this country, peat mixed with granular chalk is often used. The compost is covered with a thin layer (25 mm) of this. The transition from the vegetative to the reproductive phase takes place in the casing soil. In contrast to the compost itself, the casing soil is very low in nutrients but is completely colonized by the mycelium of the mushroom. The growing mycelium in the casing soil and the compost produces volatile metabolites such as ethanol, ethanal, ethyl ethanoate and carbon dioxide. These accumulate in the casing layer and are selective for a highly adapted micro-flora, especially the bacterium *Pseudomonas putida*, which in some obscure way triggers off basidiocarp production. An unsterilized mixture of peat and sand is essential for production of basidiocarps. They are not produced in any quantity if the casing material is steam sterilized before being applied. Bacteria, isolated from the casing layer, will also stimulate cultures of *A. bisporus* to produce basidiocarps. One suggestion is that the bacteria release ferrous iron, which is essential for the development and growth of basidiocarps, from the peat and chalk mixture where the iron is firmly complexed to organic matter. It may well be that the role of the bacteria in the casing is to remove one or more self-inhibitors of basidiocarp production rather than making any positive contribution to their induction. Reproduction in many agarics is inhibited by carbon dioxide levels in excess of atmospheric ones. The bacteria may merely remove this and other volatiles.

After casing, the environmental factors are again strictly controlled. The relative humidity is maintained at 95%, air is passed over the trays at 100–250 mm^3 min^{-1} and to favour basidiocarp production the temperature is lowered to 15–17°C. The first crop of basidiocarps is picked 15–25 days after casing and the trays may crop for several months. However, most growers pick the trays for 30–40 days, the period when production is highest, and in this way grow 6–7 crops per year.

In this rather complicated process there are three major events – the provision of a suitable substrate, the compost; the production of a suitable inoculum to take over the compost; and the induction of the switch from vegetative to reproductive growth by casing. The compost must contain adequate and appropriate sources of carbon and nitrogen as well as a supply of minerals, vitamins and ethanoate. Carbon as an energy source is provided in the straw by cellulose, hemicelluloses and lignin. Composting removes any simple carbohydrates which would not be conducive to the rapid production of cellulase, in particular, by the spawn. Cellulose is the bulk carbon source, 60% of its dry mass, in straw. Other fungi would compete more actively for simple carbon sources and the idea of spawning is to provide a massive inoculum to give the mushroom an enormous competitive advantage over other would be colonizers. *A. bisporus* specifically requires protein as its nitrogen source. Inorganic nitrogen sources added to activate the fermentation and simple ones, such as urea, present in the initial mixture are converted into microbial protein. This, after thermogenesis, provides the required protein. Minerals, such as potassium, phosphorus, magnesium, calcium, iron and micro-elements, such as zinc and manganese, are provided in the straw, manure and conditioners or added separately. *A. bisporus* is also heterotrophic for certain vitamins, including biotin and thiamine. These, too, are produced by other micro-organisms during composting. It also requires ethanoate as a building block for a number of its essential macromolecules. This is released by microbial activity in the compost and in addition the build up of fats and oils by the thermophiles provides a reservoir of acetate for its subsequent utilization.

Many variations occur in the methods of compost preparation but the aims of all are the same – to achieve the microbial conversion of a totally non-selective substrate into a highly selective one suitable for the growth and later reproduction of *A. bisporus*. An extremely similar succession of fungi occurs in mushroom composts as in wheat straw ones but at the critical stage when conditions are conducive for the establishment of species of *Coprinus*, a massive inoculum, the spawn, of the mushroom is added to outcompete the Coprini.

Pathogens, pests and competitors

Although thermogenesis itself tends to sterilize the compost and every effort is made to ensure pasteurization, the mushroom, like other crop species and toadstools in the wild, is subjected to pathogens, pests and competitors. Pathogens may be other fungi, bacteria, viruses or nematodes. The most ubiquitous and destructive of the fungal pathogens is *Verticillium malthousei*. It is endemic on many mushroom farms and can devastate crops by infecting most basidiocarps produced. Infection of very young basidiocarps produces small, deformed, undifferentiated spheres called 'dry bubbles', with a dry surface covered with dusty grey conidia. Surface infections of older and larger basidiocarps develop into localized, depressed, brown necrotic lesions. *V. malthousei* is spread both by pickers and flies to which the adhesive conidia cling and by water droplets splashed from infected basidiocarps during the application of water to the beds. It is difficult to control without resorting to chemical eradicants or protectants. The very conditions – a relative humidity of 90–95% and a temperature of 15–17°C – which are necessary for intensive production if 6–7 crops are to be grown each year are most conducive to its spread and for infection to occur. Systemic fungicides are available which affect the pathogen but not the mushroom. A dramatic decrease in the incidence of the disease can be achieved by reducing the relative humidity to 80–85% and the temperature to 14°C. These environmental changes however, lengthen the time between successive flushes of mushroom and also reduce the size of the individual basidiocarps so that most growers prefer to cope with the disease chemically.

Many so-called 'weed' moulds may grow with the mushroom mycelium. These, too, can cause extensive loss of crop but many have been virtually eliminated by refinements in the composting process. For example, *Papulospora byssina*, the brown plaster mould, also inhibits the growing mycelium of *A. bisporus*. It is a good indicator of over-composting and has virtually disappeared since the widespread introduction of the short composting system, in which, during the first phase, the compost is left for only 7–10 days. Also if over-composting occurs in the first phase, insufficient readily available carbon sources will be present for the thermophiles, which develop in the second phase to convert all the ammonia and amines present into microbial protein. Although these are toxic to the mycelium of *A. bisporus*, they are not to such fungi as *Coprinus* spp. and *Thielavia thermophile* and these subsequently develop as indicator weeds.

Cultivation of *Volvariella volvacea*

Samples of any composting material always yield abundant thermophilous fungi. *Volvariella volvacea*, the padi straw or Chinese mushroom, has been cultivated for many centuries in Southern China and neighbouring Asian countries on composts of rice straw but other types of material such as oil palm pericarp, banana leaves, sugar cane refuse and cotton waste may also be used. The preparation procedure is far simpler than that used in the cultiva-tion of *A. bisporus*. Small bundles of rice straw are soaked in ponds or tanks and stacked in long piles about 1 m high and 1 m wide which are spawned at the time of construction, usually with 'spent' compost from old beds. After a few days the temperature in the centre of the piles reaches 40–50°C. At this stage the majority of the thermophilous fungi found in wheat straw and mushroom composts are present. Basidiocarps begin to appear after 15–25 days without any casing layer being applied. One of the major problems encountered by the growers is again competition from Coprini such as *C. cinereus* in the final stages. These inhibit the growth of *Volvariella* and compete with it for both nutrients and space. The extent of the crop from any one compost very much depends upon the vigour of the spawn used as an inoculum and the intensity of the chance contamination by the Coprini. The latter can effectively be suppressed by sterilization of the straw before spawning by steam or chemical fumigants.

Garden and municipal composts

In normal garden composts and in many municipal composting systems, three factors may limit the growth of thermophilous fungi. These are exces-sively high temperatures, acidity and low oxygen levels. For example, adding abundant fresh grass cuttings to garden compost heaps may cause a rapid rise in temperature to well over 70°C and cause compaction which severely reduces the inward diffusion of oxygen. Although fungi are obligate aerobes, several species of thermophilous ones have been shown to produce significant growth in oxygen tensions as low as 0.2–1.0%. However in this particular example temperature would be the major limiting factor, 70°C being well above the temperature maximum. The response of thermophilous fungi to pH varies, as in mesophiles, with the species. Some will grow uniformly over a pH range from 4–8, but others, such as some isolates of *Chaetomium thermophile* and *Humicola insolens*, are very sensitive to pH and grow well only at pH 7–8. Most fungi isolated from composts tend to have a neutral or alkaline pH optimum and in this respect are like many coprophilous and pyrophilous fungi.

One problem associated with municipal wastes is that they contain large quantities of plastic materials which do not decay during composting. These, and inorganic materials such as glass and metals, have to be separated from the waste before composting. Many thermophilous fungi can use the plasti-cizers present in plastics as a carbon source but not in the polythene itself. However, chemical oxidation of the polythene yields substrates which can be used as a sole carbon source by several thermophilous fungi. One way being investigated to dispose of these is to follow chemical oxidation by composting.

Industrial wood chip piles are another good example of where self-heating of plant materials may occur. In Canada, Sweden and the USA, wood chips are stored outside in piles until processed for paper and pulp products. These piles, each containing millions of cubic metres of chips, or isolated pockets within them, may heat spontaneously. Well over 100 million tonnes fresh mass are used annually and more than three million tonnes of this are lost through thermogenesis and deterioration by thermophilous fungi. Additional financial losses are caused by both colour and chemical changes which increase the cost of processing and also reduce the quality of the products. The temperature within such piles increases rapidly to about 60°C within the first few weeks of storage. It remains at that level for a variable period depending upon the type of chip. It then falls slowly to ambient over a period of months. The respiration of the dying wood parenchyma cells and purely chemical oxidations release sufficient energy to cause the temperature to rise initially, but microbial thermogenesis helps to maintain the temperature of the pile and sometimes increases it sufficiently to lead to spontaneous combustion. Tansey (1971) isolated eight different thermophilous fungi and the thermo-tolerant *Aspergillus fumigatus* from chip piles of Douglas Fir, *Pseudotsuga menziesii*, in California, USA. He sampled fresh, unheated, self-heated and charred chips. The chips had spontaneously ignited and had been extinguished with water. The thermophiles were much more common on heated than on unheated chips, both in number of species and abundance. This, together with the fact that many of the fungi present, such as *Chaetomium thermophile* and *Humicola grisea*, are able to degrade cellulose, suggests that they contribute to many of the deleterious physical and chemical changes which occur in wood chip piles. Effective chemical treatment of wood chips to prevent such losses is apparently possible, but not economically feasible, under conditions of actual commercial storage and at the present, management, in terms of avoiding long term storage, seems to be the only way to prevent major losses.

Most thermophilous and thermo-tolerant fungi are dispersed by air-borne spores and they are the major source of inoculum for wood chips as they are for those which occur in self-heating and stored grain. The latter may also be subject to losses especially in quality as a result of storage. Spores of thermophiles and other storage fungi, such as the osmophilous Aspergilli, may be impacted onto grains before harvest but do not grow. At this stage they are mere casual inhabitants or superficial contaminants. Cereal grains are usually stored commercially at below 13.0% moisture content on a wet mass basis. This is sufficiently low to prevent the growth of any fungi. If even small moisture pockets occur in the stored grain giving local moisture contents of above 13.2%, the spores of the osmophilous Aspergilli germinate and grow. Their metabolic activities not only raise the water content sufficiently to enable less osmophilous fungi to grow but because of the close packing bring about heating which may be sufficient to cause rapid growth on the grain by a variety of thermophiles. The final result may be complete spoilage.

Thermophilous fungi in soils

Thermophilous fungi can be isolated from most soils but it still has to be ascertained whether they actually grow there or whether they are merely represented by spores which have fallen out or have been washed out of the atmosphere. Even in temperate climates in the summer in the sun, soil surfaces may be above 40°C for relatively long periods, but it has not been directly proven that such insolation provides sufficient heat for thermophiles to grow. Even if it does, much of their existence will be spent in co-existence with mesophiles which would outgrow them under mesic conditions. It has been suggested more often that tropical soils, with their higher insolation and temperature regimes, might well support a higher population of thermophilous fungi than temperate soils and that these fungi might be expected to play a more important role in the decomposition of plant materials in these soils. Hedger (1975) investigated this. He found very much higher populations of thermophilous fungi from mud and leaves, from pools in hot springs in West Java, than in the soil from the surrounding dense tropical rain forest and he attributed this to the marked temperature differences between the two sites. The water temperature at the hot springs was 40–50°C and the soil temperature in the forest 20°C. The population of thermophiles in the forest was no higher than in temperate soils. In East Java, which has a drier monsoon climate, he found a higher population of thermophiles in soils of the more open teak forests. This suggested the possibility that in such soils insolation caused sufficiently high temperatures to induce growth of thermophiles. To test this he investigated the microflora of two adjacent 1 m² plots, one exposed to the sun and the other shaded with palm thatch, in the Botanic Gardens at Bogor. Maximum daily temperatures reached in the open plots were 40–48°C at 10 mm depth and in the shaded plots 35–38°C. Suitable substrates, such as crushed rice, were added to the two plots. One week after such addition the population of thermophilous fungi had risen from 2.0×10^2 to 50×10^5 viable propagules g^{-1} in the open plots but only from 5.0×10^1 to 7.0×10^2 propagules g^{-1} in the shaded plots. This strongly suggests that under conditions of high insolation growth of themophilous fungi may be stimulated in tropical soils.

Nests as sources of thermophilous fungi

Birds' nests, especially those of passerine birds, are a rich source of thermophilous fungi. The nesting bird is presumed to be the source of heat for the growth of the fungi. In contrast the mallee fowl's nest is to all intents and purposes a self-heating compost heap and is a rich source of thermophiles as are the nests of the alligator. Although there is little doubt that the mallee fowl has adopted the strategy of using microbial heat for the incubation of its eggs, there is considerable doubt as to the role this heat plays in the biology of the alligator. Indeed this heat fermentation could be a hazard to the eggs and this may explain why the female selects a cool nesting site to minimize overheating. However the heat may help to counteract the cooler night time temperature in swamps, Tansey (1973) isolated five thermophilous fungi and two thermo-tolerant ones from the partially decomposed vegetation from the interior of a recently used nest of the American alligator. He also makes the point that habitats suitable for the origin and evolutionary survival of obligately thermophilous fungi have existed for long periods of time. That must be so. Alligators existed in the late Jurassic period!

PSYCHROPHILY IN FUNGI

Psychrophilous fungi, in contrast, are those which are cold-loving and which grow best at low temperatures. Since most mesophiles have a growth minimum in the range 5 to 10°C, for convenience cold environments for fungi are considered as those below such temperatures and usually taken as those below 5°C. As with thermophily, there is no widely accepted definition of psychrophily with regard to fungi. The problem that exists in defining psychrophily is that many fungi which can grow at 0°C can also grow well in the upper range at which mesophiles grow whereas for others that can grow at 0°C the maximum growth temperature may be relatively low. It is a problem of distinguishing between the true psychrophile and the cold-tolerant mesophile and appreciating that many fungi can survive, but not grow, in cold environments. They may best be considered as having cardinal temperatures for growth with optima at 15°C or lower, maxima at 20°C, not growing above, and minima at 0°C or below.

Basis of psychrophily

Very little is known of how changes of temperature affect the cells of psychrophilous fungi. A wide variety of possibilities exist. Toxic metabolites may accumulate at higher temperatures within their growth range. Key enzymes may cease to function at low temperatures. Moderate temperatures, 20–25°C, may prevent protein synthesis. The permeability of cell membranes may be altered by changes in the saturation of their fatty acids. *Sclerotinia borealis*, in culture, has an optimum temperature for growth of 0°C with a minimum below – 7°C. It fails to grow above 15°C. It is a true psychrophile being able to grow well at extremely low temperatures and unable to do so at temperatures at which mesophiles normally grow. *Typhula idahoensis* and *T. trifolii* grow at – 5°C with an optimum at 5°C and a maximum near 20°C. Neither grows at 25°C. Both of these and *S. borealis* produce survival sclerotia in larger numbers at the higher temperatures within their growth range. Cultures of these fungi grown at 10°C, and transferred to 20–30°C for varying periods of time before returning to 10°C, all develop a lag period, proportional to the duration and degree of temperature elevation, before resuming growth. A number of explanations have been offered for this phenomenon. There may be an increased accumulation of toxic metabolites at higher temperatures which are utilized or detoxified in the lag period. Some reversible thermo-sensitive system may be present. There are isolated reports of abnormally heat sensitive enzymes from a number of psychrophilous bacteria, such as malic dehydrogenase in *Vibrio marinum*, but as yet none from psychrophilous fungi. It seems more likely that in mesophiles some key enzymes cease to operate at low temperatures rather than all the enzymes of psychrophiles being especially adapted to low temperatures or particularly sensitive to high ones. However, in these species of *Sclerotinia* and *Typhula* the optimum temperature for respiration coincides with or is above the maximum permitting growth. This suggests that growth becomes completely uncoupled from respiration at 20°C and above. As a result oxygen uptake continues at these temperatures but energy ceases to be directed towards synthetic processes and growth ceases. In *S. borealis*, the optimum temperature for respiration is 25°C, considerably above the maximum temperature for growth. With glucose as a substrate, stimulation of oxygen uptake by uncoupling agents, such as dinitrophenol, is relatively greater at 5°C than at 25°C. This would be expected if growth and respiration were already partially uncoupled at 25°C. Also the degree of stimulation at 25°C varies inversely with the duration of exposure at that temperature. It thus seems probable that the mechanisms responsible for coupling growth to respiration may be abnormally heat sensitive and it is this sensitivity which determines the low maximum temperature for growth.

In the psychrophilous yeast, *Candida gelida*, the protein synthesis mechanism is heat sensitive. A positive correlation has been demonstrated between the presence of abnormally heat sensitive transfer RNA synthases and the maximum temperature for growth. Seven out of thirteen aminoacyl-*t* RNA synthases were temperature sensitive in cell-free extracts. They were completely inactivated by treatment at 35°C for 30 min. So also were some of the soluble enzymes involved in the formation of ribosomal bound polypeptide chains. The presence of heat sensitive synthase in *Sclerotinia* and *Typhula* could also explain their response to high temperature.

In psychrophiles, mesophiles and thermophiles as the growth temperature decreases it is often argued that cellular lipids are composed of an increasing proportion of unsaturated fatty acids. In psychrophiles, this could affect metabolic functioning by serving to maintain fluidity in the lipids of cell membranes. Such membranes would be able to function better at lower temperatures. However, Sumner *et al.* (1969) found that, in a range of Mucorales, the lipids of the psychrophilous and mesophilous species had similar degrees of saturation but were more unsaturated than those of thermophilous species. Thus, it seems unlikely that, as some argue, the minimum growth temperature in psychrophilous fungi is the temperature at which the capacity to produce more unsaturated fatty acid ceases. As with thermophily, psychrophily in fungi warrants further study.

Variety and distribution of psychrophilous fungi

Both polar regions support many fungi. Members of all classes exist in the Arctic Tundra. Micro-fungi occur as saprotrophs on dead leaves. The whole range of coprophilous fungi occur on herbivore dung. A number of bracket polypores decompose wood. Several agarics are mycorrhizal with species of *Salix* and *Betula*. There is also a rich lichen flora. In the Tundra, the temperature for much of the year is between − 60 and 0°C but for at least one month of the year the average temperature is between 0 and 10°C, with maxima exceeding 20°C on occasions. The temperature range over which these fungi make active growth has not been ascertained but it is quite clear that they must be able to tolerate extreme cold.

'Snow moulds'

Fungi from a number of different genera, such as *Fusarium*, *Sclerotinia*, and *Typhula*, are called snow moulds because they cause diseases of cereals, grasses and food crops grown at low temperatures. They are favoured by prolonged cold or wet conditions when there is deep snow over unfrozen ground. *Sclerotinia borealis* is particularly severe in climates as cold as those of North British Columbia and Sweden. Diseases caused by species of *Typhula* are often called snow scald because of their association with snow and the scalded appearance of the host plant after the snow melts. *Phacidium infestans*, a snow mould of pine nurseries and young plantations, is another true psychrophile. It causes serious damage at high elevations and high latitudes where snow cover is deep and prolonged. It has an optimum temperature for growth of 15°C with growth rate decreasing very rapidly above. At 20°C the growth rate is lower than at 5°C. It grows at − 3°C and endures − 20°C for prolonged periods but is killed at 27°C. In contrast, *Micronectriella nivalis*, another snow mould but of cereals, grows at temperatures between 2–32°C with an optimum at 20°C. It causes disease most rapidly at temperatures below 5°C but, with such cardinal temperatures, it is best considered as a psychro- or cold-tolerant mesophile. The majority of fungi of cold environments are probably such cold-tolerant mesophiles. Although there is a dearth of information about the activity of true psychrophiles in natural ecosystems and about the effect of temperature on their activities, the evidence available suggests that at low temperatures, 5°C and below, their rate of growth is such as to give them a competitive advantage over cold-tolerant mesophiles.

Fungi of frozen food

Freezing and cooling, using temperatures in the range of − 10 to 5°C, are extensively employed to preserve perishable materials, especially foodstuffs. Such temperature and such man-made habitats select for both psychrophiles and cold-tolerant mesophiles. But the majority of fungi isolated from such environments fall into the latter category. They are present in a survival state rather than actively growing. Gunderson (1962), using the ability of fungi to grow at 5°C, isolated 113 taxa from 21 genera from frozen food products. Of these *Aureobasidium pullulans*, an exceedingly common mould on other substrates, was the most common. Black spot of chilled and frozen meat is caused by the equally common *Cladosporium herbarum*. It grew on infected meat between – 6 and 0°C but was more profuse at 2°C. *Sporotrichum carnis* has been recorded on meat stored for up to ten weeks at temperatures from – 10 to 0°C. Although slow growth occurred at – 5°C the optimal growth temperature for this fungus in culture was 25°C. All three must be regarded as cold-tolerant mesophiles.

XEROPHILY IN FUNGI

Like all other organisms, fungi have evolved in environments where water is the solvent. It is the medium for all reactions. Although it is a universal practice to maintain potentially degradable materials at a low moisture content to prevent fungal growth, it is an equally well-known fact that many fungi can tolerate desiccation and many can actually grow on substrates with relatively low levels of available water. Fungi in the latter category are the main concern of this section.

The availability of water expressed as 'water activity'

Water moves in and out of hyphae by osmosis, a process of diffusion, in response to physical and chemical gradients across the cell membrane. Each growing colony has to come to terms with its own external aqueous environment. There are two extreme situations. In freshwater systems, aquatic fungi (Chap. 4) grow in a solution more dilute than their internal solution as do other freshwater organisms. At the other extreme there are a variety of fungi variously termed halophiles, osmophiles or xerophiles, which grow in substrates where solute concentrations are relatively high or available water is relatively scarce in relation to the internal composition. Not all the water in a substrate is available to any particular fungus growing in it; although it is usual to speak of a 'safe' moisture or water content for any particular commodity below which it will not be attacked by fungi, for example cereal grains stored at or below 13% moisture content, the moisture or water content of many substrates is not at all indicative of the availability of water to micro-organisms growing on it. Substrates with the same water content can differ markedly in their water availability. The molecular state of the water in each can be quite different. A more suitable measure of the availability of water in a substrate is 'water activity' designated a_w which is a measure of the relative escaping tendency of water from a system compared with pure water. It may be expressed as:

$$a_w = \frac{p}{p_0}$$

where p = the vapour pressure of water over the substrate and p_0 = the vapour pressure over pure water. Its relationship with relative humidity is thus obvious. The a_w is numerically equal to RH except the latter is expressed as a percentage. Thus:

$$RH = \frac{p}{p_0} \times 100$$

A substrate of a water activity of 0.85, for example, would be in equilibrium with air of a relative humidity of 85%. Thus the a_w of a system can be equated to the equilibrium relative humidity which the system could maintain. The interconversion of water activity and water potential data can be made by

$$\psi \text{ (water potential in bar)} = 1.37 \times 10^3 \ln a_w$$

Range of water activity permitting fungal growth

Fungi as a group can grow over a range of a_w values from 0.999 to 0.65. The absolute values are set by the fact that none can grow in an a_w of 1.00 or in pure water as no nutrients are available, nor below 0.55 because DNA becomes denatured. Two main methods have been used to manipulate the water activity in media being used to assess fungal growth at different a_w under experimental conditions. The media may be exposed to a constant RH and allowed to absorb or desorb water from the air until it is in equilibrium. This requires a closed chamber maintained at a constant temperature and a substantial volume of, preferentially, a saturated solution of a salt giving the required a_w. The preferred method is to adjust the solute concentration in the medium, and thus its a_w, by adding various salts, such as sodium chloride, or non-electrolytes, such as sucrose or glycerol. Both methods have their disadvantages. In the former, carbon dioxide may accumulate in sufficient quantity to inhibit growth and it is not suitable for a_w levels above 0.95 because of the nature of the water absorption isotherm. As the a_w approaches 1.00, the rate of change of water content approaches infinity. The period of equilibration may be very long and the peculiarities of dry agar films add to the difficulties. If the latter method is used, it is often found that fungi respond differently to added electrolytes than they do to non-electrolytes but, on the whole, different solutes produce the same effects on fungal growth. The addition of solutes by sugaring or salting to control the availability of water is a recognized technique in food technology.

Xerophilous and xero-tolerant fungi

Particular fungi stand out as organisms that can grow at low a_w. Various terms have been used to describe such fungi. Those that grow well on high sugar or high salt containing substrates have been called osmophiles or osmophilous and halophiles or halophilous respectively. Those that can tolerate high sugar or high salt containing substrates have been called osmo-

or halo-tolerant. These terms are all very imprecise. Griffin (1981) accepts this and states a preference for the terms xerophilic and xero-tolerant. The term xerophilous is equivalent and preferred here. He defines a xerophilic fungus as one that requires a low a_w for optimal growth, and a xero-tolerant one as one that grows at a low a_w but has an optimum growth at higher levels. The problem, as he points out, is that no term has been used to denote fungi that require a high a_w for growth nor is there any universally acceptable definition of the dividing line between a high and a low a_w. The range from 1.00 to 0.96 a_w covers freshwater and seawater habitats which might be considered the norm. It would seem more appropriate to consider those fungi with growth optima below 0.96 as xerophiles and those which grow below 0.96 but have growth optima above to be xero-tolerant. Even so, given these definitions, most fungi would be xero-tolerant, at least to a degree.

The 'osmophilous' Aspergilli

Two groups of Aspergilli, members of the *Aspergillus glaucus* and *A. restrictus* groups have always been regarded as osmophilous – indeed as classic examples in the fungi. They occur on all types of organic materials undergoing slow decay at water activities just above those at which no decomposition can occur. They have been much studied in relation to the incipient spoilage of stored products, such as grain, foodstuffs, especially jams, jellies, salted meat and fish, dried fruit, and leather. They are the green moulds that grow over the surface of jam in an improperly sealed jar or on the inner soles of shoes left in a damp cupboard. They are osmophilous rather than osmo-tolerant because they make better and faster growth and produce more conidia and ascocarps in the *A. glaucus* group on high sugar containing substrates than they do on low ones. They make very slow growth on ordinary laboratory culture media such as 2% malt extract or Czapek-Dox agar but grow very rapidly on the same media supplemented with 20 or 40% sucrose. When grown on malt extract agar with added solutes to adjust the a_w of the medium, the growth optimum of *A. amstelodami*, a member of the *A. glaucus* group, depends upon both the added solute and the pH. It is at an a_w of 0.90 at pH 4.0 with added glycerol and at 0.93 at pH 4.0 and 6.5 with added sodium chloride and at pH 6.5 with added glycerol. In contrast in the non-osmophilous *A. niger*, growth is best at 0.97 and above with both glycerol and sodium chloride at pH 4.0 and pH 6.5 (Fig. 6.4).

Members of the *A. glaucus* and *A. restrictus* groups are best known as 'storage fungi' of stored grain (Christensen, 1965). Cereal grains, and indeed many seeds, become colonized by 'field fungi' before they are harvested. These fungi include *Alternaria, Aureobasidium, Cladosporium, Epicoccum* spp., the common primary saprotrophs which also colonize grass and tree leaves and all manner of herbage (Chap. 2). On cereal grains they grow on or within the pericarp and discolor the grain. They have growth optima above an a_w of 0.96. *Cladosporium herbarum*, for instance, grows best at an a_w of 0.97 and above and does not grow at all below an a_w of 0.90. Most of these

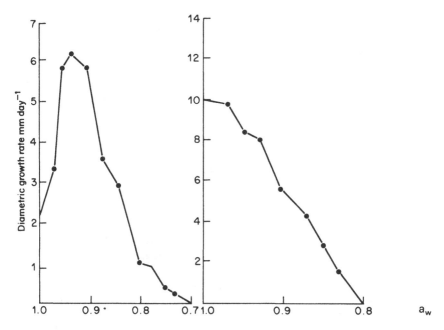

Fig. 6.4 Diametric growth rates of (**a**) *Aspergillus amstelodami* and (**b**) *A. niger* on malt extract agar with added glycerol to adjust the a_w, all at pH 6.5.

fungi require the grain to be in equilibrium with a relative humidity of over 90% before they will grow. This is equivalent to a moisture content of 22–25% on a wet mass basis. Spores of the storage fungi are air-borne and are impacted onto the grain before it is harvested but they do not invade the grain to any great extent. They remain as surface contaminants or casual or exochthonous fungi (Chap. 2). After harvest, cereal grains are usually stored at a moisture content of 13.0% or below. This is equivalent to an a_w of 0.65. No fungi can grow at this low a_w. However, should small moisture pockets occur in the stored grain giving local moisture contents of around 13.2% or just above, the most xerophilic of the Aspergilli begin to grow. For example, *A. halophilicus* and *A. restrictus* begin to grow between 13.2 and 13.5%. They grow only very slowly but by their respiration produce sufficient metabolic water to increase the overall level slowly. When the level reaches 14.0–14.2%, equivalent to a substrate a_w of above 0.70, other members of the *A. glaucus* group, such as *A. repens*, *A. chevalieri* and *A. amstelodami* likewise begin to grow. Their metabolic activities not only increase the overall moisture content still further but may also bring about heating. This increases the rate of deterioration of the grain and encourages the development of thermophilous fungi. As a result, the grain becomes discoloured and biochemical changes occur in the endosperm so that it is no longer fit for food. Toxins may be produced in sufficient quantities to constitute a health hazard. The percentage germination of the grain falls so that it can no longer be used for seed. The final result may be complete spoilage.

Spores of species in the *A. glaucus* and *A. restrictus* group are some of the most prevalent in house dust and are responsible for inhalent allergies. This is because they can grow on a great variety of household materials in equilibrium with a relative humidity of 68–75%. In most parts of the world such relative humidities are recorded in homes, especially in closed cupboards and cool basements. At such relative humidities they predominate over their competitors. Christensen (1975) states that the moisture given off by a human body as it lies on a foam rubber mattress is sufficient to allow fungi from these two groups to grow and produce spores.

Basis of xerophily

Two strategies appear to have been evolved which will allow organisms to grow in environments of reduced a_w. In environments where ionic levels are high, such as salt pans, some bacteria, the so-called halophiles, have proteins which have become modified to function optimally in high ionic solutions. Their enzymes are adapted to operate in high concentrations of sodium chloride. Other organisms can grow in concentrations of salt which would inhibit their essential enzymes. Their proteins are normal but they manufacture compatible solutes to equalize the internal a_w with the external one. The solutes produced protect their enzymes from the lower a_w. Such solutes may be amino acids, such as proline, which is also used by many seed plant halophytes. In the majority of fungi the solutes are mostly polyhydric alcohols or polyols. Twelve polyols have been reported as occurring in fungi. All of these could be osmoregulatory by lowering the osmotic potential within the fungus but their role has not been critically examined. Smith (1978) states that the hyphae of many fungi in lichens must contain a virtually saturated solution of mannitol and he believes that polyols may be important in enabling the lichen to endure environmental stresses, such as desiccation. Lichen fungi are certainly the most desiccation tolerant of all fungi (Chap. 7). The mode of action of polyols is obscure. Most are compatible with fungal enzymes and they could replace water as the medium for enzyme activity. A further suggestion is that their hydroxyl groups could replace the water of hydration of macromolecules allowing them to retain their molecular configuration as the hyphae become progressively dehydrated.

7

Fungi as mutualistic symbionts in ectomycorrhizas and lichens

A wide variety of fungi are apparently unable to exist either as free-living saprotrophs or as necrotrophic or biotrophic parasites and have developed various kinds of mutualistic associations with plants and animals. The majority of those associated with plants exist in association with roots as mycorrhizas or with algae and cyanobacteria (blue-green algae) as lichens. A very diverse array of mutualistic associations exists between fungi and the roots of plants. On morphological and anatomical features such mycorrhizas can be divided into two groups, ecto- and endo-mycorrhiza. In ectomycorrhizas, the ultimate absorbing rootlets of the root system are completely surrounded by a distinct mantle or sheath of fungal tissue from which hyphae penetrate between the outermost cell layer or layers of the root. In endomycorrhizas, there is no such well-developed sheath. Most of the fungus is within the root and intracellular, as well as intercellular, penetration is common. Intermediates between these two groups are recognizable and have been termed ectendomycorrhizas.

ECTOMYCORRHIZAS

Ectomycorrhizas, aptly called sheathing mycorrhizas, are the minority group. Some 3% of all seed plants have ectomycorrhizas. They are particularly common in northern temperate forest trees, especially in the families Fagaceae and Pinaceae but also occur in others such as the Betulaceae and Tiliaceae. However, they also occur in families well-represented in the tropics, especially some members of the Myrtaceae and Dipterocarpaceae.

Structure

The sheathing mycorrhizas of the european beech, *Fagus sylvatica* and various species of *Pinus* have been most thoroughly investigated. In *Fagus* the ultimate lateral rootlets are differentiated into 'long' and 'short' roots. The long roots are of indefinite growth. They are the main extending roots and their branches are the short roots which are of restricted growth and short lived. These are the absorbing roots and are usually modified into mycorrhizal organs. They are thicker, more brittle and coloured differently from uninfected roots. They develop many branches also of restricted growth so that a complex racemose branching system is formed. The mycorrhizas of *Pinus* are exceptional amongst sheathing types in being once or several times dichotomously branched (Fig. 7.1b). There is also á much more marked differentiation into long and short roots. The whole of these short rootlets, including the apex, are bound by a sheath of fungal pseudoparenchyma. In *Fagus*, this is some 20–40 μm thick and comprises some 20–30% of the total volume of the root (Fig. 7.1a,c). In transverse section the sheath can usually be differentiated into two layers – an outermost layer of thicker walled hyphae with no or very few interhyphal spaces and an innermost layer of loosely packed thinner walled hyphae. Anything absorbed from the soil must pass through this thick and somewhat differentiated sheath. There are no root hairs and the meristematic tissues are reduced in comparison with non-mycorrhizal rootlets. From the outside of the sheath, hyphae may spread to varying distances and in varying densities into the soil; from the inside of the sheath, hyphae penetrate between the cells of the epidermis and occasionally further between one or two cell layers of the cortex forming a hyphal network called the Hartig net. The cell layers penetrated by this Hartig net may also be radially elongated. In *Pinus* with narrower roots, the net may penetrate as far as the endodermis. The hyphae never penetrate the endodermis or stele. The only other visible change in such roots is that the epidermal layer of cells adjacent to the sheath is frequently impregnated with dark staining tannins.

Fungi involved

The majority of the fungi involved are members of the Basidiomycotina especially Hymenomycete Agaricales. Well over 100 species of common toadstools have been shown to form mycorrhizas. The majority of these are in the genera *Amanita*, *Boletus* and *Tricholoma* but there are also many species in the genera *Cortinarius*, *Lactarius* and *Russula*. Some of these are relatively specific in their associations, for example *Boletus* (*Suillus*) *elegans* and *B.* (*Suillus*) *aeruginascens* associate only with larches, *Pseudotsuga* and a few species of *Pinus*. Others may associate with 20 or more hosts. *Amanita muscaria*, for instance, is in Britain chiefly associated with birches, *Betula* spp., but is also known to be mycorrhizal with various pines, larches and Norway spruce, *Picea abies*, amongst others. Similarly many host species are capable of forming mycorrhizal associations with several fungi. Professor E. Melin, to whom we owe much of our knowledge of the identity and cultural characteristics of sheathing mycorrhizal fungi, has shown by experimental inoculation that 40 species are capable of forming mycorrhizas with *Pinus sylvestris*. It is also clear that any one tree may associate with several fungi at any one time. Other Basidiomycotina involved are Gasteromycetes in such genera as *Rhizopogon* and *Scleroderma*. Still others are Ascomycotina, such as *Gyromitra esculenta*, all species of *Tuber*, such as *T. melanosporum*, the perigord truffle, and the sterile form *Cenococcum graniforme*.

Although all sheathing mycorrhizas have a number of common characteristics, in reality they show quite wide diversity in both their morphology and anatomy. *C. graniforme*, which is perhaps the least specific of all mycorrhizal fungi, produces one extreme type of mycorrhiza. The fungus has a black mycelium which produces variable sized black sclerotia. The mycorrhizas which it forms with beeches, birches, limes, pines and spruces are also black and brittle due to dense deposits of melanins encrusting the outer walls of the sheath. Abundant very coarse, stiff, bristle-like structures grow out from the sheath into the soil. In still others, the sheath is a loose weft of easily recognizable, tangled, clamp-bearing hyphae.

Nutrition of fungi

Many of the agarics and others can be readily grown in pure culture so that they are not obligate symbionts in the sense that they cannot be grown in the absence of the host. Others, especially members of the genera *Cortinarius*, *Lactarius* and *Russula*, are more difficult to grow. Some have never been grown and are probably obligate symbionts. But in nature, virtually all, from their specialised physiology, are probably ecologically obligately dependent upon their host tree. Most are incapable of leading an independent saprotrophic existence in the soil. This is the more striking when it is appreciated that many are closely related to free-living saprotrophs inhabiting the leaf litter around the roots. In their carbon nutrition they contrast markedly with these cellulose and lignin decomposing litter-inhabiting fungi. With very few exceptions, they are incapable of utilizing polymers such as cellulose and lignin. Most are 'sugar fungi' requiring simple sugars as carbon sources. This suggests, because of the scarcity of such carbohydrates in the litter and soil in general, that they must obtain these from their host tree. All the evidence, which is very varied, points conclusively towards this. The fact that they develop better on seedlings grown in full sunlight than in the shade is suggestive. High radiant flux allows rapid photosynthesis and a surplus of soluble sugars for translocation to the roots. Explanted pine roots become mycorrhizal if supplied with sucrose via their attached hypocotyl. In culture they have a relatively slow growth rate and a very poor competitive ability. In mixed culture with soil saprotrophs, they are readily overgrown and intolerant of competition. This all indicates that they would fare badly in competition with soil micro-organisms for limited soluble carbohydrates as they become available in the litter. However there are a few exceptions. One notable one is *Boletus subtomentosus* which is mycorrhizal with a number of pines. In culture it produces extracellular phenolases as rapidly as any white-rot fungus. It has also been shown to produce cellulases and to bring about very rapid decomposition of leaf litter. A few others also produce amylases, cellulase and hemicellulases but their ability to do so is often poor compared with litter-inhabiting species such that they would be adversely affected by competition in the free-living state; as symbionts, high sugar levels in the roots would be sufficient to repress synthesis of such enzymes.

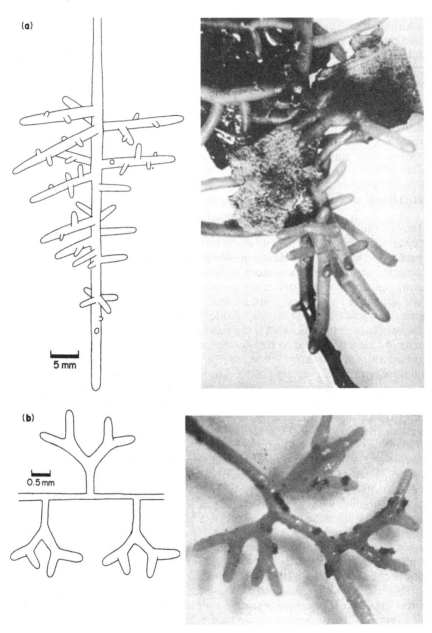

Fig. 7.1 (*Above*) (**a**) Part of a mycorrhizal root of beech, *Fagus sylvatica*, showing complex racemose branching pattern. (**b**) Three mycorrhizal rootlets of *Pinus sylvestris* showing dichotomous branching. (*Right*) (**c**) Transverse section of mycorrhiza of beech with an enlargement of the sheath and Hartig net.

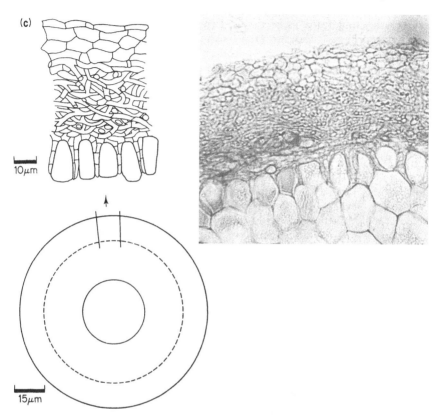

(c)

10μm

15μm

Most sheathing mycorrhizal fungi require one or more vitamins, especially thiamine or one of its moieties. Many also require, in addition, biotin, nicotinic acid, inositol and others especially of the B complex. Some require specific amino acids. These they also obtain from their host but even so, given these and a suitable carbon source, most grow better if supplied with exudates from living roots. All of these commodities would be very scarce in the litter and soil outside the rhizosphere and there would be very intense competition for them in the rhizosphere. Because the requirements of the rhizosphere micro-organisms and the sheathing mycorrhizal fungi are so very similar and because they both obtain these from the same source, the host roots, one way of looking at the latter is to consider them as a logical development from the rhizosphere where one fungus becomes dominant on the root surface; even so why a sheath develops has still to be explained.

Physiology, fungus and host benefits

Virtually all the motivation for research into the physiology of sheathing mycorrhizas has stemmed from the inspiration and dedication of Professor

J.L. Harley and for a more detailed treatment his monograph with S.E. Smith *Mycorrhizal symbiosis* (1983) should be consulted.

Dependence of the fungi on the host for their carbon sources

Direct evidence for the dependence of the fungi on their host tree for carbon sources comes from tracer experiments. When mycorrhizal seedlings of *Pinus sylvestris*, produced by inoculation with *Boletus variegatus*, were allowed to photosynthesize in ^{14}C labelled carbon dioxide, ^{14}C labelled photosynthate was translocated to the root system and it rapidly appeared in the fungal sheath. In *Fagus* mycorrhiza, host sucrose translocated to the roots is converted into the fungal carbohydrates trehalose, glycogen and mannitol; none of this can be effectively reabsorbed and utilized by the host. It is a remarkable feature that in all symbioses involving an autotrophic and a heterotrophic partner the heterotroph always converts the products of the autotroph's photosynthesis to compounds which the autotroph cannot utilize. This is as true for the invertebrates which associate with algae as it is for the fungi of lichens and for the fungi of mycorrhizas as it is for biotrophic pathogens such as the rust fungi. This conversion obviously creates a sink for host sugars in the fungus and maintains a concentration gradient in favour of the fungus. It also prevents reciprocal movement. The rate of translocation to the roots of mycorrhizal systems is several magnitudes more rapid than to uninfected ones. The sheath may accumulate as much as 70% of the sugars supplied to the roots. There is thus not only conversion of sugars but also a massive diversion of these into mycorrhizal roots. Because of this it is often argued that mycorrhizal roots are a considerable drain on host photosynthate and that any selective advantage that such mycorrhizas confer on their host must outweigh this large drain. This should be considered in terms of the concept of source to sink ratios. It is probably merely a question of the source, the photosynthesizing leaves, being more efficient in the presence of a large sink. To achieve such efficiency they would probably require only an enhanced mineral supply. This is the contribution of the fungus to the symbiosis.

Enhanced mineral ion uptake

Ectomycorrhizal infection tends to be most intense under conditions of mineral deficiencies or imbalance in the soil; the greater the availability of mineral nutrients, the less infected the root system tends to be. It is in such conditions as relative mineral deficiency or where there is intense competition from other plants, indeed in most natural habitats, that mycorrhizal infection enhances growth and increases the mineral status of the host. The fact is that most northern temperate trees are normally infected and benefit from infection under natural conditions. The evidence for this is again very varied. Foresters and others who attempt to introduce exotic trees via seed into new localities often find that the seedlings fail to make adequate growth. Conifer seed imported into Western Australia in the late 1920s germinated well and initially grew into healthy seedlings which soon began to show symptoms of mineral deficiency. The introduction into the nursery beds of small quantities of soil in which the same conifers had grown successfully elsewhere was more effective in causing recovery than was adding the deficient minerals. The recovery was associated with mycorrhizal development and failure was due to the absence of the mycorrhizal fungus in the soil. There is a large volume of evidence of this nature. Other evidence comes from a comparison of the growth of inoculated and uninoculated seedlings. Hatch's early work (1937) is often quoted. He compared the growth of inoculated and uninoculated seedlings of *Pinus strobus* in prairie soil from Cheyenne, Wyoming, USA. He chose this soil because it was treeless and assumed, from the sort of evidence dicussed above, that it would not contain the appropriate mycorrhizal fungus. After one year's growth, inoculated, and thus mycorrhizal, plants were healthier, taller and heavier. This is indicated in the differences in their dry mass (Table 7.1). Inoculated plants also showed greater absorption of nitrogen, phosphorus and potassium expressed both as a percentage of their dry mass and per seedling. Most striking of all was the increased phosphorus uptake. There was an almost three-fold increase over uninoculated plants whereas nitrogen uptake was less than doubled. In data from other workers, phosphate absorption was always conspicuously and consistently increased. In such studies, it has been invariably found that the increased uptake of the three mineral nutrients, nitrogen, phosphorus and potassium, was proportionately higher than the increase in dry mass of the mycorrhizal seedlings. It can be argued from this that the mycorrhizal fungi increase nutrient uptake in general. This can be explained on the increase in absorbing area on mycorrhizal infection, achieved by increase in diameter of the absorbing roots, increase in their number by stimulation of branching and the extension of the hyphae well into the soil. This is offset to some extent by the absence of root hairs on mycorrhizal roots and it assumes that host and fungal surfaces have equivalent absorptive capacity. Mycorrhizal roots also live longer. The absorptive efficiency of mycorrhizal and infected roots is not the same on a unit area basis. By using excised mycorrhizal and uninfected roots in ^{32}P-labelled orthophosphate solution, it has been shown that uptake by mycorrhizal roots is between 2–5 times greater than that of uninfected roots per unit area per unit time.

It is often argued that the enhanced growth of mycorrhizal seedlings is primarily due to increased phosphorus uptake, increase in its uptake leading to increased growth rate and as a consequence increased uptake of other elements. This is a fair argument but it should not be allowed to obscure the

Table 7.1 Results of experiments by Hatch (1937) on the effect of mycorrhizal infection on growth and nutrient uptake by *Pinus strobus*.

Treatment	Dry mass (mg)	Nutrients absorbed per cent of dry mass		
		Nitrogen	Phosphorus	Potassium
Mycorrhizal	404.6	1.24	0.196	0.744
Non-mycorrhizal	302.7	0.85	0.074	0.425

fact that phosphates are either deficient or at least in an unavailable form in most soils and in any case are very immobile. It would thus be expected that any increase in absorbing area would markedly increase their uptake.

The sheath as a reservoir of mineral ions

In short term experiments carried out with attached and excised roots at various times during the year, the primary site of absorption in the mycorrhizal roots is the fungal sheath. About 90% of the phosphate taken up remains in the sheath (Table 7.2). Even in experiments lasting a few days and in which phosphate uptake was linear with time, a constant proportion of about 10% passed into the host. Only if the mycorrhizas were swamped with unrealistically high levels of external phosphate did more than about one tenth pass to the host. The fungal sheath thus acts as a reservoir of accumulated mineral ions but it then releases these to the host in periods of deficiency. This has been demonstrated in a number of experimental situations. One of these has been to grow mycorrhizal and uninfected pine seedlings in ^{32}P- labelled orthophosphate solutions and then transfer them to phosphate-free solutions and to follow movement of the label to the shoot. This is steady in both initially while in the phosphate, but both uptake into the plant and movement to the shoot are greater in mycorrhizal plants. The movement continued·at a steady rate for 21 days in mycorrhizal plants when placed in the phosphate free solutions but fell rapidly in infected plants. Phosphate appears to be accumulated as polyphosphate granules. These are chains of covalently linked orthophosphate molecules complexed with calcium. These increase in both quantity and size during phosphate accumulation. This is a convenient way to store phosphate as the polymerized form is osmotically inactive. Polyphosphate granules also occur in other types of mycorrhizas and in some lichens. They may be a common characteristic of phosphate storage in such symbiotic systems.

Table 7.2 Accumulation of phosphate in the sheath of attached and excised mycorrhizas of mature beech trees at three different seasons. (After Harley and McCready, 1952.)

Condition of roots	Attached			Excised		
Date	31 Mar	1 May	23 June	31 Mar	1 May	23 June
Condition of leaves	In bud	·Expanding	Fully expanded	In bud	Expanding	Fully expanded
Conc. H_2PO_4 supplied (mM)	0.074	0.32	0.16 1.6	0.074	0.32	0.16 1.6
Mean percentage phosphate in sheath	88	88	89.8 86.8	91	89	84.9 91.3

Organic phosphorus and phosphatases

It has often been suggested that sheathing mycorrhizas may actually mobilize insoluble mineral nutrients which are unavailable to uninfected roots thus explaining their efficiency as absorbers. Much of the phosphate in the soil is in an organic form. The average value for mineral soils lies between 50–65%. Of this organic phosphorus, various inositol containing salts of iron, aluminium and particularly calcium, such as calcium inositol hexaphosphate, commonly called calcium phytate, are by far the most important components.

Sheathing mycorrhizas develop best in raw humus where as much as 90% of the phosphorus may be in this organic form. Both mycorrhizal and uninfected roots of beech possess surface bound acid phosphatases which will hydrolyse calcium phytate into free inositol and orthophosphate. The activity of these phosphatases is 2–8 times higher on mycorrhizal roots than on uninfected ones. These enzymes can be found throughout the sheath. Under experimental conditions no phosphatase activity occurs in the external medium. Phosphatases thus appear to be firmly bound to the plasmalemma or cell wall. The ability to hydrolyse such insoluble organic phosphates has been clearly demonstrated experimentally and it is often assumed that they contribute to the effectiveness of phosphate uptake. However, rather more information is required before the role of these enzymes in the mineral nutrition of the host can become more than a matter of pure conjecture. An important question still to be answered is whether or not these phosphatases contribute significantly to phosphate availability in the soil. Obviously any hydrolysis by these enzymes can take place only when the substrate is in contact with them so it would depend upon such factors as the degree of exploitation of the soil by the mycorrhizal roots and their hyphae, hyphal density and the mobility of the substrate. Some information would also be needed as to what proportion of the total phosphate taken up is derived from organic sources before the ecological significance of the possession of such enzymes can be evaluated. Surprisingly the experiment of labelling the labile pool of phosphate in the soil has not been carried out with sheathing mycorrhizas.

The dual role of the sheath

The large and organized sheath is a conspicuous feature of ectomycorrhizas. In *Fagus*, it comprises some 35–40% of the dry mass of the mycorrhizas and is responsible for about half of their carbon dioxide emission. Harley (1975) argues that it must be an expensive structure to upkeep and so must have some selective value. It is a storage structure for fungal carbohydrates. Such a storage structure may be of selective advantage to fungi which produce a perennating mycelium and seasonally produce large reproductive structures such as basidiocarps. They store up carbohydrates when the host is most active in photosynthesis in the spring and summer and utilize this store to produce basidiocarps in the autumn as and after the leaves fall. Most fleshy agaric-type basidiocarps produced by mycorrhizal fungi and indeed the litter-inhabiting Basidiomycotina are ephemeral structures. The main flush of these in northern temperate climates starts towards the end of August. A heavy rainfall at that time and subsequent warm humid weather will be followed in about a week by the appearance of basidiocarps of various species of *Boletus*, *Lactarius* and *Russula*. The main season lasts from early September to late October when the litter is moist and the atmosphere humid. Most are frost sensitive and their disappearance is correlated with the first frosts of autumn. They cannot withstand drought. Spore production requires an ample supply of water and spore discharge involves the activity of turgid cells. They also rely upon the turgidity of their hyphae for support rather than upon any distinct mechanical tissues. In dry conditions they lose water very easily to the atmosphere and shrivel. Spore discharge ceases. The majority do not recover on return to more humid conditions. Thus the development and appearance of basidiocarps is restricted mainly to the autumn. This periodicity of reproduction is thus a reflection of a periodic climate. But the sheath is also a storage tissue for mineral nutrients which in a way benefits the host. The availability of these nutrients in the litter is also seasonal. After leaf-fall the litter is moist and microbial and micro-faunal activity is high. Minerals will be made available for absorption by leaching and mineralization. However, in the dry summer both of these processes will be depressed. Efficient absorption by the fungi in times of plenty and storage for release to the host in times of scarcity has obvious merits.

Ectomycorrhizas thus appear to be adaptations shown by trees to seasonal climates, particularly the northern temperate one, where host and fungus grow and reproduce out of phase with each other. Such a system would be less well adapted to herbaceous plants where in particular new roots as well as shoots are produced each year from reserves in some special storage organs or seed. In the wet tropics the trees grow and the fungi grow and reproduce throughout the year and in the dry tropics in the dry season everything is virtually dormant. Most trees in such climates have endomycorrhizas of the vesicular-arbuscular type.

Other beneficial effects on the host

It is becoming apparent that the beneficial effects on the host cannot be ascribed solely to the increased efficiency of the mycorrhizal roots in mineral nutrient absorption. The fungal symbiont also supplies the host with growth substances such as auxin, indole acetic acid, and probably cytokinins and gibberellins. Since these growth substances are also formed endogenously by the host, it may possess above normal levels of these and they must greatly influence its growth and development; in all probability they change the physiological and biochemical processes going on in the roots (Slankis, 1973).

It has been known for a long time that cell-free extracts from cultures of mycorrhizal fungi induce the development of structural features, such as branching pattern, swollen appearance and lack of root hairs, associated with mycorrhiza in non-mycorrhizal roots. Such changes can also be brought about by the application to roots of above optimal levels of auxin. Many mycorrhizal fungi produce indole acetic acid and several other indole compounds in culture and it has been suggested that the mycorrhizal fungi produce extracellular auxin on infection and that it produces the characteristic morphology of ectomycorrhizas. Certainly high levels of auxin are characteristic of mycorrhizal roots and support for the idea that it is the fungus which induces the changes comes from the observation that, if the association terminates for any reason, any new roots assume the appearance of non-mycorrhizal ones.

High auxin levels induced by the fungus may thus help to explain, perhaps too naively, the increase in absorbing surface provided by the fungus. It may also go some way to explaining why mycorrhizas develop only in mineral deficient soils. High available minerals can terminate the association. The synthesis of auxin is influenced by the concentration of available nitrogen. An ample supply of easily available nitrogen inhibits extracellular auxin production by mycorrhizal fungi in culture. If the fungi cannot produce sufficient auxin, successful infection is not established. If the same applies to the established mycorrhizal association, high external nitrogen levels would mean an increase in endogenous nitrogen in the roots leading to lower levels of fungal produced auxin and eventual breakdown of the association.

More recent investigations have provided evidence that several ecto-mycorrhizal fungi also produce cytokinins and substances related to gibberellins. Their role in the symbiosis is as yet even less well determined. These and the high levels of auxin could help to explain the diversion of host photosynthate to mycorrhizal roots and its conversion to fungal carbohydrates. They might enhance the synthesis of particular RNA and enzymatic proteins such as those necessary to convert sucrose to mannitol. Such an increase in metabolic activity would create the necessary sink for host metabolites. Cytokinins have also been implicated in directing both short and long distance transport of organic nutrients in the phloem in comparable situations in biotrophic parasitic symbioses, such as in 'green island' formation by leaf-infecting rust fungi. However, at the moment we can only speculate as to whether or not they actually do so in ectomycorrhizas. They could also help to explain why mycorrhizal roots live longer, and hence function longer than non-mycorrhizal ones.

There is some evidence that mycorrhizal roots are less susceptible to fungal root pathogens. The sheath may act as a mechanical barrier to such fungi or may produce antibiotics which deter them. Root pathogens would no longer be attracted by root exudates. Ross and Marx (1972) found that non-mycorrhizal roots of *Pinus clausa* infected by *Phytophthora cinnamomi* exhibited a massive root necrosis and after two months 60% had died. On comparable mycorrhizal seedlings, 25% of the absorbing roots were mycorrhizal, thus reducing the amount of susceptible root tissue exposed to the pathogen. This resulted in nearly 70% of the test seedlings surviving. The cortical tissues of mycorrhizal roots were free of *P. cinnamomi*, verifying their resistance to attack by the pathogen. This protection may be achieved physically by the sheath acting as a purely mechanical barrier to infection or chemically by the production of antifungal compounds. These, such as diatretyne nitrile, a polyacetylenic antibiotic, may be produced by the fungus. This compound is produced by mycorrhiza formed between *Pinus echinata* and *Leucopaxillus cerealis* and has been shown to contribute to the resistance of roots to *P. cinnamomi*. Others, such as tannins, may be produced by the host in response to infection by the mycorrhizal fungus and subsequently act as a chemical deterrent against pathogens. Mycorrhizal roots may produce different exudates from non-mycorrhizal roots and may also support a different rhizosphere population. These may also affect the competitiveness of any would be pathogen in the rhizosphere.

LICHENS

The ectomycorrhizal root system with its distinctive sheath to all intents constitutes a dual organism similar in many ways to that found when an alga and a fungus combine to form a lichen. But in a lichen the physiological interactions of the alga or cyanobacterium and fungus result in a new distinct morphological entity which is a self-reproducing functional unit behaving as a single organism with no obvious similarities to either of its components. For convenience they are named and studied as if they were a single organism.

Fungi involved and structure

There are some 13 500 known species of lichens representing about one sixth of all the known fungi. The fungi involved are mainly Ascomycotina with a few Basidiomycotina. A very wide variety of Ascomycete fungi are involved. The majority are Discomycetes producing abundant apothecia. Others are Pyrenomycetes or Loculoascomycetes producing perithecia and pseudothecia respectively. In addition there are some 14 Basidiomycete lichens, such as species of the genus *Cora*. None of the fungi involved have been given generic names as none have been encountered free-living and none have sufficiently similar characters to any known free-living terrestrial genera to indicate that they are co-generic with any of these. It appears that they cannot survive independently in nature. In lichens, one particular alga or cyanobacterium is associated with one particular fungus. Unlike the fungi, the majority of the algae and cyanobacteria belong to well-known free-living genera. Some 21 genera of algae are involved in lichen associations. Twenty of these are green Chlorophyceae. Of these, members of the genus *Trebouxia* are most common and they are also the only ones which are rarely, if ever, encountered free-living. Other common genera of green algae include *Trentepohlia* and *Coccomyxa*. The remaining genus is *Heterococcus*, a member of the yellow-green Xanthophyceae. Twelve genera of cyanobacteria are involved. *Nostoc* and *Scytonema* are the most commonly encountered ones.

Lichens are usually separated morphologically into three major growth forms – crustose, foliose and fruticose (Fig. 7.3). Crustose lichens form crusts in such intimate contact with their substrate that often they cannot be readily separated from it. Foliose lichens (Fig. 7.2 and 3) are leafy dorsiventral structures often attached to their substrate by hair-like rhizinae and are easily peeled off from it. Some show remarkable parallels with the conventional dicotyledonous leaf. Fruticose lichens are erect, shrubby or pendent, hair- or tassel-like structures attached only at their base and usually circular in transverse section. Two major types can be recognized on internal structure (Fig. 7.4). In a few lichens the alga and autotrophs are distributed at random throughout the vegetative body or thallus. Such undifferentiated mixtures are called homoiomerous. The majority are heteromerous or stratified with the alga confined to a well-defined layer just below the upper surface. The alga is covered and protected by a thin, tough pseudoparenchymatous cortical layer composed of thick-walled, heavily gelatinized hyphae firmly cemented together. In the orange and yellow pigmented forms, the pigments, such as usnic acid, are deposited in this layer. Below this is the so-called algal layer. It is usually 10–15 μm thick and occupies some 5–10% by

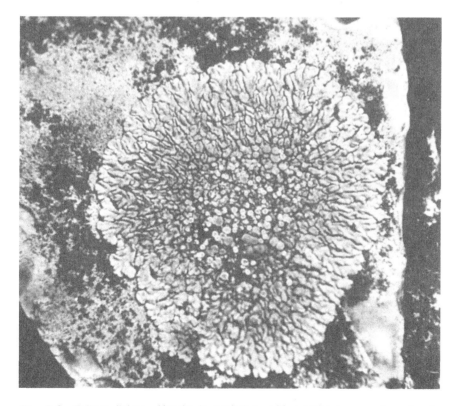

Fig. 7.2 Foliose lichen, *Xanthoria parietina*, with apothecia, growing closely applied to a rock.

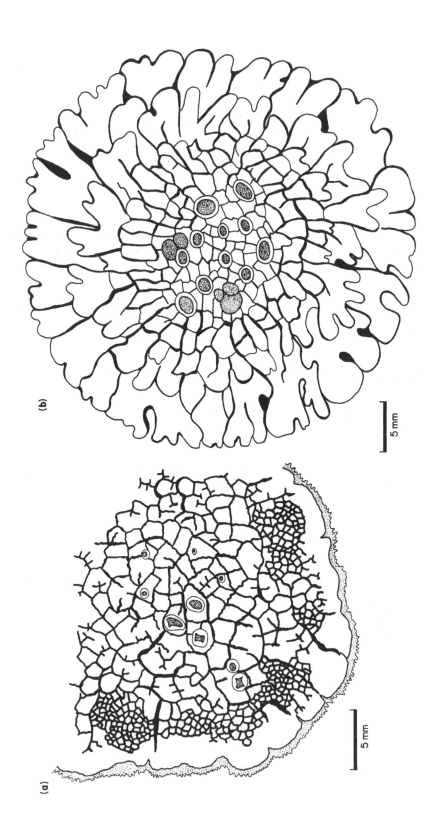

(b)

5 mm

(a)

5 mm

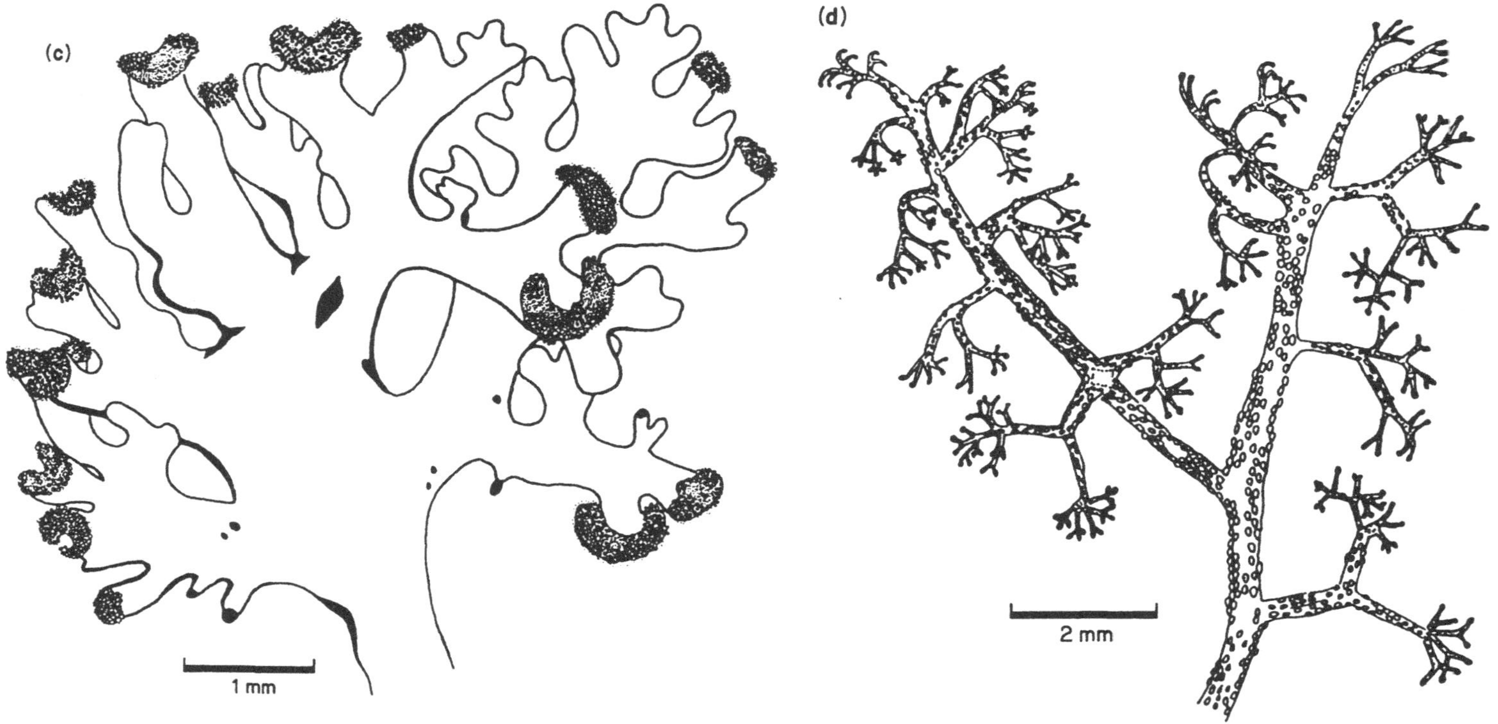

Fig. 7.3 Three major growth forms of lichens. (**a**) Crustose: part of a thallus of *Ochrolechia tartarea* dried out on a rock. (**b**), (**c**) Foliose: (**b**) *Xanthoria parietina*, with apothecia, growing closely applied to a rock; (**c**) part of a thallus of *Hypogymnia physodes* with reflexed tips of branches bearing powdery soredia underneath. (**d**) Fruticose: part of an erect thallus of *Cladonia rangiformis*.

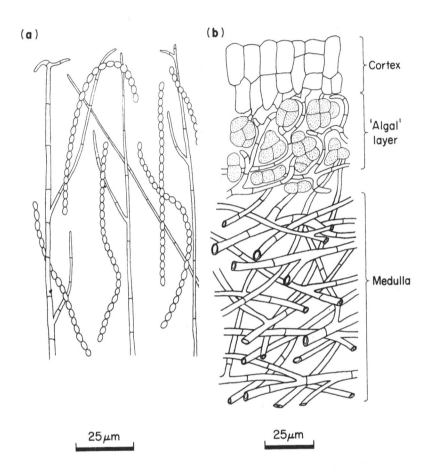

Fig. 7.4   Internal structure of lichens. (a) Homoiomerous type: *Collema* with the chains of the cyanobacterium scattered at random through the widely spaced fungal filaments. (b) Heteromerous or stratified: *Peltigera* with the cyanobacterium confined to a distinct layer just below the upper surface, the cortex.

volume of the thallus. In this layer the algal cells are surrounded by thin-walled, loosely packed hyphae. Below this layer is the medulla with again loosely packed hyphae but this time thick-walled. It makes up the bulk of the thallus and may be up to 0.5 mm thick. In some, especially foliose forms, there may be a lower cortex with or without attaching and presumably absorbing rhizinae. The fungus thus makes up the bulk of the thallus. As in ectomycorrhizal associations, both partners can reproduce separately, the fungi by ascospores or by basidiospores, and the algae by non-motile spores. Many of the green algae involved will produce zoospores in culture as their free-living relatives do but do not do so in the lichen. The spores of the fungi, at least, are actively discharged from the thallus and it is assumed that they unite with a suitable alga after dispersal to form a composite plant. This must be especially so in those, such as *Peltigera*, with no vegetative means of dispersal of both. Unlike ectomycorrhizal assocations, the majority produce some combined reproductive structure, such as soredia (Fig. 7.5). These are separable clumps of algal cells closely enveloped by fungal hyphae. Individual soredia range from 25 to 100 μm diameter well within the range of wind dispersed propagules. Although most are dispersed by wind, some are dispersed by rain splash or animals. Aggregations of soredia occur on the surfaces of lichens as powdery areas. Their size, shape and position on the thallus are characteristic for each lichen species.

Physiology, carbon and mineral metabolism

The physiological relationships of the symbionts have been the object of a great deal of study since lichens were first discovered to consist of two organisms in the nineteenth century. As in ectomycorrhizal associations, the autotrophic host, in this case the alga or cyanobacterium, supplies the fungus with all its carbon requirements and with vitamins, such as biotin and thiamine. Movement of photosynthetically fixed carbon from the algal layer to the fungal medulla has been demonstrated in several lichens. In the foliose lichen *Peltigera polydactyla*, the thallus can be dissected horizontally into the algal layer plus the upper cortex and the fungal medulla. Smith and Drew (1965) exposed discs cut from the thallus of *P. polydactyla* to ^{14}C labelled carbon dioxide for 4 h in the light. The discs were then dissected and they showed that about 40% of the ^{14}C fixed by the cyanobacterial symbiont *Nostoc* had passed into the medulla, indicating a rapid movement of photosynthate to the fungus. In other lichens containing other algae this movement has been shown to occur but it may be slower. The algae of lichens can be cultured; immediately after isolation into culture solutions, they release appreciable amounts of carbohydrates from their cells as they photosynthesize. Each alga usually releases a single, simple carbohydrate. In *Nostoc* and other cyanobacteria, it appears to be glucose. In the Chlorophyceae it is always a polyhydric alcohol. *Trebouxia* always releases ribitol and *Trentepohlia* erythritol. These are also the forms in which carbon moves from alga to fungus in the lichen. The carbohydrates, in whatever form they are released, become converted in the fungus into fungal carbohydrates, usually mannitol but sometimes into arabitol as well. *Nostoc* and lichen algae are unable to utilize these and as in sheathing mycorrhizas there is again a one way flow of carbon from autotroph to heterotroph. It is not known how the fungus induces its partner to release carbohydrates. Lichens, but not its two components alone, produce a wide variety of chemicals called lichen

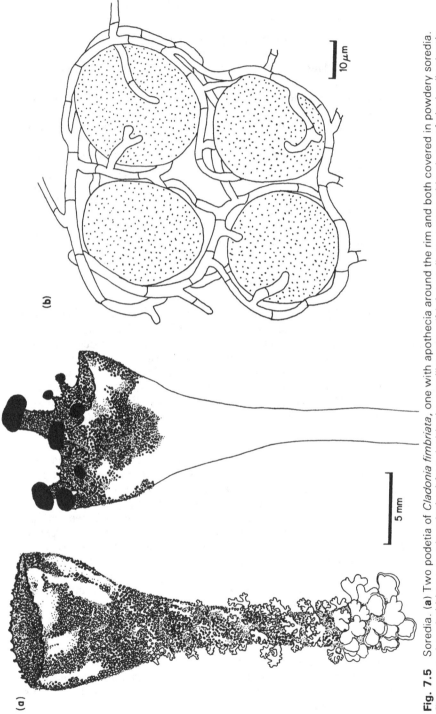

Fig. 7.5 Soredia. (**a**) Two podetia of *Cladonia fimbriata*, one with apothecia around the rim and both covered in powdery soredia. These may be wind blown or rain splashed from inside the cup-like apex. (**b**) A soredium, a separable clump of algal cells closely enveloped by fungal hyphae.

substances or lichen acids. It has been suggested, with very little supporting evidence, that these may in some way alter the selective permeability of the cell membranes of the autotroph. Alternatively it may be a special feature of physical contact between the symbionts. A further possibility is that the fungus interferes with the metabolism of the alga or cyanobacterium so that the carbohydrate intended for cell wall or sheath synthesis is released instead. For example, in *Nostoc*, immediately after isolation into culture from *Peltigera*, 50% of the carbon fixed in a 3 h period was released as glucose. After two days in culture, the amount released had dropped to 12% and glucose was no longer released but polysaccharides were. Free-living *Nostoc* and symbiotic ones grown in culture for some time have thick mucilaginous sheaths surrounding their filaments. This sheath is lacking or greatly reduced in *Nostoc* in the lichen. The loss in ability to release glucose after a period in culture is correlated with an increase in thickness of the mucilaginous sheath. Thus glucose intended for sheath synthesis may be diverted by the fungus and converted to mannitol. Hill (1972) has provided evidence that before the glucose is released the fixed carbon is converted to sugar phosphates and then a glucan. It is suggested that the glucan is produced outside the cyanobacterial plasmalemma as part of the wall or sheath and that it is hydrolysed to glucose by an extracellular fungal glucanase, the free glucose formed being converted to mannitol (Fig. 7.6). The green alga *Trebouxia* isolated from the lichen *Xanthoria aureola* behaves very similarly to *Nostoc*. Initially in culture it releases about 40% of the carbon fixed in photosynthesis but in this case as ribitol. Only about 2% is converted into insoluble materials such as storage products, cell walls, proteins and lipids. It also has a very thin wall. With time in culture, the cell wall becomes much thicker and some 50% of the carbon fixed is incorporated into insoluble materials and only 2% is released. However in this case ribitol is not a component of any algal polysaccharides and a different mechanism for its release must operate.

The fungi then can derive their carbon requirements from their algal partners but they must also do this. In culture they grow very slowly and they are poor competitors. Most can utilize only simple carbohydrates and as most lichens live in environments where these are certainly in very short supply or even non-existent, they, like the fungi of sheathing mycorrhizas, must be regarded as ecologically obligately dependent for their carbon on their autotrophic partner. They may also obtain other materials from them. Most lichen fungi are deficient for biotin and thiamine. It is assumed that these are made available by the autotrophic partners. Some lichens, containing cyanobacteria, such as *Nostoc*, are known to fix atmospheric nitrogen. Most of the nitrogen fixed passes to the fungal partner. Many lichens with green algae possess cephalodia, small outgrowths on the surface, which usually contain a cyanobacterium. These fix nitrogen and constantly release it to the thallus where it is utilized almost exclusively by the fungus. For example *Peltigera aphthosa*, which unlike other species of *Peltigera* has *Coccomyxa* as its algal component, has cephalodia containing *Nostoc*, normally found as the autotroph in other species of *Peltigera*. In this case only 3% of the nitrogen fixed by *Nostoc* is eventually used by *Coccomyxa*.

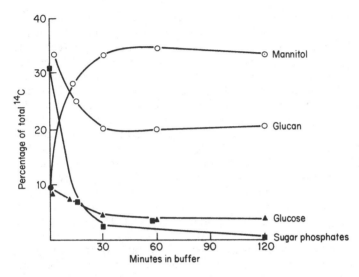

Fig. 7.6 Redistribution of ^{14}C in *Peltigera polydactyla* in various fractions after pulse feeding with labelled CO_2 followed by incubation in buffer. (After Hill, 1972.)

Lichens have very efficient mechanisms for the accumulation of substances from any watery solution with which they are in contact. In nature supplies of nutrients, especially mineral ions, are again likely to be both very low and very erratic. Such efficient absorption mechanisms would be advantageous. In laboratory experiments, the absorption of mineral ions, such as phosphates, and other nutrients, such as amino acids and sugars, is very rapid and metabolically dependent. For example, *Peltigera polydactyla* may take up the equivalent of 3% of its dry mass of phosphate in 24 h and the equivalent of 25% of glucose. It is also clear that these are accumulated in periods of relative abundance and utilized at a slow rate in periods of prolonged scarcity in between. There is no direct evidence that the fungus absorbs mineral ions and passes them on to the autotroph but, because the fungi are such efficient absorbers in general, it is often assumed that they do so. In the lichen it is the algal layer with its fungal hyphae which is the most active in ion uptake but it is not known whether uptake by the autotroph is direct or via the fungus. The medulla does not play the role of the sheath in ectomycorrhizal associations. There is in fact no experimental evidence that anything passes from the fungus to the autotroph. This is a major point of contrast between lichens and ectomycorrhizas. From the evidence presented for nitrogen fixation by cephalodia in green alga lichens, it could even be argued that the fungus actually reduces the amount of minerals reaching the alga.

Detrimental consequences of efficient absorption mechanisms

Such efficient mechanisms for absorption would also help to account for the well-documented reports of relatively high accumulation by lichens of radioactive fall-out and their extreme sensitivity to atmospheric pollutants (Hawksworth and Rose, 1976; Ferry, Baddeley and Hawksworth, 1973).

Accumulation of strontium and caesium

The so-called 'reindeer moss', *Cladonia rangiferina*, and 'Iceland moss', *Cetraria islandica*, are fruticose types which have covered much of the tundra. They have accumulated large quantities of radioactive strontium (^{90}Sr) and caesium (^{137}Cs) from atomic fall-outs. These lichens are eaten by reindeer in northern Sweden and Russia and by caribou in Canada and Alaska. They supplement their diet of sedges and willows with the lichens during the winter. If the winter is very severe, lichens may form up to 95% of their food. Their nutritive value is due to such storage carbohydrates as lichenin, a β-1,3 and β-1,4 linked glucose polymer and the less common isolichenin, with α-1,3 and α-1,4, linked glucose. The lichens are eaten and the radioactivity built up in the animal and passed on in the food chain to especially the Eskimo and Lapp. Eskimos who consumed caribou have been found to contain over four times as much ^{90}Sr as the average populations of the northern temperate zone. This can be attributed to three factors: the rapid and efficient uptake by the lichens, their slow growth rate and great age so that they persist, and the reindeer and caribou grazing over very wide areas.

Sensitivity to sulphur dioxide

Their extreme sensitivity to atmospheric pollutants, such as sulphur dioxide, means that they are markedly absent from towns and cities and around industrial sites, such as oil refineries and brickworks. This sensitivity is no doubt enhanced by their efficient absorption mechanisms. The sulphur content rises markedly as the amount of sulphur dioxide in the atmosphere increases. As a group they show differential sensitivity and can thus be used as reliable biological indicators of pollution. There appears to be a direct relationship between the zonation of lichens and the mean annual, but more especially the mean winter, sulphur dioxide levels. A better correlation exists between zonation and winter means because lichens are more sensitive to sulphur dioxide at high humidities. Relative humidities are on average higher in winter. The sulphur dioxide penetrates the lichen tissues more easily if they are moist. Lichens are also physiologically more active in the winter months.

Lichens vary to such a degree in their tolerance to sulphur dioxide that around a city or an oil refinery one finds concentric patterns of exclusion of particular species. The zones are related to the mean winter sulphur dioxide levels. All except *Lecanora conizaeoides* show these patterns but it occurs even in city centres. It is the most tolerant of all lichens to sulphur dioxide. This tolerance can be explained in part by the fact that it is virtually un-wettable in the natural state. This is probably due to heavy incrustations of non-wettable lichen substances. In this sense it avoids the pollutant by reducing the amount that it takes up dissolved in water. It is also a very poor competitor with other lichens; it seems probable that the elimination of its competitors by levels of sulphur dioxide which it can tolerate leads to its abundance in polluted areas. Although it was unknown before the middle of the last century, it is now extremely common and very abundant, in terms of density, on tree bark and wood throughout Britain where mean winter sulphur dioxide levels are in the range 55–150 μg m^{-3}. Hawksworth and Rose (1970) have devised a qualitative scale for the estimation of mean winter sulphur dioxide levels using epiphytic lichens growing on tree bark and in a later publication (Hawksworth and Rose, 1976) have discussed in much broader terms lichens as pollution monitors.

It is possible to oversupply nutrients to the symbionts so that the association breaks down. This is exactly the situation that occurs in ectomycorrhizal associations where, given high levels of mineral nutrients in the soil, the host is non-mycorrhizal. However, in both cases the fungi are incapable of an independent existence in nature.

General biology and possible benefits to the autotroph

The benefits which the autotroph derives from the association are much more difficult to assess. There is no unequivocal evidence that any specific compounds move from fungus to autotroph. Although for some algae the association may not be an obligatory one, once in an association, constraints are applied on the alga by the fungal partner. Green algae no longer reproduce sexually or produce their motile stages in the lichen. The alga makes up only 5–10% of the dry mass of the lichen yet must supply the whole with all its carbon requirements. It is often argued that it can produce only sufficient organic carbon photosynthetically to support very slow growth of itself and the fungus. The fungus would thus appear to be too large a sink for the alga to grow quickly and the alga equally too small a source for the fungus to grow quickly. Thus growth of both is restricted. There is, however, some evidence now that the algae photosynthesize more efficiently in the lichen than in culture at least and that the supply of photosynthate from the alga in a lichen is substantially greater than that needed to sustain the fungus (Table 7.3). The massive movement of photosynthate to the fungus and its conversion to the soluble mannitol may be the key. The cell contents of many lichen fungi must contain a near saturated solution of mannitol. Its function may not be so much for growth and metabolism as seen in other symbiotic systems but its high concentration may be part of the mechanism enabling lichens to withstand environmental extremes. For example, under conditions of moisture stress a low internal osmotic potential is obviously of great advantage.

The algae and cyanobacteria benefit to the extent that they are physically supported and effectively displayed to the light by the fungi, protected to

Table 7.3 Rates of growth and photosynthesis in lichens and lichen symbionts. (After Farrar, 1978.)

Maximum relative growth rates
Fungi in culture   = > 0.33 d^{-1}
Algae in culture   = > 0.39 d^{-1}
Thalli in nature   = > 0.01 d^{-1}
Thalli in laboratory   = > 0.015 d^{-1}

Photosynthetic rates (mg CO_2 g^{-1} algae h^{-1})
Algae in culture   = 3–14
Algae in thallus   = 2–28

Growth and photosynthesis in *Hypogymnia physodes*
Maximum growth rate   = 0.005 g g^{-1} dry wt in 24 h
Net carbon fixation   = 0.02 g (CH_2O) g^{-1} dry mass (12 h light in 24 h)

some extent from environmental extremes and may have their habitat range extended. These benefits can best be considered in relation to the general biology of lichens.

Growth rates

Lichens are exceedingly slow growing plants. Foliose types usually grow the fastest but rarely grow more than 10 mm per year. Crustose types often grow less than 1 mm per year. Individual plants may be very old. Some have been estimated to be able to maintain themselves as individuals for 4500 years. Such age indicates a very well-balanced and stable association. Their slow growth rate may be attributed to a number of factors including the inherently slow growth rate of the fungi, their inability to conserve water and their restriction to habitats where the supply of minerals is very low. When dry they are exceedingly resistant to environmental extremes. These features enable them to grow where neither partner can alone. As such they tend to be pioneer colonizers of such inhospitable habitats as rock faces, even as far as 7300 m up Mt Everest, tree bark and man-made habitats such as stone walls, tombstones, roofing tiles, asbestos and concrete. Their resistance, their low demand for minerals and their highly efficient absorption mechanisms enable them to grow usually where no other organisms can. They are favoured by high humidities and low light intensities.

Water relations, photosynthesis and respiration

Lichens are in many ways a paradox. They are able to survive in drier habitats than either of their two components. Fungi as a group are somewhat tolerant of water stress (Chap. 6) while most algae are very susceptible to desiccation. Brock (1975) infers that the fungus confers desiccation resistance upon the lichen. The lichen autotrophs which he tested were able to photosynthesize down to a lower a_w within the lichen than when liberated from it. The lowest a_w permitting photosynthesis of the whole lichen ranged from 0.96 to 0.80 and the lowest of the autotrophs separated from these ranged from

0.995 to 0.90. He suggested that one function of the stored fungal carbo-hydrates, such as mannitol, is to act as osmotica to lower the osmotic potential so that water accumulation could occur in environments of low water availability. Because of the intimate association of the fungus and the autotroph some of the water becomes available to the autotroph, allowing it to photosynthesize at a lower a_w. This may be so but the polyols may also have other major roles.

Lichens are metabolically active only when moist. One reason for their low productivity is that they spend the major part of their time dry and thus meta-bolically inactive. The fungal medulla, although it does not appear to have an important role to play in mineral uptake and storage has two important functions which are beneficial to the autotroph. It provides a supportive skeleton to the whole over and around which the cells of the autotroph are displayed in a thin layer to the light. However, more relevant to the present discussion is that it also has a higher saturated water content than the remainder of the thallus. In *Peltigera polydactyla*, for example, the medulla contains 25% more water per unit dry mass than the algal layer and upper cortex. It thus has an important role to play in the water relations of the whole thallus. Since the medulla is either beneath or internal to the algal layer, it provides water for it when the thallus is drying out. Lichens have no physio-logical control over their water content. Water loss and uptake are purely physical phenomena. Loss is governed by the physical factors that affect transpiration. Thus lichens can show large and rapid fluctuations of water content during the course of the day. They dry out rapidly in the sun with the water content falling as low as 5% of their dry mass and they become satu-rated immediately it rains as water is taken up in the extra hyphal capillaries and into the thick hyphal walls. The water content of the thallus has profound effects on the rates of respiration and photosynthesis and as such affects the carbon balance of the lichen. The saturated water content of lichen thalli lies between 100–300% of their dry mass. The maximum rate of photosynthesis occurs at between 65–95% of complete saturation, depending upon the species, and the rate decreases above and below this, declining parti-cularly rapidly at water contents higher than those for maximum photosyn-thesis. The maximum rate of photosynthesis is also relatively low per unit area compared with leaves. The autotroph occupies only 3–10% by mass of the lichen yet has to supply the fungus with all its carbon. The chlorophyll content may be less than one quarter of that of leaves and the several layered and often pigmented fungal cortex absorbs far more of the incident light than does the epidermis of a leaf. The maximum rate of respiration is reached somewhere between 40–95% of complete saturation. The rate of respiration increases approximately linearly with water content up to this point but does not decrease in rate above or in saturated thalli. Respiration also continues at around 6–7% of the maximum rate even in air-dry thalli in which photosyn-thesis has ceased. Thus it is clear that under many natural conditions there may be only a few hours each day, as in the early morning following wetting by dew and before they dry out in the sun, when there is a positive carbon balance. This is even more likely in extreme environments. Lange, Schulze and Koch (1970) have shown that in a typical day in the Negev Desert *Ramalina maciformis* fixes 1.32 mg CO_2 g^{-1} (dry mass) in the period between sunrise and drying out. It then remains dry until nightfall when it is re-wetted by dew. It subsequently loses 0.78 mg CO_2 g^{-1} (dry mass) before dawn by basal respiration. Thus some 60% of the fixed carbon is lost within 24 h of fixation. Net production is therefore a relatively low percentage of gross primary production. Obviously any means of delaying drying out, for however brief a period, would be beneficial to the carbon balance. The water storage capacity of the medulla should be evaluated in this light.

The decreased rate of photosynthesis which occurs below the optimum water content of the thallus can be attributed to two factors, reduced light transmission and eventual dehydration of the alga. The water-holding capacity of the medulla delays the latter. The fungal cortex acts as a light screen. Air-dry *Peltigera polydactyla* is quite grey but on wetting with water turns blue green in under a minute as a result of the cyanobacterial cells showing through the more translucent moist fungal cortex. When saturated, the cells of the fungal cortex are fully expanded and allow maximum transmission of light. With loss of water they contract so that there are more vertical walls per unit area and the whole becomes opaque as less light is transmitted. The cyanobacterial cells also contract so that there are more of them per unit area (Fig. 7.7). Thus a fixed number of cyanobacterial cells are exposed to less radiant energy than they are when the thallus is saturated.

Significance of fluxes in water content

Fluxes in water content, i.e. alternating periods of wetting up and drying out,

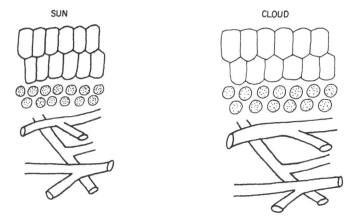

Fig. 7.7 Control of light transmission by the cortical cells. In the sun the cortex contracts as it dries out so that there are more cells per unit area and thus less light is transmitted through the thick walls. There are also more algal cells per unit area because they contract as well so each receives less radiant energy.

may be far more important to the lichen than its water content at any particular moment of time and may be absolutely critical in maintaining the lichen in a healthy state. Most attempts to maintain lichens in the laboratory have been successful only when the thalli were alternately wetted and dried. Slow alternating phases of wetting and drying also appear to be one of the factors that favour the synthesis of a lichen from its components. Constant saturation has a marked deleterious effect. Farrar (1976a) has shown in *Hypogymnia physodes* that the capacity to photosynthesize, the total polyol content and the ability to absorb phosphate all declined over 7 days at complete saturation. He suggested that this might be due to anoxia consequent on reduced gas exchange through the water saturated hyphal walls. He also concluded that lichens may actually require alternating periods of wetting and drying when he investigated the effects of various wetting and drying periods on *H. physodes* (Farrar, 1976b). It maintained high photosynthetic activity and high polyol content when so subjected but not when maintained continuously saturated. The anomaly of the situation is that subjection to wetting and drying cycles leads not only to restricted periods of photosynthesis but to high losses in the fixed carbon. A substantial proportion of the carbon fixed when photosynthesis is possible under such conditions is converted to polyols to maintain the pool of these at a high level. *H. physodes* commonly contains as much as 10 mg g^{-1} ribitol in the alga and 10 mg g^{-1} mannitol and 50 mg g^{-1} arabitol in the fungus on a dry mass basis (Farrar, 1973). The physiological response of the lichen to such cycles of stress was characterized by both a depletion and turnover of the polyol pool. Levels of insoluble materials such as proteins remained unchanged. Farrar suggests that the polyols are used to buffer the latter against change. They could, for example, protect such macro-molecules by direct substitution for water molecules in their hydration shells. Sudden re-wetting by rain after a dry spell results in a burst of carbon dioxide emission. This lasts less than two minutes and is probably a purely physical effect. This is followed by a period of several hours of resaturation respiration which is well above the basal rate. The intensity of this respiration is less in lichens from dry habitats than in those from wet ones and it may be related to drought tolerance. It may be due to an expenditure of energy in repairing membranes damaged as a consequence of desiccation. There is also a period of one or two minutes when organic solutes are lost. There is a loss of polyols of both algal and fungal origin and sucrose. The amount of carbon lost by leakage may be as much as that lost by resaturation respiration. During the initial stages of resaturation the rate of photosynthesis is low and recovery is slow; rapid cyclic wetting and drying could thus be disastrous to the lichen in that carbon losses could exceed gains.

   The factors discussed above go some way to explaining the slow growth rates of lichens but this tolerance of marked fluctuations in their water content, or drought resistance, enables them to persist in such severe habitats as rock faces where no other plants, certainly no other fungi or algae, can, and thus extends the habitat range or ecological tolerance of either component. When dry, lichens are extremely resistant to high temperatures.

Thallus temperatures of 53–69°C, some 20–40°C above ambient, have been recorded in lichens exposed to the sun on rocks. Such lichens experience higher temperatures than the vegetative phase of any other organisms except those living in hot springs. But such lichens are thermoduric – they survive but do not grow at these temperatures. High temperatures on rocks are invariably the result of high solar insolation. Yet many lichen algae are quite intolerant of strong sunlight. The contraction of the cortical cells helps to decrease the intensity of the exposure. This is also aided in many by deposition in the fungal cortex of coloured lichen substances, such as the deep orange anthraquinone parietin of *Xanthoria parietina*. Many lichens are normally deeply pigmented when growing in sunny areas and weakly or not pigmented in shade. This can be verified in the case of *X. parietina* by comparing its colour on a north and a south facing rocky cliff. Parietin has maximum absorption at 436 nm filtering out part of the blue region of the spectrum. Such pigments thus also change the quality of the light reaching the alga.

Lichen substances

Lichens thus have a number of quite different properties from those of either of its components, especially in their ability to withstand desiccation and high temperatures. These series of peculiar characteristics are further illustrated by their production of lichen substances, often also called lichen acids but not all are acids. A wide variety of these are formed as granules or crystalline encrustations on the surfaces of hyphae. The commonest fall into two major groups – the depsides and depsidones (Fig. 7.8). These are usually formed from two simple phenylcarboxylic acids, mainly orsellinic acid, held by an ester linkage in the depside and with an additional diphenyl oxygen linkage in the depsidone. Neither the alga nor the fungus produces these substances in culture. Many monocyclic phenyl carboxylic acids, especially orsellinic acid (methyl-dihydroxy-benzoic acid), have been identified in pure cultures of lichen fungi. It is thought that the alga is responsible for converting these units to the dimers. Because they do not appear in cultures of isolated symbionts or in cultures of non-lichenized algae, cyanobacteria and fungi and because similar ones are produced by different lichens, it is often assumed that they have some role to play in the symbiotic state. They may have a varied protective role. Their function in increasing the opacity of the fungal cortex has already been mentioned. The general resistance of lichens to insect and microbial attack is well-known and it has been suggested that lichen substances may be responsible for this. Many have antibiotic properties which could protect them from bacterial and fungal attack. Most have a bitter taste which would make them less palatable to slugs and other herbivores. Reindeers, however, appear to enjoy the very bitter 'reindeer moss'. Simple phenolics are toxic and it may be that depside and depsidone formation is a protective response of the algae to decrease their toxicity. A further suggestion, already made, is that they may increase the selective

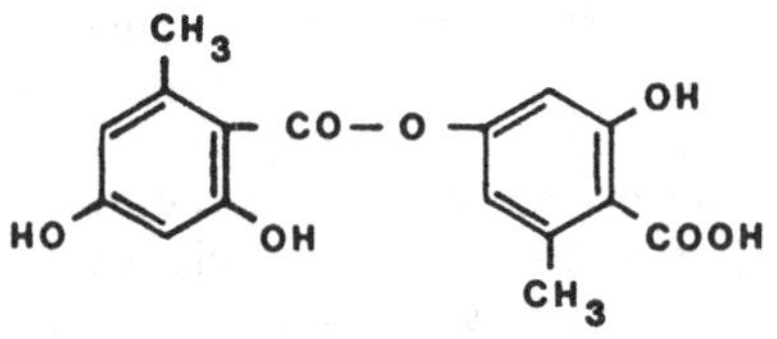

LECANORIC ACID-DEPSIDE (ester of phenylcarboxylic acids, often <u>orsellinic acid</u>, methyl-dihydroxy-benzoic acid)

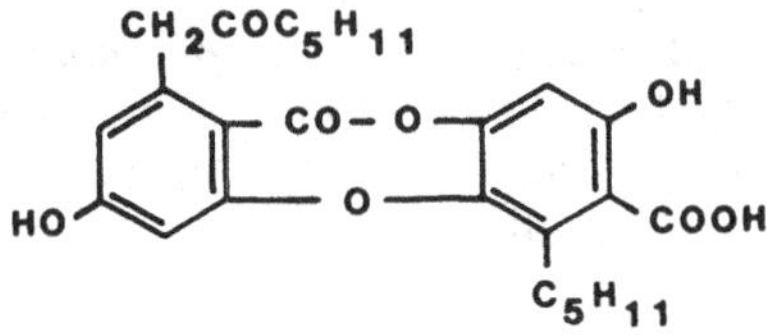

PHYSODIC ACID-DEPSIDONE (like depside but with additional diphenyl linkage)

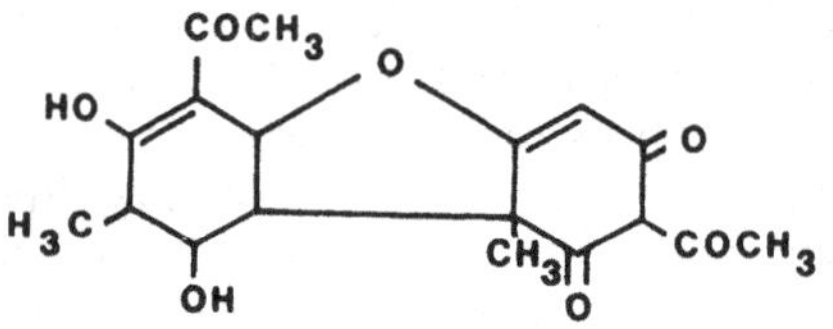

USNIC ACID-DIBENZOFURANE derivative

Fig. 7.8 Lichen substances. (Reproduced by permission of The Systematics Association.)

permeability of the algal membranes and thus be important in the flow of carbohydrates to the fungus. They may also be important in two respects in the chemical weathering of substrates such as rocks. Any acids may bring ions into solution and some lichen substances are known to chelate these. Both of these processes may be important in maintaining an adequate supply of minerals when growing on a substrate very low in available mineral ions. Whatever their role, they are not produced by all lichens and so cannot be essential to the symbiotic association.

Lichen chimeras

The association has conferred ecological versatility on the two partners by changing a number of their physiological attributes, enhancing in particular their resilience to extreme environmental conditions. Both partners are also morphologically changed. Lichenized green algae have thinner cell walls than their free-living relatives and cyanobacteria lack a typical mucilaginous sheath in the lichen. The fungus shows tissue differentiation often with a pseudoparenchymatous cortex and an organized medulla, features not found in the vegetative phase of non-lichenized fungi. In ectomycorrhizal associations, it appears that the fungus induces both morphological and anatomical changes in the root system of the host thus determining its form but it is not known what controls the development and form of the sheath. Since in the lichen the fungus forms the bulk of the thallus, it is often assumed that it determines the form, the size and the rate of growth of the whole.

The processes by which lichens develop their complex form and high degree of internal organization remains at the moment as just another fascinating unsolved problem. Recent evidence suggests that the algal partner plays a vital role in determining both the pattern of development and the final form of the lichen in spite of the fact that it usually makes up no more than about 5% of the whole. This evidence comes from the discovery of lichen chimeras. Several physically united pairs of lichens containing two different algae but the same fungus have been found. Each member of the pair was known previously as an entirely distinct species.

Many green algal lichens have developed a subsidiary association with a cyanobacterium. The latter develops in small, special outgrowths of the surface, a millimetre or so in diameter, called cephalodia. This is so in *Peltigera*, a genus easily recognizable on morphological and anatomical features, in which there are some species with cyanobacteria and others with green algae and cephalodia. However in these chimeras or composite morphotypes the two algal fungus associations have developed more or less to the same extent. This has led to marked divergence in morphology in the two parts. For example, chimeras have been found in which shrubby species of *Dendriscocaulon* with cyanobacteria have united with a leafy dorsiventral species containing green algae belonging to the genus *Sticta*. In all cases, clear continuity of fungal tissue between the two parts has been established, eliminating the possibility that the association may be epiphytic or parasitic. The difference in morphology between the two components of these lichen chimeras plainly demonstrates the importance of the alga in controlling form. This is particularly obvious in *Dendriscocaulon-Sticta* chimeras where the radial arrangement of the internal tissues around a central axis of hyphae in the cylindrical branches of the *Dendriscocaulon* part containing cyanobacteria changes in the immediate vicinity of the green algal cell of the *Sticta* part to a layered dorsiventral structure. The growth pattern is also different. In the part with cyanobacteria, growth is confined to the tips of the branches whereas in the green algal part growth is essentially marginal. The mechanisms by which the growth of the fungus is controlled to form the lichen thallus are not known.

Environmental factors appear to play an important part in determining which alga is selected by a particular fungus and by virtue of being associated with two algae the range of ecological tolerance of the fungus may be

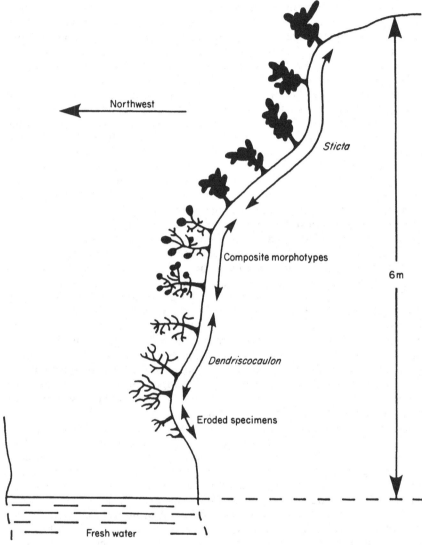

Fig. 7.9   Distribution of green algal morphotype (*Sticta filix*) and cyanobacterial morphotype (*Dendriscocaulon* sp.) with composite morphotypes in relation to aspect and associated environmental factors at a site in a forest in New Zealand. The fungus is the same in all states. (After James and Henssen, 1976.)

considerably increased as was shown by James (James and Henssen, 1976). He studied the distribution of composite morphotypes between *Dendris-cocaulon* and *Sticta filix* and normal ones in New Zealand. *S. filix* is widespread in many well-wooded areas in New Zealand. It requires a variable regime of relatively high illumination and fluctuating humidities whereas *Dendriscocaulon* prefers very sheltered and shaded habitats of unvarying high humidities and very low illumination. At one site consisting of two large northwest facing boulders, some 6 m tall and influenced by spray from a small water-fall, there were some 90 separate thalli of *Sticta* and *Dendris-cocaulon* and almost one-third of these were composite ones (Fig. 7.9). On the upper parts of the boulder, which although partially shaded by the tree canopy received some 5 hours of sunshine per day through October to May, only *Sticta filix* occurred. At the base of the boulders which were in permanent shade only *Dendriscocaulon* occurred. However in the middle zone only composite plants were present. Even so there was a gradation; those at the top had many broad leaflets of the green algal morphotype and those at the bottom had only a few small leaflets. Thus the fungus, by virtue of its association with the two symbionts, has increased its range of tolerance to both light and humidity.

Lichens show striking parallels with ectomycorrhizal associations in the substantial movement of carbohydrates from the autotrophic to the hetero-trophic partner, for example, but they also have many distinctive features, not least amongst these being the formation of an entirely new entity, the lichen thallus. The outcome of a very complex series of interactions between the partners is that in each case the symbiotic association leads to ecological success enabling the symbionts to exploit habitats in which neither would be successful alone. Perhaps the major difference between the two is that in ectomycorrhizal associations the growth of the autotrophic partner is markedly stimulated so that it is a more effective competitor; however, in the lichen association the growth of the autotrophic partner is not so stimulated, but at least it is sufficient to allow slow growth of the whole. A slow growth rate may even be an advantage for any plant growing in extreme and severe habitats. This is true for many xeric communities. As primary colonizers of such habitats, lichens are not subject to competition from others and growth rate becomes less critical as long as they can maintain a high reproductive capacity.

8

Fungi as mutualistic symbionts in endomycorrhizas

In ectomycorrhizal associations, the fungus exerts a demand on the host's photosynthate for its carbon supplies but offsets this demand by increasing the efficiency of absorption and temporary storage of mineral ions. A wide variety of endomycorrhizal associations exist and some of these, vesicular-arbuscular and ericoid mycorrhizas in particular, afford interesting parallels and also contrasts with ectomycorrhizal ones.

VESICULAR-ARBUSCULAR MYCORRHIZAS

Occurrence

Vesicular-arbuscular mycorrhizas are produced by aseptate mycelial fungi and are so-called because of the two characteristic structures – vesicles and arbuscules – found in roots with this type of infection. They are by far the commonest of all mycorrhizas and are found in Bryophytes, Pteridophytes, Gymnosperms, excluding the Pinaceae which have sheathing mycorrhizas, and in virtually all families of Angiosperms. They are of general occurrence in the Gramineae, Palmae, Rosaceae and Leguminosae, which all include many crop plants. Indeed most crop plants, including herbs, shrubs and some trees, possess this type of mycorrhiza. In the Angiosperms, apart from such families as the Ericaceae and the Orchidaceae which have other types of mycorrhiza, the Chenopodiaceae, the Cruciferae, the Cyperaceae and the Resedaceae are odd exceptions in that most species from these families appear to be either non-mycorrhizal or at least only very sparsely infected. However, the striking fact is that the majority of all plants that are mycorrhizal have this type of infection.

Associations resembling modern day vesicular-arbuscular mycorrhizas were present very early in the evolution of land plants. Kidston and Lang (1921) examined the 'thallophyte' flora of the petrified fossil genera *Rhynia* and *Asteroxylon* which are now believed to be some 370 million years old. Illustrations and slides made from Kidston and Lang's material show aseptate hyphae bearing structures with striking resemblance to vesicles and spores which are of the size and with a wall complexity similar to those seen belonging to these fungi in present day soils. It is not known whether vesicular-arbuscular mycorrhizas existed before *Rhynia*. Earlier plants, such as

Cooksonia, have not been examined for mycorrhiza because their fossil remains are mainly in the form of compression material.

Features of infection and fungi involved

In these types of mycorrhizas there is no sheath of fungus around the host root only a loose and very sparse network of rather thick-walled, irregular and aseptate hyphae. These are often aligned more or less parallel to the length of the root with numerous cross connections. Hyphae grow out from this network into the soil and may spread from one to several centimetres away from the root. These hyphae, depending on the genera of fungi involved, produce one or more of a number of spore types in the soil. These may be very thick-walled, brown to black, balloon-like chlamydospores, some 100–250 μm in diameter. They are produced singly or in clusters as the blown-out ends of hyphae (Fig. 8.1a). Loose aggregations of spores may also be formed within a poorly differentiated reproductive structure called a sporocarp (Fig. 8.1b) or thick-walled zygospores may be formed from the fusion of two hyphae. Chlamydospores are the most common.

The hyphae which establish the infection penetrate the epidermal cells after forming an irregular disc- or lozenge-shaped appressorium on the surface and then grow intercellularly, often forming an intracellular coil in the first cell penetrated. Inside the root the parallel alignment of the aseptate hyphae is again most noticeable. The meristem of the root is not infected. The intercellular hyphae completely permeate the entire cortex but do not invade the endodermis or stele. Arbuscules are formed inside the parenchymatous cortical cells by the repeated dichotomous branching of the penetrating trunk hyphae. The ultimate branches may be less than 1 μm in diameter and are very difficult to resolve. They often virtually fill the host cell in which they are produced and resemble little trees – hence their name. The life span of an arbuscule is about four days. They then quickly disintegrate and it is often assumed that they are 'digested' or lysed by the host. It may be however that the arbuscules are merely ageing and undergoing autolysis just as fungal hyphae do in culture. During this process the host's nucleus enlarges considerably and the arbuscules release oil as droplets into the host cell. The hyphal branches in the arbuscule and the hyphae between the cells often become septate at this stage. The hypha or hyphal branch producing the arbuscule penetrates the wall but only invaginates the host's plasmalemma so that the ultimate branches of the arbuscules are surrounded by it. Arbuscules are usually regarded as complex haustoria through which a two-way transfer of materials may occur. If they are haustoria, they are unusual in that they often lyse. This may be a consequence of a host defence mechanism comparable with the laying down of callose around the haustoria of some obligate biotrophic parasites. The hyphae may also form vesicles (Fig. 8.2 and 8.3). They are readily discernible in roots, using suitable stains such as trypan blue in lactophenol, as they are usually in excess of 100 μm diameter. This is sufficiently large often to distort the cell or intercellular space in which they occur.

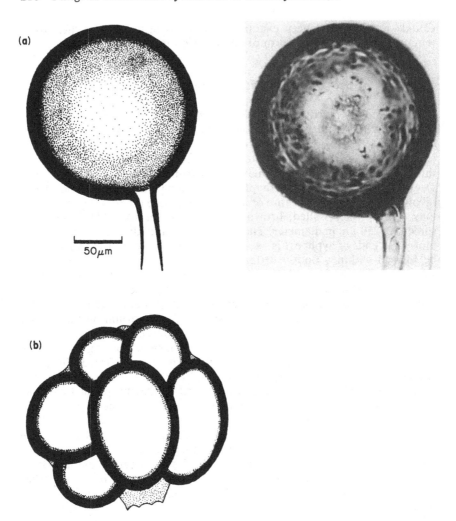

Fig. 8.1 (a) Chlamydospore of *Glomus* sp. (b) Sporocarp of *Sclerocystis rubiformis.*

Vesicles arise as swellings terminating hyphae or in an intercalary position along a hypha. They may be thin- or thick-walled, ovoid or spherical. Most contain one or more large oil droplets suggesting that they are involved in food storage (Fig. 8.3).

In marked contrast to ectomycorrhizal associations, infected roots show little or no morphological modifications. Most look superficially like uninfected roots, undistorted and with root hairs. Some infected roots lack root hairs; in maize, onion and some other plants infected roots are coloured a bright yellow by a light-sensitive pigment. However considerable

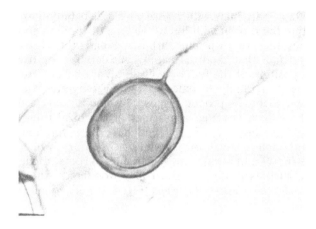

Fig. 8.2 Vesicle of vesicular-arbuscular mycorrhiza.

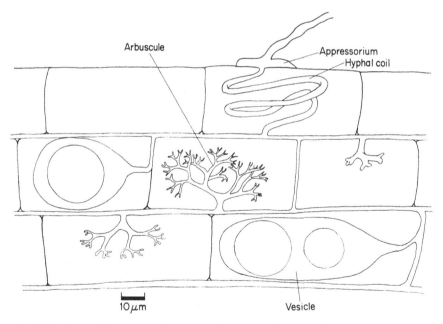

Fig. 8.3 Vesicular-arbuscular mycorrhiza, showing irregular external hyphae, appressorium, hyphal coil in the first cell penetrated, intercellular hyphae, vesicles and arbuscules in various stages of lysis.

differences do occur between the vesicular-arbuscular mycorrhizas of different plants in particular with regard to the distribution and form of the fungus within the root in addition to the type of spore or spores produced. Some, for instance, rarely produce arbuscules and others vesicles. In spite of this and the fact that there has been considerable controversy over the taxonomic position of the fungi involved, all are believed to belong to the Endogonaceae, a family included in the Zygomycete Mucorales largely because some members produce zygospores and all have aseptate hyphae. Most of the British vesicular-arbuscular mycorrhizal fungi only produce chlamydospores singly in the soil and have been referred to the genus *Glomus*, although in some soils species of *Gigaspora*, which produce relatively large spores and usually only arbuscules in roots, occur with these. In other soils, such as those of boulder clay woodlands, the blackberry-like sporocarps of *Sclerocystis rubiformis* (Fig. 8.1b) are equally common.

Spores in the soil

Until the late 1950s little was known about these members of the Endogonaceae and they were believed to be fairly rare but the evidence available now would suggest that they are some of the commonest soil-borne fungi. Because their spores are so large, 100–250 μm in diameter, they can easily be extracted from soil by decanting and wet sieving or by using a flotation technique. In the former technique about 250 g of soil are suspended in a litre of water, the heavier particles are allowed to settle out for a few seconds and the liquid decanted and washed through a series of size-graded sieves. The fraction passing through a 250 μm mesh sieve but retained by a 100 μm one contains the majority of the spores. A simple flotation technique is to add soil to a beaker some two thirds full of water, stir thoroughly to dislodge the spores from the soil crevices and allow to settle. The balloon-like spores will float to the surface and collect in the superficial scum. Most of these will adhere to the surface of a filter paper if it is partially submersed at a very acute angle to the surface and withdrawn. By such techniques, large numbers of spores have been recorded from soils, for example 1–8 g^{-1} under wheat at Rothamsted, 20–70 g^{-1} under wheat in Ontario and over 400 g^{-1} under *Mercurialis perennis*, Dog's Mercury, in a Cambridgeshire boulder clay woodland. These numbers may be deceptive. High spore numbers are not necessarily indicative of high spore productivity. Because the spores have such thick melanized walls, they are very resistant to decay and may persist for very long periods so accumulating and appearing abundant. A large number of spores encountered in the soil are devoid of contents and are no longer viable. Be this as it may, these large spores which appear to be poorly adapted for widespread dissemination are present in all soils whether they support pioneer or climax plant communities.

Spores extracted by such techniques have been used to infect plants and have produced typical vesicular-arbuscular mycorrhizas but so far these fungi have not been grown in pure culture. Chlamydospores will germinate on nutrient agar but the hyphae stop growing when the food supply in the spore is used up so they cannot be sub-cultured. They are thus obligate biotrophs but are unusual in displaying a surprising lack of host specificity, contrasting markedly with other obligate biotrophs such as the parasitic rusts and powdery mildews. Spores produced from an infected maize plant have been used to produce typical vesicular-arbuscular mycorrhiza in straw-berries, onions, red clover and soybeans.

Benefits to host

Enhanced host growth and increased phosphate uptake

The inability to culture vesicular-arbuscular mycorrhizal fungi has made them more difficult to work with in terms of understanding their physiology and conducting experiments involving inoculating plants with these fungi. Isolates are maintained in 'pot cultures', i.e. on the roots of living plants in pots. Plants are inoculated with single spores and grown in pots of sterile soil. The spores produced are collected by decanting and wet sieving the soil in the pots and used to inoculate others. Observations on the effect of infection on a wide range of host plants have been variable but in general enhanced plant growth occurs and this is associated with increased mineral uptake, especially phosphate. The·extent of the enhanced growth depends on the level of infection in the root system and on the nutrient status of the soil. There is a much more marked stimulation of growth with low phosphate availability in the soil. Other factors may also be involved; increased growth will not occur if other mineral nutrients are limiting in the soil. The degree of root hair development on the host roots may also affect the response. Plants with many and very long root hairs are less dependent upon vesicular-arbuscular fungi for their mineral supplies than are those with very few, short ones. This may be one explanation why members of families, such as the Resedaceae, are rarely, if ever, infected, their root hairs possessing sufficient surface area to exploit a large enough volume of soil to satisfy their mineral requirements.

There are a wealth of published reports which now provide quite indis-putable evidence that infection of plant roots by vesicular-arbuscular mycorrhizal fungi increase phosphate uptake, especially under conditions of low phosphate availability. For example, Hayman and Mosse (1971) found that in some phosphate deficient soils they obtained up to a 20-fold increase in shoot mass of non-mycorrhizal onions merely by adding phosphate in the order of 1 ton superphosphate acre^{-1} (2.5 Mg 10^4 m^{-2}). Comparable increases in shoot mass were obtained by growing mycorrhizal onions in the same soil but without the added phosphate. It appears, as with sheathing mycorrhizas, that the host obtains maximum benefit when the mineral nutrient regime is least favourable for growth. Most attention has been paid to the effect of vesicular-arbuscular mycorrhizas on crop plants such as apples, maize, onions and tomatoes but one of the earliest reports was of their effect on plants of the mixed rain forests of New Zealand (Baylis, 1959,

1967). Much more recently attention has been paid to their role in other natural or semi-natural ecosystems, such as grasslands in the Pennines, tussock grassland in New Zealand, lowland tropical rain forest trees in Costa Rica and in man-made habitats such as plant communities establishing on coal tip spoils in Pennsylvania and Scotland.

Data from Khan (1975) using maize and wheat in Lahore, W. Pakistan may be used to illustrate some of the effects which these fungi are reported to have on crops grown in the field. Mycorrhizal plants were obtained by growing seedlings on steam sterilized sand mixed with spores of *Glomus mosseae*. After 3 weeks mycorrhizal and non-mycorrhizal control plants were transplanted into field plots with both a very low phosphate content and indigenous spore population. Half of the plots containing mycorrhizal and control plants were treated with triple superphosphate at 280 Kg 10^4 m^{-2}. Thus there were plots containing mycorrhizal plants without phosphate (M) and with phosphate (MP) and similarly non-mycorrhizal without phosphate (NM) and with phosphate (NMP) but with time the non-mycorrhizal plants gradually become infected from the indigenous spores in the soil. At harvest, after 90 days, the mycorrhizal plants without phosphate had grown the most as indicated by the fact that they had the greatest dry mass (Table 8.1). They had also taken up the most phosphate. Furthermore they produced the heaviest grains and more grains per cob. The non-mycorrhizal plants fared badly all around. The trends were the same in wheat except that the dry mass of 1000 grains was similar in all treatments. However mycorrhizal plants without phosphate produced more ears per plant and ear bearing stems per square metre, thus giving a greater yield. A number of workers have found that, in some soils, non-mycorrhizal plants grown with

Table 8.1 Effects of mycorrhizal infection with and without added phosphate on maize and wheat. (After Khan, 1975.)

Host	Character	Treatment			
		NM	NMP	M	MP
Maize	No. of grains per cob	31	279	354	321
	100-grain mass (g)	2.4	19.8	23.7	20.9
	Total dry mass (g)	119	175	264	244
	% P	0.18	0.48	0.61	0.54
Wheat	Ear bearing stems m^{-2}	20	26	32	28
	Ears per plant	3	5	8	4
	No. of grains per ear	36	69	72	70
	1000-grain mass (g)	38.5	39.0	38.0	37.5
	Calculated yield (g m^{-2})	85.8	215.3	275.5	235.2

NM non-mycorrhizal without added phosphate
NMP non-mycorrhizal with added phosphate
M mycorrhizal without added phosphate
MP mycorrhizal with added phosphate
Phosphate added as superphosphate at 280 Kg 10^4 m^{-2}
Seedlings (N and NM) planted out into field plots (+ or − P) after 21 days
Figures at harvest 90 days later

added phosphate were smaller than those with mycorrhiza in the same soil but without added phosphate. This is true of Khan's plants. In such soils, the added phosphate rapidly becomes unavailable. It is 'fixed' so that, even if added at high levels, it is still less available to non-mycorrhizal plants than the small amount initially available in the unamended soil is to mycorrhizal plants. Only about 20–30% of any added phosphate is ever available to the standing crop. The rest becomes slowly available over the next 10–15 years. The mycorrhizal plants in plots with added phosphate performed on the whole less well than mycorrhizal plants in plots without added phosphate. With a high soil phosphate content, infection is often suppressed. Khan noted this in his maize plants. The incidence of infection was much lower than in mycorrhizal plants without phosphate and fewer arbuscules and vesicles were formed. This is also reflected in the numbers of spores produced. Initially there were 550–600 spores Kg^{-1} soil. At harvest in plots with mycorrhizal maize without added phosphate the number had risen to about 2800 spores Kg^{-1} soil whereas in plots with mycorrhizal maize and with added phosphate there were only about 1800 spores Kg^{-1} soil. In addition the external hyphae attached to the roots did not extend so far into the soil so that they exploited a smaller volume of soil. Others have noticed in such circumstances that the growth of the fungus along the surface of the roots may be quite extensive even although actual infection is less. This might lead to competition for phosphate between fungus and host in the rhizosphere. Uptake by the host may thus be reduced and its growth rate decreased.

Possible use as 'biological fertilizers'

Because virtually all crop plants possess vesicular-arbuscular mycorrhizas and because they benefit so much from these in their mineral nutrition, the research effort which has been put into these is quite enormous. Much emphasis is being placed on their use as 'biological fertilizers'. Techniques as to how this might be achieved vary. Attempts at field inoculation with the most effective fungus have been made by placing suitable infected roots at sowing depth, decanting and wet sieving spores from plots where a particular crop has grown well and adding these to fields and pre-inoculating plants before planting out. The ultimate might be to incorporate spores into pellets around seeds. The results of Azcon – G. de Aguilar, Azcon and Barea (1979)

Table 8.2 Effects of *Glomus* and *Rhizobium* on the yield of *Medicago sativa* growing under normal cultivation in an arable field soil. (After Azcon-G. de Aguilar, Azcon and Barea, 1979.) Reprinted by permission from *Nature*, Vol. 279, No. 5711, pp. 325–7. Copyright © 1979 Macmillan Journals Limited.

Inoculation treatments	*Shoot dry mass g^{-1}
Uninoculated controls	15.2 ± 1.0
Rhizobium	15.9 ± 1.2
Glomus	22.6 ± 1.1
Rhizobium + Glomus	32.1 ± 1.8

* Mean of four replicates

illustrate some of the possibilities (Table 8.2 and Fig. 8.4). They grew alfalfa, *Medicago sativa*, in an arable field soil with very low available phosphate in Granada, Spain and investigated the effects on yield of prior inoculation with *Glomus*, the symbiotic nitrogen-fixing bacterium *Rhizobium* and the two together. In terms of shoot dry mass, an important criterion of yield as a fodder plant, *Glomus* improved yield but *Rhizobium*

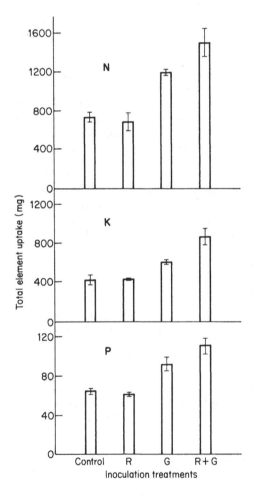

Fig. 8.4   Effects of *Glomus* (**G**), *Rhizobium* (**R**) and both (**R + G**) on nitrogen, phosphorus and potassium uptake by *Medicago sativa* in an arable field under normal cultivation. The total N, P and K uptake by the plants was calculated from the data of shoot dry masses and the percentage of the element. The figures are the mean of four replicates. Standard errors are given. (After Azcon, -G. de Aguilar, Azcon and Barea (1979). Reprinted by permission. Copyright © 1979 Macmillan Journals Limited.)

did not. But the synergistic effect of both was most obvious. It was argued that *Rhizobium*-inoculated plants were growth limited by phosphate so that yield was no greater than controls. *Glomus* increased the phosphate uptake of the plants so that inoculation with both greatly improved the yield. This argument is borne out by analysis of the shoots for nitrogen, phosphorus and potassium. *Glomus* improved the uptake of all three but inoculation by *Rhizobium* did not unless *Glomus* was present.

Vesicular-arbuscular mycorrhizas may be of equal importance in natural communities in assisting plants to obtain sufficient phosphates and in the recycling of phosphate. In such communities the available phosphate levels are often exceedingly low. In the New Zealand forests in which Baylis worked it was 4–8 p.p.m., as compared with what he considered as a minimum agricultural standard level of 28 p.p.m. All the trees in these forests possess vesicular-arbuscular mycorrhizas and infection by appropriate mycorrhizal fungi was essential if seedlings were to become established and grow. Agricultural and horticultural crops are grown in soils usually with much higher phosphate level and mycorrhizal infection may be of less significance but even so all the available evidence indicates that mineral nutrient uptake by mycorrhizal roots is the norm rather than the exception. There is also a need for recycling because there are no natural processes, such as there are for nitrogen, by which phosphorus can be increased after its depletion. Phosphorus from plant remains, unless rapidly reabsorbed, is quickly converted to various insoluble organic compounds, phytates, in the soil and, although as such it is less likely to be leached than are nitrogenous compounds, it is unavailable to plants. The role of the mycorrhiza can thus be seen as ensuring efficient and rapid transfer of phosphate from the available pool in the soil into the root, thus decreasing the time period in which it might be subjected to conversion to organic forms and to leaching.

The main interest in vesicular-arbuscular mycorrhizas has centred around the enhanced phosphorus nutrition of the host. Their effect is thus seen to be much more marked in phosphate deficient soils. Evidence is also available that they may increase the nitrogen, potassium, calcium, magnesium and iron uptake. There are two possible ways, or a combination of two possible ways, in which this increased uptake can occur. The fungus could increase the effective absorbing area of the root so that a much larger volume of soil is exploited for its labile phosphate or it could make soluble insoluble or organic forms of phosphate which are not available to the uninfected plant. They may do both but there is not much evidence for the latter and so are exactly comparable with the sheathing mycorrhizas.

Source of phosphate taken up

If various phosphates are added to the phosphate deficient soils it can be shown that mycorrhizal and non-mycorrhizal plants of the same species grow equally well on available phosphate such as calcium monophosphate ($CaHPO_4$) and that mycorrhizal plants grow better on tricalcium phosphate ($Ca_3(PO_4)_2$) which is only slowly available. However, this may be because they have a larger surface area to absorb what little phosphate is available rather than making it more soluble and hence more available. If the labile phosphate pool in the soil is labelled with ^{32}P and the non-labile pool is left unlabelled the specific activity (c.p.m. mol P^{-1}) of phosphorus compounds in plants grown in the soil is the same in both mycorrhizal and non-mycorrhizal ones, indicating that they both use the same pool of phosphate. If the mycorrhizal plants used some of the non-labile insoluble soil phosphates, the specific activity of the absorbed phosphate would be less than that of the non-mycorrhizal plants.

Phosphate ions have a very slow diffusion coefficient in soils and there is likely to be a marked depletion of phosphate in solution immediately around the root. This depletion zone usually extends 1–2 mm around the root up to the tips of the root hairs. The hyphae of the mycorrhizal fungus may grow well outside the depletion zone and considerably extend the volume of soil explored. In mycorrhizal onions, hyphae have been demonstrated to absorb at 70 mm from the root surface. The enhanced phosphate uptake by the mycorrhizal plants thus depends on a more complete exploration of the soil by hyphae outside the depletion zone of the root hairs. There is also some evidence that there is a more complete exhaustion of the soil explored by the hyphae including the root hair depletion zone. It appears that non-mycorrhizal plants of different species differ in their ability to utilize the low levels of phosphates which occur in phosphate deficient soils. They may not take up phosphate until it reaches a certain miminum or threshold level. This level is different for different species. Mycorrhizal plants either have no threshold value or a very much lower one and can thus more completely exhaust a soil of its labile phosphate.

Inter-plant transfer of nutrients via fungus

In many natural plant communities endomycorrhizal infection arises when uninfected roots make contact with hyphae spreading from infected roots. Since any one fungus may be capable of infecting a number of different plant species, a series of plants may become connected via the hyphae of their common mycorrhizal fungus. This could lead to inter-plant transfer of carbon and mineral nutrients. Recent autoradiographic studies by Francis and Read (1984) have shown that such inter-connections provide a direct pathway for the transfer of carbon compounds between different plants. They used seedlings of *Plantago lanceolata* as 'donors' of both infection and ^{14}C labelled carbon. After their establishment, uninfected seedlings of *Festuca ovina* were planted amongst them and they became mycorrhizal. The 'donor' plants were then allowed to photosynthesize in ^{14}C labelled CO_2 and the 'receiver' plant of *F. ovina* were kept in full light, half light or the dark. After two days, autoradiography showed that the label moved from 'donor' to 'receiver' roots only if they were connected to one another by hyphae. At that stage uninfected 'receiver' roots contained no label. Quantitative analysis of the radioactivity in the plants confirmed the transfer of assimilate between the two plants but the quantity of label moved was strongly influenced by the light regime. The levels of activity were six times higher in the *F. ovina* plants kept in the dark than in those fully illuminated. Thus the movement appears to be governed by source-sink relationships. The ecological and physiological significance of this may be far-reaching. In numerous plant communities many young seedlings may spend part of their early life in poorly lit environments and even in full shade. To such seedlings early mycorrhizal infection would be vital on the one hand in providing a necessary enhanced mineral nutrient supply but the carbon drain imposed by the fungus, to provide such enhancement, could be critically disadvantageous on the other hand with regard to growth. Direct transfer of assimilate from more fully illuminated neighbouring and connected mature plants may sustain seedling infection. Furthermore during the critical establishment phase the whole seedling may live for a period at least as a partial heterotroph depending upon its neighbours to which it is connected by its bridging fungus.

Benefits to fungus

Virtually all the evidence indicating that the fungus benefits from this association is at the moment based on assumptions. It is assumed that since these fungi have never been grown in culture they are like all other biotrophs in being dependent on their hosts for their carbon supplies. It is also assumed that they are like all other Zygomycotina in being unable to utilize polymers such as cellulose and are thus 'sugar fungi' requiring simple carbon compounds. Again, as with sheathing mycorrhizal fungi, because of the scarcity of such carbohydrates in the soil, they must obtain these from their host. Potted plants have been allowed to photosynthesize in ^{14}C labelled CO_2 and the label has appeared in spores formed around the roots. This indicates, but does not necessarily prove, direct translocation of photosynthate to spores. The carbon metabolism may be quite different from that of sheathing mycorrhizal fungi but again there is no clear indication of this. There are generally more soluble sugars present in mycorrhizal than in non-mycorrhizal roots. Something like 75% of the labelled photosynthate reaching mycorrhizal roots is in the form of soluble sugars. Although there is diversion to mycorrhizal roots they do not, unlike sheathing mycorrhiza, contain higher levels of trehalose and mannitol but some glycogen and lipids are present. These may be storage compounds but very little carbohydrate appears to be stored. This may be because storage occurs in spores and vesicles and these are scattered. There is no obvious bulk storage tissue, such as a sheath. The absence of polyhydric alcohols such as mannitol is to be expected as they have not been recorded from other aseptate fungi.

Mosse and Phillips (1971) grew *Trifolium parviflorum* aseptically on agar slopes and inoculated them with spores of *Glomus mosseae*. The addition of calcium phytate to the agar medium greatly stimulated the growth of the hyphae out from the root. Growth of the hyphae was virtually as good when inositol was added and they argued that the stimulation caused by the addition of calcium phytate was due to the fungus using its inositol component as a carbon source. If this is so the fungus has a potential carbon source in the soil for which there may be very little competition.

In this mycorrhizal system, the host depends on the fungus for soil-derived nutrients and the host has appeared to adopt the strategy that it is less costly in carbon and more flexible in operation that the absorbing surface should be an extensive hyphal system rather than a much branched root system of a much larger biomass.

ERICOID MYCORRHIZAS

Occurrence, structure and fungi involved

Ericoid mycorrhizas are another type of endomycorrhiza but here the fungi involved are septate ones. Ericaceous plants, such as those of *Erica* (heather), *Calluna* (ling) and *Vaccinium* (bilberry), are mostly calcifuges associated with acid, peaty, mor-humus soils of very low mineral status. These soils are particularly low in exchangeable nitrogen and phosphorus, most being bound as insoluble inorganic or organic compounds. They are plants which you would expect to benefit from mycorrhizal infection.

They have a dense root system ending in very fine, branched absorbing rootlets, often unfortunately called 'hair roots'. These rootlets have a very narrow central stele, surrounded by a cortex of 1–3 layers, often only one, of parenchymatous cells. There are no root hairs. The rootlets are covered by a very sparse, loose weft of dark brown, septate hyphae. From this weft branches penetrate the cortical cells forming compact intracellular coils invaginating and enclosed by the host's plasmalemma. They do not penetrate the stele or the apical meristem. After a time, some 3–4 weeks, the cell contents of the host begin to degenerate with the plasmalemma losing its integrity. The intracellular hyphal coils then collapse, although they may persist intact for some time in the degenerate host cell. They lose their filamentous structure and the coils become an amorphous mass of fungal tissue (Fig. 8.5). This process may be compared with the degeneration of arbuscules in vesicular-arbuscular mycorrhizas but is much more comparable with similar phases in orchidaceous mycorrhizas.

A number of sterile, slow-growing, dark, septate fungi have been obtained in culture from root segments and have been shown to form mycorrhiza with various ericaceous plants. Most work on these has been done by Read and his associates. They first obtained their own isolates by thoroughly washing hair roots in 20–25 changes of sterile distilled water. They then macerated the roots and plated out onto distilled water agar separate cortical cells containing hyphal coils. They isolated only those fungi which were seen to grow out from the hyphal coils within the cells. These fungi were all slow growing, dark and septate. Most proved to be sterile but one isolate produced ascocarps, apothecia, and they described these as belonging to a new species *Pezizella*

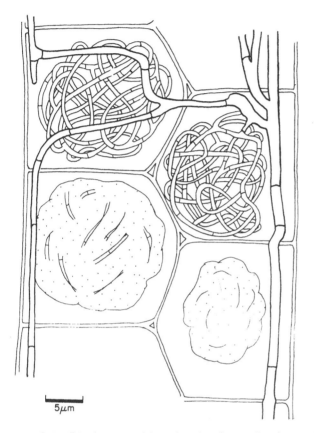

5μm

Fig. 8.5 Ericoid mycorrhiza, showing four cells of a rootlet with a loose weft of septate hyphae and hyphal coils, some undergoing lysis, in the cortical cells.

ericae. Most of their work has been done with this Ascomycete which does form typical mycorrhiza with several ericaceous genera. Other isolates may yet be shown to belong to this species but other fungi may also be involved. For example, a group of Australian workers have produced evidence that *Clavaria vermicularis* may be mycorrhizal with *Azalea indica* and various rhododendrons and ericas. Basidiocarps of *C. vermicularis* are almost always found in close association with ericaceous plants. Their attempts to grow the fungus from basidiocarps failed so that they could not synthesize mycorrhiza directly. They used a fluorescent antibody technique and claimed that the mycorrhizal fungus within the root cells was serologically the same as the *Clavaria* growing in the neighbourhood of the plants. In this country *Clavaria argillacea* is always associated with *Calluna vulgaris* and may be mycorrhizal with it.

Benefits to host and fungus

It has become clear from the work of Read and his associates that ericoid mycorrhizas behave like sheathing and vesicular-arbuscular mycorrhizas. They improve the mineral nutrition, especially the nitrogen and phosphorus uptake, and growth of their hosts while the fungus is dependent upon its autotrophic host for its carbon supply. It is important to appreciate that the nitrogen supply to plant roots or mycorrhizas may be in the form of ammonium ions, nitrate ions or both and that organic nitrogen in the soil may or may not be available to the absorbing systems. Nitrate and ammonium ions have different mobilities in soils. Ericaceous plants, such as *Calluna* and *Erica*, typically grow in acid soils rich in organic matter in which nitrification rates are very low and the available mineral nitrogen source is ammonium, in very low concentrations. Ammonium ions, in contrast to nitrate ions, are relatively immobile in soil although they are not as firmly bound as phosphate ions. In acid peaty soils only a small fraction, less than 0.5% of the total nitrogen, is present as ammonium and not all of this is exchangeable and thus available to plant roots. Most of the nitrogen, over 70%, is present in the organic form, often as unidentified complex compounds and again unavailable to roots. The story is very similar to that for phosphorus. It has been argued, without substantive proof, on many occasions that mycorrhizal plants can make use of these compounds. The ability to utilize these would be very important to ericaceous plants growing in such acid moorland soils.

Source of nitrogenous compounds utilized by mycorrhizal plants

Mycorrhizal ericaceous plants grown in such soils are larger and healthier and their nitrogen, and phosphorus, content are significantly higher than in non-mycorrhizal ones. For example, *Vaccinium* seedlings grown on sterile moorland soil for six months after inoculation with a mycorrhizal fungus contained over twice as much nitrogen on a dry mass basis as did non-mycorrhizal seedlings (Table 8.3). Stribley and Read (1974) investigated this further and labelled moorland soil with ^{15}N by long incubation with ^{15}N ammonium sulphate. After removing excess ammonium ions, they found that the labelling pattern of the nitrogen in the soil was complex but basically there

Table 8.3 Nitrogen content and dry mass yield of mycorrhizal and non-mycorrhizal seedlings of *Vaccinium macrocarpon*. (After Read and Stribley, 1973.)

Growth stage	Total nitrogen content (mg per plant)	Total dry mass yield per plant (mg)
Seedlings at time of inoculation	0.11	18.70
Mycorrhizal	0.88	124.60
Non-mycorrhizal after three months	0.62	121.0
Mycorrhizal	1.82	235.0
Non-mycorrhizal after six months	0.81	184.0

were two fractions. One was very small with high labelling. This was the exchangeable or labile ammonium fraction. It was highly labelled as the pool was small. The other was relatively large with low labelling. This was the organic unavailable nitrogen. After sterilizing this soil by gamma irradiation, they grew mycorrhizal and non-mycorrhizal seedlings of cranberry, *Vaccinium macrocarpon*, in it in pots for six months before analysing their shoots for their nitrogen content. Mycorrhizal plants were heavier, larger and healthier. They contained more nitrogen and a greater concentration of nitrogen on a dry mass basis. But what they called ^{15}N-excess (labelled atoms % N) of mycorrhizal plants was lower than that of the shoots of non-mycorrhizal plants – that is to say that the non-mycorrhizal plants contained a high proportion of ^{15}N which suggests that they mainly utilized the exchangeable ammonium-N. Mycorrhizal plants, although they contained more total nitrogen, had a lower ^{15}N enrichment and had thus absorbed more unlabelled nitrogen (Table 8.4). This implies that they had a source of nitrogen which was unavailable to non-mycorrhizal plants and Stribley and Read argue that it seems likely that the mycorrhizal fungus is able to assimilate nitrogen from organic compounds in the soil and conduct these directly to the host plant. This seems a fair argument as they have also shown that the fungus does not cause net mineralization in the soil. They grew the fungus in sterile soil for six months and found that it did not bring about an increase in exchangeable ammonium nor did it stimulate the growth of other species such as the grass *Festuca ovina* grown in that soil.

Not all the organic nitrogen in soils is in the form of complex compounds. Free amino acids are present in low concentrations in organic soils but, because they are not in equilibrium with a large source of supply, it is generally believed that they cannot be an important source of nitrogen for plants. They are released from two sources – from humified organic materials and as leachates from freshly fallen leaf litter. Stribley and Read (1980) investigated the relationship between mycorrhizal infection and the capacity to utilize simple organic nitrogen sources, such as amino acids, and complex ones, such as humic and fulvic acid. They grew mycorrhizal and non-mycorrhizal seedlings of *Vaccinium macrocarpon* in sterile sand which was saturated with 125 cm^3 of filter-sterilized mineral nutrient solution at pH 5.8 and which contained as the nitrogen source one of the amino compounds glycine, glutamic acid, glutamine, aspartic acid or alanine at a nitrogen concentration of 20.5 mg dm^{-3}. Control plants received either ammonium nitrogen at the same concentration or no nitrogen. They were grown in a controlled

Table 8.4 Nitrogen content, dry mass and ^{15}N-excess of shoots of mycorrhizal and non-mycorrhizal plants of *Vaccinium macrocarpon* after six months' growth in ^{15}N-labelled soil. (After Stribley and Read, 1974.)

	N content % oven dry mass	Yield mg oven dry mass	Total N mg plant^{-1}	^{15}N-excess (atom %)
Mycorrhizal	1.20	30.32	0.36	15.38
Non-mycorrhizal	0.98	20.97	0.21	20.03

Table 8.5 The effects of different nitrogen sources on shoot dry mass (mg pot^{-1}) of mycorrhizal (M) and non-mycorrhizal (NM) *Vaccinium macrocarpon* seedlings. (After Stribley and Read, 1980.)

Nitrogen source (20.5 mg dm^{-3})	M	NM
Ammonium	43.4	41.4
Glycine	43.9*	11.6
Alanine	47.6*	22.5
Aspartic acid	29.5*	7.0
Glutamic acid	39.4*	8.2
Glutamine	44.6*	18.2
No nitrogen	6.0*	8.3

* Significant difference in dry mass between M and NM at P 0.05.

environment room for 8 weeks. The shoots were then removed and dried. The results are given in Table 8.5. Mycorrhizal plants of *V. macrocarpon* utilized all the amino acids except aspartic as readily as they utilized ammonium. Even so growth on aspartic acid was still some 5 times greater than in the controls without nitrogen. The non-mycorrhizal plants had restricted ability to utilize these nitrogen sources, growth being better on ammonium than on any of the amino acids. They also tested the ability of the mycorrhizal fungus to degrade humic and fulvic acids extracted from soils under ericaceous plants; the fungus was grown in liquid culture containing 0.1% (w/v) of the acids as sole nitrogen sources and after 40 days at 20°C both the percentage loss of the acids (i.e. amounts utilized) and the dry mass of the mycelium produced were measured. The fungus had very little capacity to utilize these acids as nitrogen sources. Stribley and Read suggest that it is the capacity of the fungus to utilize simple organic compounds, such as amino acids, in the soil which explains the enhanced nitrogen uptake by mycorrhizal plants.

Read and his associates have done much less work on phosphate uptake but have shown that the fungus shows acid phosphatase activity and that it can use phytates at least in the form of the sodium salt of myo-inositol hexaphosphate. They grew their isolate on a glucose-mineral medium amended either with inorganic or organic phosphorus. For unit phosphorus in the medium, the yield of the fungus, in terms of dry mass mycelium, was about the same. This would help to explain the enhanced phosphorus content of mycorrhizal plants in such acid nutrient poor soils but, as in sheathing mycorrhizas, this aspect of their nutrition requires further investigation.

Nutrition of fungi

There is a direct evidence that the fungi of sheathing mycorrhizas absorb carbohydrates from their hosts. It is assumed that the fungi of vesicular-arbuscular mycorrhizas also do this. The fungi of ericoid mycorrhizas are again mainly intracellular and as with vesicular-arbuscular mycorrhizas it is impossible to separate fungus from host. This makes any assessment difficult. When ^{14}C labelled carbon dioxide is supplied to photosynthesizing shoots, sucrose becomes the most strongly labelled sugar in both mycorrhizal and non-mycorrhizal roots of *Vaccinium*. But mycorrhizal roots contain both trehalose and mannitol which are not present in non-mycorrhizal roots as they are fungal compounds and both of these are labelled. Thus it appears that, in their carbon nutrition, ericaceous mycorrhizal fungi are more like those of sheathing mycorrhizas than those of vesicular-arbuscular ones. Mycorrhizal roots also show a striking increase in a mannose polymer. This is a cell wall component of the fungus. Such a component is also present in sheathing mycorrhizas and in many rust-infected tissues. Although isolates of *Pezizella ericae* can utilize such polymers as pectins and carboxymethyl-cellulose, but not native cellulose, in culture, their relative slow growth rates are indicative of low competitive saprotrophic ability, which would suggest that this fungus too is an ecologically obligate biotroph.

ECTENDOMYCORRHIZAS

A number of genera in the Ericaceae and members of some other families in the Ericales possess mycorrhizas which are different from those discussed above. They are intermediate in form between ecto- and endomycorrhizas and have been called ectendomycorrhizas. For example, in the Ericaceae *Arbutus* and *Arctostaphylos* possess such mycorrhizas. The root system of *Arbutus* is differentiated into long and short roots. The short roots are swollen and are invested with a hyphal sheath. There is no Hartig net as in sheathing mycorrhizas but intracellular coils develop in the outer cortical cells and these are eventually lysed. Virtually nothing is known of the fungi involved or the physiology of these arbutoid mycorrhizas.

Monotropa hypopitys

Mycorrhizas somewhat resembling these are found in the Monotropaceae, a family in the Ericales. A good example is *Monotropa hypopitys*, the yellow Bird's nest or Pinesap. It is a non-green or achlorophyllous herb and its short, fleshy roots are invested by a sheath of hyphae with some of these forming a network between the epidermal cells comparable with the Hartig net of sheathing mycorrhizas. But such mycorrhizas must be physiologically different as the non-photosynthetic *Monotropa* cannot possibly supply carbohydrates for its fungus. The hyphae of the sheath and net however also form structures resembling simple haustoria, usually only one per cell, within the outermost cells of the host root. These so-called haustoria are not true ones but consist of fungal intrusions from the sheath into the outer cortical cells. They cause the host cell wall to invaginate, inducing the formation of structures reminiscent of transfer cells. Duddridge and Read (1982) have called these 'fungal pegs' and suggest that they facilitate enhanced nutrient supply during periods of peak demand. They could be involved in the rapid transfer of carbohydrates from fungus to plant in the period of expansion of the inflorescence. Such cellular structures are common in healthy plants and in some other symbiotic associations. They have been recorded recently in the ectomycorrhiza of *Pisonia grandis* which have a poorly developed Hartig net.

Björkman (1960) isolated a fungus, a *Boletus*, a genus well-known to be mycorrhizal with conifers, from mycorrhizal roots of *Monotropa* growing amongst pines and found that it also formed typical mycorrhiza with pine seedlings. He injected ^{14}C glucose into the phloem of mature pine trees with *Monotropa* growing amongst them and showed that after 5 days low but significant levels of radioactivity occurred in young *Monotropa* plants 1–2 metres away but not in other herbs close by, indicating that labelled compounds had been translocated from *Pinus* to *Monotropa*. Support for this idea also comes from physically separating *Monotropa* from the roots of its associated trees with metal plates driven into the soil. Its subsequent growth is very poor. It thus appears that photosynthesis by the pine provides carbon compounds not only for its symbiotic fungus, as one would expect but also for *Monotropa*, so there is a three tier system involving a chlorophyllous symbiont and a non-chlorophyllous one and a fungus as a bridge. The remarkable feature is the dual role of the fungus, extracting carbohydrates from the one and contributing carbohydrates to the other. This state of affairs may be relatively more common than is generally credited. The movement of nutrients between the pine and *Monotropa* may not be all one way. ^{32}P labelled inorganic phosphate injected into *Monotropa* plants has been detected in associated pine roots within 2 h of injection. It is not at all easy to see what advantage the fungus gets out of this sort of association.

Sequestration of heavy metals by ericoid mycorrhizas

Ericaceous plants are normally mycorrhizal and are very succesful colonizers of natural and man-made environments which contain very high levels of metallic elements. Members of the family colonize soils of high mineral composition such as those on serpentine and even natural heathlands have potentially toxic levels of metals such as aluminium, because of its availability increasing at low pH. *Calluna vulgaris* is frequently the dominant and only colonizer of coal tip spoils and *Vaccinium macrocarpon* is prominent on soils contaminated by metal smelting. Enhanced uptake under these circumstances would be physiologically damaging rather than beneficial.

Bradley, Burt and Read (1982) investigated the relationship between mycorrhizal infection and resistance to metal toxicity in *Calluna vulgaris*, *Vaccinium macrocarpon* and *Rhododendron ponticum*. Plants were grown from seed and some inoculated with mycorrhizal fungi. Mycorrhizal and non-mycorrhizal plants were grown in acid washed sand containing a dilute nutrient solution supplemented with copper over the range 0–75 mg dm^{-3} and zinc over the range 0–150 mg dm^{-3} for 12 weeks. The growth of non-mycorrhizal plants was severely inhibited in all copper treatments and all but the lowest level of zinc. Mycorrhizal infection provided considerable resistance to the effects of the metals. Some growth occurred in all concentrations of copper and zinc. Mycorrhizal plants showed larger increases in shoot:root ratios in the metal treatments compared to control plants. The shoot:root ratios were higher at higher levels of metals, reflecting a greater diminution of root relative to shoot growth with increasing concentration of metals. Compared with non-mycorrhizal plants, mycorrhizal plants had a lower concentration of metals in their shoots. The concentration of metals in the roots of mycorrhizal plants was considerably higher than that in their shoots which suggests that metals are being complexed in the roots. They propose that hyphal complexes of the fungus in the cortical cells of the root provide absorptive surfaces which facilitate the exclusion of metals from the shoot and avoids metal toxicity. The surface for heavy metal absorption is present in mycorrhizal plants but non-mycorrhizal ones can rely only on root extension to provide such a surface; in the presence of heavy metals this is reduced by restriction of root growth. Thus as in plants, such as *Agrostis capillaris (tenuis)*, heavy metal tolerance depends upon sequestration of the metal in the roots, the normal site of complexing being the wall, i.e. the wall provides a series of catchment surfaces upon which metallic ions can be fixed. Resistance is thus based on an exclusion mechanism and stress is avoided rather than tolerated. If such a mechanism is to operate, it is clear that the fungus itself must be resistant to heavy metal contamination. The mycorrhizal fungi were cultured on the same levels of copper and zinc. They grew without any significant reduction up to 50 mg dm^{-3} of copper and 100 mg dm^{-3} of zinc. Thus in Ericaceous plants in addition to increased nutrient uptake, a further benefit of the mycorrhizal association is the exclusion of undesirable elements.

Mycorrhiza of the Gentianaceae

A number of other flowering plant families have mycorrhizas in the form of hyphal coils. Two good examples are the Gentianaceae and Orchidaceae. Little is known about the mycorrhiza of the former family. Gay, Grubb and Hudson (1982) investigated the mycorrhiza of *Blackstonia perfoliata*, *Centaurium erythraea* and *Gentianella amarella*, three turf compatible biennial members of the Gentianaceae found in chalk grasslands. All three plants are usually mycorrhizal. Their seedlings become infected within two weeks of germination. The cortical cells of the roots become full of irregular coils of aseptate hyphae. The coils are eventually lysed. Vesicles can occasionally be seen attached to these coils (Fig. 8.6). This and the nature of the surface hyphae suggest that they are a peculiar form of vesicular-arbuscular mycorrhiza. Indeed Gay, Grubb and Hudson infected seedlings of these three species by growing them in pots of sterile soil in which plants of *Sanguisorba minor* and *Leontodon hispidus* infected with typical vesicular-arbuscular mycorrhizal fungi were growing. Such a view has been confirmed by Jaquelinet-Jeanmougin and Gianinazzi-Pearson (1983). They grew various species of *Glomus* and *Gigaspora margarita* in roots of onions or raspberries

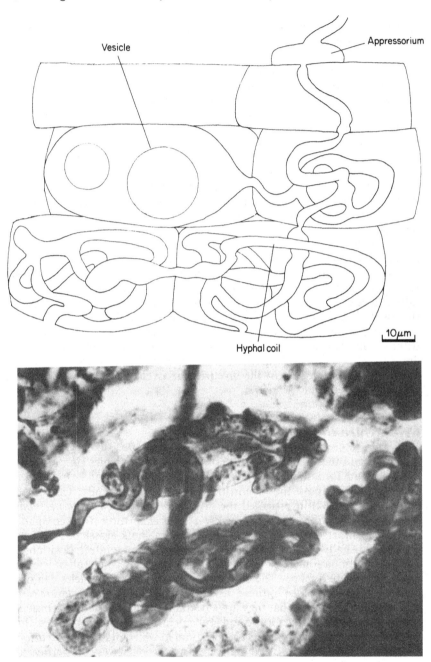

Fig. 8.6 Mycorrhiza of the Gentianaceae. Cortical cells of a root of *Blackstonia per-foliata*, showing appressorium, aseptate hyphal coils and a vesicle.

in pots of γ irradiated soil and used chopped roots from these to inoculate seedlings of *Gentiana lutea*. All produced typical intracellular coils in the roots of the gentian. The turf compatible biennials have two periods of maximum growth and these are associated with maxima in the accumulation of mineral nutrients. These are in the mid-summers of their first and second years. These periods also coincide with the periods of greatest density of mycorrhizal infection. At these times the hyphal coils are intact and thus presumably active whereas they become lysed in the autumn periods. The assumption is that they are behaving as typical vesicular-arbuscular mycorrhizas in enhancing mineral uptake while being dependent upon their hosts for their carbon supplies.

Orchidaceous mycorrhizas

Orchidaceous mycorrhizas, although very similar to ericoid ones, are unusual in a number of respects, especially in their carbon nutrition. At the onset it must be appreciated that most of the work on these has been done with heterotrophic orchid seedlings rather than with the autotrophic adult plant. For most green orchids the association with a particular fungus is obligatory in nature if their seeds are to germinate and become successfully established. For the non-green or achlorophyllous orchids, this association is obligatory for all their lives.

Orchids produce vast numbers of tiny seeds, often millions per capsule. Each seed weighs only 0.3–14.0 μg. The embryo consists of 10–100 cells and there is no, or virtually no, food store. The embryo is merely enclosed in a loose-fitting, thick-walled, net-like testa which is essentially a dispersal structure (Fig. 8.7a). The majority of seeds are incapable of successful germination unless assisted by an external source of carbohydrate. For this, in nature, they rely upon fungal infection. Any orchid until it is photosynthetic relies upon its fungus for its carbon supply. It may be some considerable time after germination before the first photosynthetic leaf is produced. In many orchids, such as species of *Dactylorchis*, the first green leaf may arise as early as in the second year of growth. In *Spiranthes spiralis*, it does not arise until the eleventh year and in others it may take even longer before any photosynthesis occurs. So all have a prolonged and some a very prolonged phase in the seedling stage when they are non-green, subterranean and obligatorily mycorrhizal.

Fungi involved

By far the commonest of the fungal associates are members of the Basidiomycete genus *Rhizoctonia*, a genus recognized on hyphal characteristics. Many of these have been demonstrated also to produce ill-defined crust-like basidiocarps in such genera as *Corticium*, *Ceratobasidium* and *Tulasnella*. Unlike the Basidiomycotina of sheathing mycorrhizas, these fungi are not

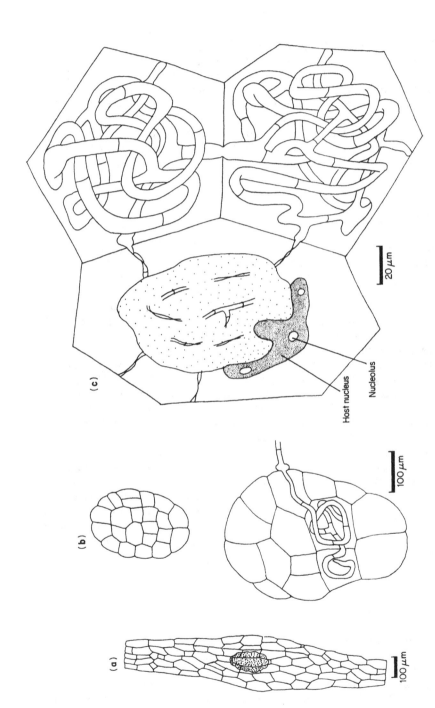

Fig. 8.7 (a) An orchid seed. (b) Embryo swelling and initial stages of infection. (c) Three cortical cells from the root of *Dactylorchis purpurella*, two with intact hyphal coils and one with coils undergoing lysis.

Host nucleus

Nucleolus

20 μm

100 μm

100 μm

(a)

(b)

(c)

members of the Agaricales but of the Aphyllophorales. They contrast markedly with the Agaricales. The majority are known to be capable of existing as free-living saprotrophs in the soil. They are relatively fast-growing and can compete successfully with other soil saprotrophs. Furthermore they can utilize cellulose so that they are self-supporting for carbon compounds. Many have been demonstrated to be aggressive necrotrophic parasites of other hosts. They cause root diseases of cereals, tomatoes, lettuce and other crop plants and 'damping-off' diseases of seedlings, such as cress and swede. Other fungi which have been less frequently isolated as orchid mycorrhizal fungi include such Agarics as *Armillaria mellea*, better known as a devastating root parasite of trees. It is known to form mycorrhizal associations with the achlorophyllous orchids *Gastrodia elata* and *Galeola septentrionalis*.

Features of infection

Initially the fungus enters the embryo via the suspensor and this is followed by an immediate marked stimulation of growth of the embryo. The central region of the embryo initially becomes completely colonized (Fig. 8.7b). The protocorm, or seedling axis, then develops. It differentiates a conducting strand, an apical group of leaves and roots from the axis near the leaf bases. The fungus then becomes restricted to the outer parenchymatous cells of the axis with hyphae passing out into the soil. As the young roots develop, they apparently become infected by the fungus in the soil and not from that in the protocorm.

The fungus enters only the parenchymatous cortical cells of the root and it becomes restricted to these cells where it forms dense intracellular coils or tangles of septate hyphae (Fig. 8.7c and 8.8). As in the protocorm, with time these hyphae swell, lose their contents, collapse and eventually degenerate into a structureless, disorganized mass as in ericaceous mycorrhizas.

Carbon nutrition of fungi and heterotrophic growth of the orchid

There is good evidence that the fungi concerned utilize complex carbon sources in the soil and make the products of the hydrolysis available to the young orchid. If seeds are placed on moist filter paper, they do not germinate, but if the filter paper is inoculated with an appropriate *Rhizoctonia*, the seeds will grow. If pads of filter paper on which the fungus has been growing are washed, the washings contain hydrolytic products of cellulose such as glucose and cellobiose as well as the fungal carbohydrates trehalose and mannitol. From this it can be argued that the fungi either release, by hydrolysis, sugars in excess of their needs from polymers, such as cellulose, so that the sugars are available in the external medium for direct absorption by the orchid, i.e. the orchid is behaving as a commensal, or the fungi take these up and translocate them in their hyphae to the orchid, i.e. there is a

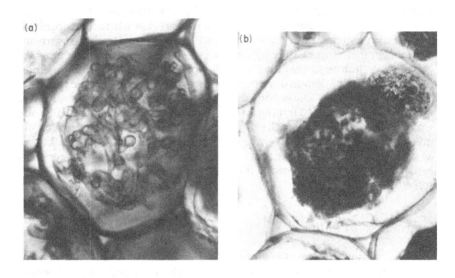

Fig. 8.8 Two cortical cells from the root of *Dactylorchis purpurella*. (**a**) Intact hyphal coils in cell; (**b**) amorphous mass of lysed hyphae with enlarged nucleus and nucleolus.

transfer across a living interface. The former is most unlikely as there would be too much competition from micro-organisms in the soil for these free sugars for the orchid to obtain sufficient. Also Hadley and Williamson (1971) have shown that not all protocorms in a population of *Dactylorchis purpurella* become successfully infected. Uninfected individuals in the presence of the fungus do not show the same increase in growth rate as do infected ones. In addition to these two possibilities the host must gain something in the way of nutrients from the periodic degeneration of the hyphae. This must release the contents of the hyphae into the host cell. One view is that hyphal collapse and degeneration is caused by the host. Such a view has led some to consider the orchid as being a necrotrophic parasite of the fungus. If host mediated, this action by the orchid is more likely to be a defence mechanism against complete parasitic invasion. There is some sort of balance in this association between the aggressiveness of the fungus and the host's defence mechanisms. This is certainly a much more precarious balance than is evident in other mycorrhizal associations. This can be seen in germination experiments with the fungus present. Three types of seedlings can be observed. There are seedlings which are growing vigorously – that is where a balanced association has become successfully established. There are seedlings which have germinated but which have subsequently been killed. In this case, the fungus has become, after initial successful establishment, wholly parasitic and incapable of being controlled by the host. Finally, there are seedlings which have germinated but which are no longer growing. In these, after successful establishment of the infection, degeneration of the fungus has been so effective that the fungus has been eliminated.

In this association both biotrophic transfer across orchid-fungus membranes and any nectrotrophic activity by the host may be involved in the movement of carbon sources from fungus to orchid. The extent of the movement would indicate that the former is the major pathway. Translocation to the developing orchid seedling by its fungus has been demonstrated by Smith (1966, 1967). She used what is known as the double dish technique and inoculated the inner dish which contained full mineral nutrients and cellulose as a sole carbon source with the *Rhizoctonia* fungus of *Dactylorchis purpurella*. In the outer dish she used a carbon free medium, silica gel, and sowed this with seeds of *D. purpurella*. The fungus grew out from the inner dish and spread across the outer and stimulated the seeds to grow. Over such a distance translocation of the products of cellulolysis in the fungal hyphae to the orchid seed could be the only feasible explanation of this stimulation. She then followed the uptake and incorporation of ^{14}C labelled glucose by the fungus growing out from the infected seedlings. Uninfected tissues of *D. purpurella* contain glucose, fructose and sucrose and the hyphae of *Rhizoctonia* glucose, trehalose, little mannitol but no sucrose. After feeding the hyphae of *Rhizoctonia* with labelled glucose, the label appeared in trehalose in the hyphae. Carbohydrates are usually translocated as trehalose in fungi. In infected parts of the seedling, the label appeared in trehalose and sucrose, i.e. in both host and fungal sugars, so that there must be a transfer between the fungal and host cells. In a time course experiment after a short feeding period with ^{14}C glucose trehalose contained the highest proportion of the label in the infected host. This proportion decline with time and was accompanied by a rise in sucrose (Fig. 8.9). This suggests that trehalose was being utilized by the seedling and thus maintaining a concentration gradient in the system. What is not clear is whether the trehalose is transferred directly or whether it is externally hydrolysed and the glucose formed rapidly absorbed by the orchid tissue.

In similar experiments, Smith showed that ^{32}P labelled orthophosphate supplied to the fungus could later be detected in the infected seedlings. Thus in the heterotrophic seedling phase there is a net movement of carbon and minerals from fungus to host orchid. At this stage this association must be regarded as a parasitic symbiosis with the orchid parasitizing the fungus. The ultimate source of carbon supplied to the orchid may be cellulose from dead plant remains in the soil or simple carbon compounds from another seed plant on which the fungus is a necrotrophic parasite in a second symbiotic relationship. As in the *Monotropa-Pinus* association, the fungus acts as a bridge.

Relationship between the adult green orchid and the fungus

In adult orchids only the absorbing roots ever become infected. Root tubers

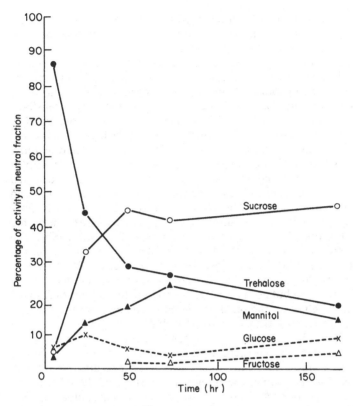

Fig. 8.9 Distribution of ^{14}C activity in a neutral extract of mycorrhizal seedlings of *Dactylorchis purpurella* after the fungus, *Rhizoctonia* sp. had been fed with ^{14}C glucose. (After Smith, 1967.)

or photosynthetic roots never become infected nor does infection spread to the stem and leaves. These parts are thought to produce a phytoalexin, orchinol, as a defence reaction on infection by mycorrhizal and other fungi. The roots may produce orchinol on infection but in insufficient quantity to inhibit the fungus. An odd feature of this association is that infection of the seedling does not initiate a permanent state of infection in the adult. The overwintering storage root tubers are free of infection and, in orchids such as *Dactylorchis fuchsii*, new roots are infected from the soil as they arise each year in the spring. As adults, some orchids are completely free of the fungus.

At the present very little is known about the relationship between the adult green orchid and the fungus. Reciprocal movement of carbon compounds to the fungus from the autotrophic orchid has not been demonstrated although suitable experiments to test the possibility have been carried out; however, at the moment it can only be concluded that this association is not a mutualistic symbiotic one. There is much which still needs to be explained in this association, especially the significance of the re-infection of the new roots of the adult green orchid every growing season. It could be argued that this is inevitable because of the aggressiveness of the fungi as root parasites but the possibility of there being a reciprocal exchange of carbon from the orchid to the fungus at this stage cannot be definitely ruled out or that the fungus does not still supply the orchid with minerals. *Dactylorchis fuchsii* and *Listera ovata* often grow side-by-side in the same woodland. The former has three relatively short roots arising from the leaf bases above the tuber. The latter has very numerous and very long roots. The roots of the former are virtually always very heavily infected and those of the latter only spasmodically infected. It could well be that *D. fuchsii* is still relying on its fungus for mineral supply. For some orchids at least one can see the possible advantages of maintaining the association. Some, for instance, by reverting to the association and becoming heterotrophic or partially heterotrophic can persist over one or more years without bearing leaves or sufficient leaves to maintain themselves fully. This would be advantageous in particularly adverse seasons or under conditions of very heavy shade or when their leaves are removed by grazing.

9

Fungi as symbionts with insects

A very wide range of fungi parasitize insects although perhaps the two best known genera are *Cordyceps* and *Entomophthora*. Such entomogenous fungi attack all stages of insects – eggs, larvae, pupae and adults. Some of these are not obviously antagonistic. This is especially true of members of the Laboulbeniales, a very little known group of the Ascomycotina which are obligate ectoparasites of Coleoptera in particular, and the Trichomycetes, a group with close affinities to the Zygomycotina, which inhabit the midgut, hindgut and rectum of many insects and other Arthropods. The Trichomycetes obtain all their nutrients from the aqueous fluids of the digestive tract in which they are bathed. They are attached to the lining of the digestive tract by a distinct device – a cellular or non-cellular holdfast and the thallus consists of a simple, cylindrical filament or a sparingly branched one, with or without septa. There is no penetration of the host tissues and perhaps these fungi may best be regarded as endocommensals (Fig. 9.1a).

In contrast, members of the genus *Coelomomyces*, one of the Mastigomycotina, as the name suggests, develop in the coelomic cavity, especially of the larvae of mosquitoes. A branched coenocytic hyphal system is produced, attached at one end to the host's tissues. All members of the genus are obligate parasites utilizing the adipose tissue and killing larvae before they pupate. Hyphal bodies, rounded off fragments of hyphae, are also produced and these circulate in the haemolymph (Fig. 9.1b). Similar bodies are found in the Entomophthorales. These are members of the Zygomycotina, which parasitize Hemiptera and Diptera although some are saprotrophic in soil and dung. On infection either a coenocytic mycelium develops or more usually this segments into rounded hyphal bodies which are produced in such profusion that the whole insect becomes more or less filled with them. The various tissues of the insect's body disintegrate at different rates – first the abdominal contents, then tissues of the head, thorax and legs and finally the trachea and tracheoles. Eventually the hyphal bodies develop conidiophores which protrude to the exterior and produce multinucleate conidia which are violently discharged. In houseflies infected with *Entomophthora muscae* the infected dying flies crawl towards the light – the top of a grass stem or a window pane. Their abdomens become distended and white bands of conidiophores project between the segments of the exoskeleton. The dead flies soon become surrounded by a white halo of discharged conidia. Some of

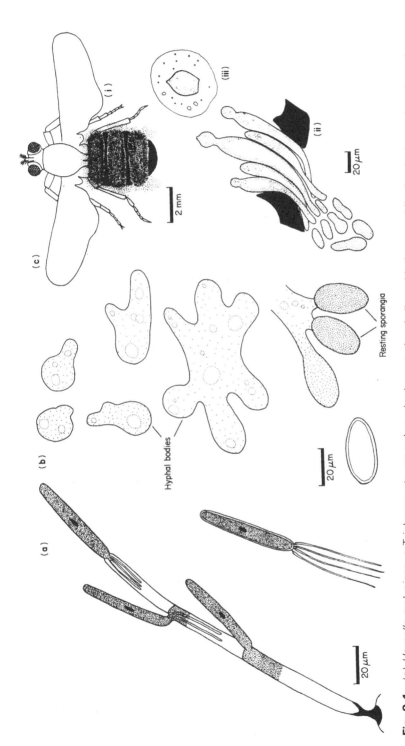

Fig. 9.1 (a) *Harpella melusinae*, a Trichomycete, an unbranched vegetative thallus with three conidia developing and a released conidium. (b) *Coelomomyces pentangulatus*. Hyphal bodies, vegetative thallus, developing resting sporangia and a released resting sporangium. (c) *Entomophthora muscae*. (i) Dead fly with distended abdomen and bands of conidiophores between the segments. (ii) Hyphal bodies giving rise to unbranched conidiophores. (iii) A discharged conidium.

these conidia will adhere to other flies as they crawl over them. Others, which have become air-borne, will impact upon flies in flight. The cycle of infection thus continues. At death, the flies virtually become mummified. They are no more than a sclerotial mass of hyphal bodies contained only by their own chitinous exoskeletons (Fig. 9.1c).

Another genus of fungi which similarly fills insects with a mass of mycelium is *Cordyceps*, one of the Ascomycotina. Most species of *Cordyceps* are parasitic on insects, especially Lepidoptera, but some are found on spiders and others on the subterranean ascocarps of *Elaphomyces*. *Cordyceps militaris* produces orange-coloured, club-shaped stromata, 1–4 cm tall, bearing numerous ascocarps, perithecia. These develop in the autumn and arise from buried mummified pupae of Lepidoptera. On infection from ascospores, cylindrical hyphal bodies develop in the haemocoel of the pupa. They multiply by budding. On death, mycelial growth occurs and the pupa becomes converted into a sclerotium. Its tissues are extremely resistant to decay due to the production of an antimicrobial nucleotide cordycepin by the fungus. Thus it does not rot before the fungus has produced its reproductive structures.

Mutualistic associations between fungi and insects

Not all associations between fungi and insects are so one-sided. The ability of many insects to exploit potential food resources to their best advantage, or perhaps even at all, may be dependent upon a reciprocally beneficial association with fungi. Some insect species are always found to be associated with a particular fungus. The insect may feed on the fungus or the fungus may digest or partly digest food for the insect. In many cases complete interdependence exists. In ectosymbiotic associations the fungus lives outside the insect but it becomes the food supply of the insect or it renders food digestible to the insect or it creates a favourable environment for the insect. It may, for example, protect the insect. The fungus is disseminated by the insect which may even have developed special pouches, mycetangia or mycangia, for keeping and carrying the fungus.

ECTOSYMBIOTIC ASSOCIATIONS

Ambrosia fungi and wood-boring beetles

Wood, although a favourable habitat for many insects, is a most unpromising food source for them. It is resistant to degradation by normal animal enzymes. It is deficient in B vitamins and sterols which insects require. For some the role of the fungi is to grow on the wood and thus provide the insect with something nutritious and palatable to gnaw on. Ambrosia beetles, either Scolytidae or Platypodidae, are the most successful of wood-inhabiting beetles. They show a specialization of the wood-boring habit by boring deep into the wood, the Platypodidae even into the heartwood. Although wood borers, they are not wood feeders. They introduce into their tunnels ambrosia fungi which grow on and line the walls of the entire tunnel system and serve as the chief source of food for all stages of the beetles. The fungi are constantly associated with the beetles and are transmitted by them frequently in specialized organs, mycetangia. The role of the fungi here is seen in their ability to convert a virtually indigestible substrate into a digestible form resulting in the conservation of time and energy for the beetle. These ambrosia beetles should be distinguished from bark beetles which feed on living tissues especially the phloem. Many of these are constantly associated with a specific fungal flora in their brood chambers at the interface of wood and bark. The fungi are transmitted from tree to tree, not in specialized mycetangia, but usually accidentally by their slimy spores adhering to the integuments of the beetles. Although the fungi may have a nutritional value to the beetles, especially the young larvae, they are not essential. Some such associations have proved disastrous to the host tree, for example, the one between elm bark beetles and *Ceratocystis ulmi*, whereas others cause blue-stain of coniferous sapwood (Chap. 3).

In northern temperate forests the Scolytidae are best known from the numerous species of bark beetles; the ambrosia beetles are comparatively uncommon. In tropical forests the position is reversed and Platypodidae are more common. In the world as a whole species of ambrosia beetles far outnumber species of bark beetle. Although the adults of the two groups differ greatly in their appearance, their larval forms are very similar. They prefer dead, dying or weakened trees as breeding material, such as trees weakened by drought, age, fungal attack, fire, competition or defoliation and as pests are referred to as secondary insects. This preference is because the moisture content of the tree is important in fungal establishment. As with wood decay fungi, the moisture content of healthy trees is too high. Most attack a very wide range of hosts. They destroy a great deal of timber. The tunnels to their brood chambers are called 'shot holes' and extend deep into the wood. In addition they may be surrounded by streaks of fungal stains.

The adult female Scolytid beetles which do all the boring are very odd. They fly to and attack the wood of a new host when their guts are empty and do not feed while excavating the entrance tunnel to their new gallery systems. Once this is achieved they mate, sow their ambrosial fungi and wait for the spores to become established. They then feed on the fungi, resume boring and lay eggs in the newly constructed brood chambers. If the fungi do not develop in the entrance tunnel, they eventually starve. In any case considerable resorption of the flight muscles occurs during this initial period of boring.

Mycetangia are usually found only in female Scolytids but are present in the males, which do all the boring, in the Platypodidae. The location and structure of the mycetangia vary enormously. In *Xyleborus*, they may be a pair of sacs alongside the brain opening to the inside of the mouth, membranous pouches at the base of the mandibles or sclerotized pouches in the base of the elytra. More usually they consist of depressions, flask-shaped invaginations, or pouches on the body surface, such as the prothorax, where oil, secreted from specialized hypodermal cells or glandular hairs, accumulates (Fig. 9.2a). The fungi are sown as the tunnels are constructed when the output of the secretions is increased, possibly for lubrication during boring, and slowly washes out oidia or yeast-like fungal cells. These germinate on the walls using the oil as an initial carbon source. The fungi were called ambrosia by Schmidberger (1836). This was at the time when spontaneous generation was still acceptable and as the material on which the beetles fed appeared to have no earthly source, he called it ambrosia, the food of the gods. But it was Hartig (1844) who recognized it as fungal and he called it *Monilia candida*. It is now apparent that there are more than one ambrosial fungus and that there are specific associations between beetles and fungi. They are conidial states of Ascomycotina. Their taxonomy is still in a state of confusion. Almost all superficially resemble species of *Monilia*, a genus including a heterogenous assemblage of fungi bearing their conidia in chains. The majority have now been placed in the genus *Ambrosiella*. They produce a greyish brown to dark brown mycelium bearing branched or unbranched conidiophores which

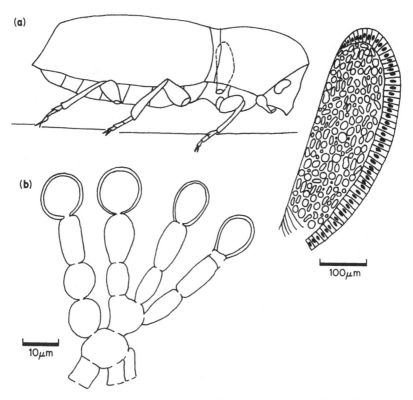

Fig. 9.2 (a) *Xyloterus lineatus*, adult female, showing position of mycetangium and mycetangium in vertical section. (b) Conidiophores and conidia of *Ambrosiella xylebori*.

produce acropetal chains of thin-walled, colourless blastoconidia or a thick-walled, solitary, terminal conidium (Fig. 9.2b).

Other ambrosial fungi have been placed in the genera *Monacrosporium*, *Phialophoropsis* and *Raffaelea*. The fungi penetrate only a few millimetres into the wood, especially the parenchyma around the tunnels. They stain the wood a dark brown and form a hyphal mat on the tunnel walls from which the closely packed conidiophores project into the tunnels. The beetles eat these. In many species of *Xyleborus* the head capsules of the larvae are poorly chitinized and their mouthparts are not sufficiently strong to be used to gnaw wood; they rely solely on the fungi as food. Very often there is insufficient fungal growth to satiate all the beetles in a particular brood. The last adults to develop in a brood are often much smaller due to a deficiency of food. Platypodid larvae feed exclusively on the fungi during the first three instars but the last two instars have a well-chitinized head capsule and much stronger mandibles. They bore deeper into the wood but it is not clear whether they feed directly on the wood or are merely excavating more tunnels to produce more fungus.

The fungi are nutritionally highly specialized. They are unable to utilize inorganic nitrogen but can use both fats and oils as sole carbon sources. When grown on a medium containing large quantities of amino acids or in pure olive oil, they produce ambrosial conidial mats as in the tunnels; however, when grown on ordinary laboratory media, such as malt extract agar, they produce a sterile aerial mycelium which produces conidia only when flooded with olive oil. The environment in which they grow clearly has a marked effect on their form. This is further borne out by the fact that they are oidial or yeast-like in the mycetangia. While the larvae inhabit the tunnels, they keep the fungus under control and in more or less pure culture. Foreign or weed fungi do not spoil the ambrosial mats. This is probably due to some form of secretion of the beetles and larvae but, once they leave the tunnels, the mats rapidly become contaminated with all sorts of fungi.

The presence of specialized organs of transmission, the mycetangia, suggests that the relationship between the fungus and the beetle is highly significant in the life of the two and the wide variation in their location and structure points to a polyphyletic origin of the relationship. Ambrosia beetles thus live in an obligatory symbiosis with special domesticated ambrosial fungi. Many details of this intriguing symbiosis need to be worked out. The beetles appear to gain the main benefits in that the fungi provide them with a rich and easily available source of food, vitamins and perhaps sterols, such as ergosterol, which is necessary for the development of the adult from the larva. Although the fungi are disseminated, protected from desiccation within the mycetangia, and inoculated directly onto a suitable, moist substrate by the beetles, they are strictly controlled by them. Readily available sources of carbon and nitrogen are scarce in wood and the quantity of nutrients in the form of fats and oils, as carbon sources, and urea and uric acid, as nitrogen sources, provided by the beetles may determine the extent to which the fungi develop. Constant grazing by the beetles also keeps them in check. This may be bondage by preference as the fungi are never found free-living.

Plant galls and fungi

Some insect plant galls may also contain fungi but, although there is a wide variety of interrelationships between the insects and fungi, they are usually all casual ones. In galls caused by Itonid midges, the fungus grows on the gall tissue and forms a thick layer of hyphae on the inside of the gall. It appears unlikely that the fungus acts as a direct source of food for the larvae but it may assist the insect indirectly by partially breaking down the gall tissue so that the larvae can digest it.

A single fungus is usually associated with each midge species and its spores are probably implanted when the female lays her eggs. The stem galls of *Rubus fruticosus* caused by the midge *Lasioptera rubi* are called ambrosia galls. The midge larvae, instead of feeding directly on the gall tissues, feed on the hyphae of the fungus lining the gall in a manner reminiscent of the wood-boring ambrosia beetles. However, in another ambrosia gall, that caused by the midge *Asphondylia sarothamni*, on the buds and pods of broom, *Cytisus scoparius*, the larvae derive their nourishment from the gall tissues and not from the fungus. As long as the larvae are actively feeding the gall continues to grow and the fungus feeds saprotrophically on the waste products of the larvae. But as soon as the larvae stop eating prior to pupating, the gall ceases to grow and the fungus fills the cavity with its hyphae so that the insect pupates in a dense mass of mycelium. The fungus then parasitizes the gall tissues. The fungi concerned have been very little studied.

The *Sirex/Amylostereum* association

Wood wasps, especially species of *Sirex*, also attack weakened or suppressed trees or freshly felled logs as do many ambrosia beetles. They are also associated with a fungus but the association is a very different one. There is specialized transmission of the fungus but not strict culture. The female wasp does not bore out brood chambers but inserts her eggs several centimetres into the wood and it is the larvae which tunnel. The adults are on the wing all their lives. All females carry *Amylostereum*, a genus of the Basidiomycotina, which causes a white rot of wood (Chap. 3). The females possess a pair of small invaginated intersegmental sacs protruding into the body and connecting by ducts to the anterior end of the ovipositor (Fig. 9.3a,b). These mycetangia contain large amounts of mucus, secreted by a pair of glands adjacent to the mycetangia, and the fungal mycelium within fragments into oidia or arthrospores consisting of 1–4 short cells with clamp connections at the septa (Fig. 9.3c,d). The female larvae, from the second instar onwards, also carry the fungus in hypodermal organs situated on both sides of the body in deep skin folds between the first and second abdominal segments. During

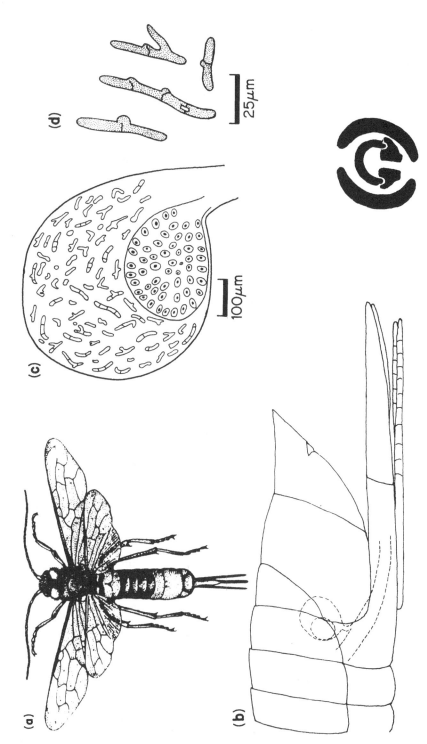

Fig. 9.3 *Sirex noctilio*. (**a**) Adult female. (**b**) Posterior end of adult showing the ovipositor, position of the mycetangia and the ovipositor in section. (**c**) Vertical section through mycetangium and mucus gland. (**d**) Arthrospores from mycetangium.

egg-laying the mycetangia contract and the oidia are squeezed out and deposited in the wood with the eggs. They germinate and grow in the wood and rapidly break down both the cellulose and lignin. One effect of this is that it enables the larvae to bore more freely in the wood. Unlike the ambrosia beetle larvae, they do not feed solely on the fungus as its hyphae are too sparse, nor do they actually ingest wood but they may be able to digest the decaying wood. It seems probable that both the fungus and the decaying wood are utilized by the larvae. The hyphae may be broken down externally by larval enzymes secreted in saliva discharged into their cupped mandibles. Larvae are able to live and grow for at least three months on pure cultures of the fungus.

Benefits to the fungus and wasp

The fungus clearly benefits from the association not only,in being effectively dispersed but by being inoculated, free of competition from other wood decay fungi, into a suitable host without having to penetrate any protective tissues, such as bark. With such an efficient means of dispersal, the production of basidiocarps becomes almost superfluous. They are very rarely found in nature although isolates of the fungi will produce them in culture. The wood wasp would appear to benefit by having a more readily available food source presented to it. This is a much less strictly defined association than the ambrosia beetle/fungus one. The fungus is not under such strict cultural control. There is no close association with the insect in the wood. It grows indiscriminately but it does again convert a relatively non-nutritious substrate into a nutritious one for the insect and the larvae feed on the softened wood. The importance of the association to the wasp may be inferred from the complicated procedure which has evolved to ensure that the oidia are established in the mycetangia. The adult wasp, unlike the ambrosia beetle, spends all its life on the wing. It does not come in contact with the fungus in the wood and must obtain it from the pupa via the larva. Mycetangia first develop in the fully-grown female larva as a series of pits on either side of the posterior end of the first abdominal segment. Just before the last larval moult, glandular cells in these temporary mycetangia secrete a copious waxy fluid containing oidia. This wax hardens into platelets which are shed and remain in the pupal chamber. When the adult female emerges from the pupal chamber, the wax platelets are transported, by retractive movements of the components of the ovipositor, to the permanent mycetangia at its anterior end. Once within, the oidia are freed from the wax. They then produce hyphae which fragment into more oidia. Thus the future inoculum multiplies in the mycetangia as indeed it does in the ambrosia beetles.

Problems for the plant pathologist

The association of wood wasps and their fungi may cause problems for the plant pathologist. *Sirex* species are confined to conifers. They are usually regarded as pests of secondary importance, losses from which are small in relation to the total volume of timber produced. Only in New Zealand and Australia has *Sirex* been viewed as a major pest, especially of *Pinus*. The association between *Sirex noctilio* and *Amylostereum areolatum* has caused serious losses in *Pinus radiata* there. Improved silvicultural practice and the introduction of parasites of *S. noctilio* now keep the *Sirex* population in New Zealand at a level at which it is no longer a serious forestry problem. In Australia it is still a problem and it is generally accepted that about a 10% absolute loss in volume of timber produced may be caused under average climatic conditions. In prolonged drought losses may be far more severe.

A. *areolatum* is a relatively weak wound parasite and most trees resist attack by it alone but, in a common attack with *S. noctilio*, it is able to kill pines. Mass attacks by the wood wasp provide a large number of the inoculation points which are more effective in enabling the fungus to establish and spread than in a single inoculation from a natural wound. *S. noctilio* is attracted to physiologically stressed trees such as weakened, suppressed, over-crowded, fire damaged or drought stricken ones. Injured trees exuding resin are particularly attractive. Volatiles in resin act as specific attractants; even so a high resin content is important in restricting growth of the fungus and it may also depress the survival rate of the larvae. Resistance of *P. radiata* to attack by *S. noctilio* is related to the ability of the tree to obtain adequate water from the soil and the moisture content of the wood. The optimum moisture content for egg-laying is 40–70% on a dry mass basis. Dominant trees with a high moisture content tend to be rejected by *S. noctilio* and the survival rate in such trees is often low. *A. areolatum*, when deposited with the eggs, spreads slowly in wood with a moisture content of 70% saturation or more. As it grows it reduces the moisture content of the wood locally, with the result that the eggs hatch and the larvae bore into relatively dry wood. The moisture content of heavily infected wood averages about 35% of the dry mass. Eggs and young larvae become desiccated in wood below 20% but fairly mature larvae can survive in very dry wood.

The spread of *A. areolatum* in wood of *P. radiata* is restricted by the formation of phenolics, such as the stilbene pinosylvin, and resins in the sapwood near oviposition holes as a reaction of the tree to attack. The killing of trees after attack by *A. areolatum* was formerly attributed to the pathogenic effects of the fungus. It was suggested that it cuts off the supply of sap to the crown. Although the fungus eventually dries the wood so much as to render it non-conducting, the restriction of sap flow is quite gradual and cannot explain the very marked physiological changes that occur in the tree within the first two weeks after attack by *S. noctilio*, The mucus, a protein mucopolysaccharide complex, secreted by *Sirex* and injected into the tree with its eggs and the fungus is phytotoxic. Mucus injections alone will induce the rapid early physiological changes typical of a *Sirex* attack. These include increased respiration in the stem, accumulation of starch in the leaves, decrease of starch in the bark of the stem, implying that translocation of photosynthate is inhibited and also some impairment of normal water relations of the needles causing desiccation, distortion and collapse of the translocating cells of the phloem. This is followed by yellowing, wilting and abscission of many needles. Such symptoms are reminiscent of premature senescence. This conditions the tree making it more susceptible to the fungus. It can then invade the sapwood and kill the tree.

Neither the fungus nor the mucus alone is capable of killing the tree but the combination of the two is lethal. The tree, having been conditioned by the mucus, is subsequently killed by the fungus invading the sapwood and cutting off the sap flow to the crown, death being preceded by a sudden increase in leaf water deficit. The mucus can be seen as a 'conditioning agent' which brings about a reduction in the effectiveness of the tree's natural resistance mechanism.

The *Septobasidium/Aspidiotus* association

A rather different mutualistic symbiotic relationship exists between scale insects and certain fungi. All species of the Basidiomycete genus *Septobasidium* are invariably found living in association with scale insects, such as species of *Aspidiotus*. In this association the fungus provides individual or communal prefabricated housing accommodation as well as protection from physical and biotic environmental factors, in particular protection from birds and wasps, for the scale insect. In return for being dispersed and being provided with the opportunity of deriving nutrients from insect secretions and excretions, the fungus also parasitizes some insects by sending out specialized, often coiled hyphal branches into the haemocoel of the insects (Fig. 9.4b). The insects can live unassociated with the fungus but probably never do so. The fungus is also obligately associated with scale insects in nature. This is a very delicately balanced mutualistic symbiosis as the scale insects may be killed by the fungus in which case the fungus also dies.

Septobasidium is a member of the Auriculariales and is found as an epiphyte on the bark of trees, especially on the lower sides of branches, in the tropics. Individual colonies form flat, radially ridged growths almost circular in outline. They may reach several centimetres in diameter and, being perennial, show concentric rings of growth. They often resemble a thick crustose lichen. Within, under the roof as it were, are a number of vaulted chambers and interconnecting tunnels or passages, opening to the outside but partially closed by a fungal flap. Uninfected female scale insects inhabit the tunnels and are free to move about but most chambers contain a single sedentary infected female (Fig. 9.4a). Only the uninfected ones are fertile. They project their abdomens to the outside through one of the openings and are fertilized by wandering males on the outside. They then remain closeted within the colony and reproduce.

Basidiospores, which may themselves bud, are formed over the surface of the colony and they adhere to the young scale insects as they emerge from the parent colony (Fig. 9.4c). After they settle, the spores germinate and surround the insect with hyphae, some of which penetrate the setal pores and enter the haemocoel where they eventually form numerous hyphal coils. Inside the haemocoel the hyphae are composed of chains of wide spindle-

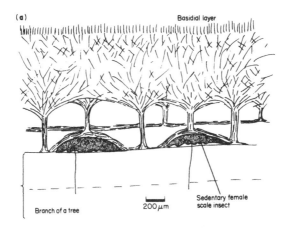

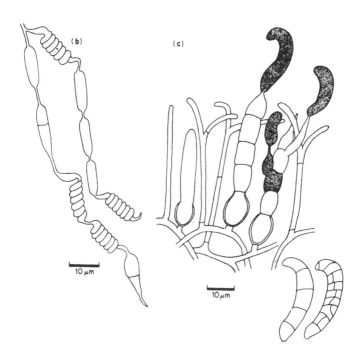

Fig. 9.4 (a) Vertical section through a colony of *Septobasidium sabalis*, showing surface basidial layer, vaulted chambers and sedentary female scale insects. (b) Spindle-shaped segments and hyphal coils from the haemocoel of a scale insect. (c) Part of basidial layer showing paraphyses, phragmobasidia and basidiospores.

shaped segments connected together by very fine hyphae, the whole resembling a string of cocktail sausages (Fig. 9.4b). The hyphal coils, each consisting of 4–8 symmetrical turns, like a coil of a hot water heater, arise on these hyphae connected end to end by fine hyphae or in clusters of two to three as lateral branches. The scales are sucking insects and the stylets of these females tap the phloem of the tree. Christensen (1965), in his rather splendid account of these, sees the insect at this stage merely as a pumping station transferring sap from the tree to the fungus. The insect converts sucrose from the phloem to its blood sugar trehalose which the fungus uses directly. A new colony soon becomes established around such a female.

In addition to shelter, gained at the expense of parasitism and sterility of some individuals, the fungus affords protection from predator birds and parasitic wasps. The scale insects are hidden from the view of birds and the particular parasitic wasps have only short ovipositors so that they can reach and lay eggs only in the superficial scale insects and not those in the vaulted chambers and tunnels. The thick mat of hyphae above the chambers also to some extent insulates the scale insects from temperature extremes and prevents their desiccation. In return, through its parasitism the fungus obtains its organic nutrients, especially its carbon source, from the insect and is also dispersed.

The Attine ants and their fungi

The fungus-growing ants and termites provide further examples of mutualistic symbioses between insects and fungi. The Attine ants culture a saprotrophic fungus in their nests or so-called 'fungus gardens' and use the fungus as their primary and probably sole source of food. The Attini are a tribe of New World Myrmicine ants which, although presently distributed between latitudes 40°N and 44°S of the equator, are best represented in the tropics, especially in Central and South America, the southern and eastern United States of America and the Caribbean. They were probably a significant part of the fauna of the Americas even before the time of the American Indians. The two most common genera, *Atta* and *Acromyrmex* are leaf-cutting ants and use the fungi in their subterranean gardens. *Atta texana*, for example, because of its relentless leaf-cutting, causes considerable damage to *Citrus* crops in eastern Texas and southern Louisiana. Because of their destructive activities, they are regarded as plague insects. There are several other genera of Attine ants such as *Trachymyrmex* and *Sericomyrmex*. Some of these use insect excrement and carcases, often of ants, and a variety of plant debris on which to cultivate their fungus. Others supplement these with leaves and flowers.

The *Atta* colony and their 'fungus gardens'

In *Atta* there are three castes – workers, females and males. The workers are polymorphic. They range considerably in size and show a gradual and continual change in size from smallest to largest. The largest are the 'soldiers', whose prime function is to guard the colony. The medium sized workers forage for leaves, cut out discs and carry them back to their nests on their backs and hence have earned themselves the name of 'parasol' ants. In foraging they make distinct well-worn tracks up to 300 mm wide through the vegetation. They also construct the nests which are underground; to build these they excavate. The excavation craters of a single *Atta* colony may cover some 8 m^2 of ground and be about 1 m or so deep. An average sized colony has some 2000 chambers, each 200–300 mm in diameter. There would be some 1–2.5 million ants weighing about 10 kg in the colony. For every 2000 chambers, they excavate about 80 000 Kg of soil and during the life of the colony about six times this mass of fresh leaf material is used in the gardens. Thus, in addition to their destructive activities, they are of some ecological value in creating fertile, organic-rich pockets in soil and in mixing the soil. They are of importance especially in soils where few roots penetrate to any depth and where there are few burrowing animals.

At the nest the leaf-cutting ants pass the discs over to the smaller workers. These clean the discs, scrape them and cut them into 1–2 mm pieces. They then pinch them to a pulp with their mandibles. Saliva and faecal droplets are added and they are deposited in the fungal gardens. Actively growing tufts of mycelium from other parts of the garden are planted on the surface of the newly added leaves. They are, in the mushroom growers' parlance 'spawned'. The garden is a speckled, brown, sponge-like or honeycombed mass of withered pieces of leaves, loosely connected by a weft of white mycelium. Exhausted or spent gardens may be deposited in certain chambers, in specially constructed larger ones or cast out by the medium-sized workers several metres from the colony.

These activities of the ants produce a flourishing growth of one fungus although the substrate is suitable for many. This pure culture is maintained in spite of the possibilities of contamination from the surrounding soil or from spores carried in by the ants on themselves or on leaves. When the ants are removed from or leave their gardens, as in the ambrosia beetle tunnels, the fungus is quickly replaced by alien fungi and bacteria. Only when the association is maintained is a healthy monoculture garden maintained.

The garden is the abode of the queen and her brood. Before swarming, a virgin queen tucks a small pellet from the fungus garden into a pocket, the infrabuccal cavity, below her mouthparts. After her mating flight, she settles, breaks off her wings and starts to excavate in order to establish a new colony. She digs a small tunnel, 150–250 mm deep, and near the end creates a chamber and plants the pellet in it. She continually adds faecal droplets and broken infertile eggs to it but also goes out and forages for fresh leaves. It may be several months before she lays fertile eggs. The offspring then attend to the garden and a new colony is formed.

The fungi concerned are Basidiomycete Agaricales which do not normally produce basidiocarps if the nest is attended, probably because the mycelium is grazed too much. A number of generic names have been given to these fungi. Most appear to be members of the white-spored genus *Lepiota* or of the macroscopically rather similar genus *Leucocoprinus*. A number of isolates have been cultured. They can be easily grown on ordinary laboratory media and are typically mycelial. When grown on acid media, rich in amino acids, and exposed to 0.5% carbon dioxide, as they are in the nests, they develop into more velvety growths producing clusters of hyphae with very swollen tips rich in cytoplasm. Such clusters are called staphyla or bromatia (Fig. 9.5). These are particularly abundant in the nests. The ants crop these. The hyphal tips are fed to the larvae and they are also the staple food of the adults.

Neither the ant nor the fungus is found in nature without the other. The association is an obligate symbiosis for both. This is probably the most advanced of all the insect/fungus mutualistic associations. One intriguing aspect of this relationship which is not understood is how the ant maintains the growth, form and purity of the fungus in its garden as each species of ant cultures only a single species of fungus. One theory was that the ants' salivary and anal secretions, which the ants regularly apply to the gardens, contain antibiotics which prevent the growth of alien fungi. Although an appealing theory, it is an untenable one. Extracts from ants' fungal gardens and cultures of the fungus have no antibiotic activity.

The biochemical basis of the symbiosis

The biochemical basis of the symbiosis in the *Atta colombica tonsipes*/fungus association has been extensively investigated by Martin and Martin (1970) and they have clarified some of the principles and mechanisms underlying this symbiosis. The fungus grows very poorly on culture media containing polypeptide nitrogen and they initially suggested that the fungus lacks the necessary complement of proteolytic enzymes to make full use of

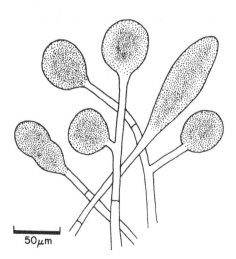

50μm

Fig. 9.5 A cluster of staphyla.

protein nitrogen. Fresh leaf material would thus seem tó be a singularly inappropriate substrate for the fungus as most of the nitrogen in leaves is in the form of protein. The application of faecal material to the leaves compensates for this. The two major nitrogenous excretory products of the ant are allantion and allantoic acid which would be available for the fungus to use as amino nitrogen. But, in addition, faecal material contains significant quantities of ammonia and some 21 amino acids. Glutamic acid, histidine, arginine, proline, lysine and leucine make up 80% by mass of the total amino acids excreted. These supplements would clearly be beneficial to the growth of the fungus but the benefit would be short lasting. They would be rapidly absorbed and utilized as they are excreted in very small volume in relation to the quantity of leaves used in the nests. In addition, Martin and Martin found that although the ants' salivary excretions showed no protease activity, the faecal droplets had significant activity. They contain three proteolytic enzymes, serine proteinase and two metallo-peptidases, which are highly stable and possess broad catalytic activity. The faecal material thus serves the additional function of catalysing the hydrolysis of leaf protein, thus further facilitating fungal growth. At first Martin and Martin thought that these enzymes were the ants' contribution to the symbiosis. But because of their stability and general resistance to denaturation and deactivation and the fact that the enzymes are only slowly degraded by other proteolytic enzymes, Martin and Martin suspected that these enyzmes were microbial in origin and have shown that they are derived from the cytoplasmic contents of the hyphae upon which the ants feed. Thus the ants do not contribute bio-chemically to the association but they are vehicles. They simply acquire, accumulate and move fungal enzymes from a region in the nest where the fungus is in a state of rapid growth and the enzymes are in ample supply to a site of inoculation, the newly added leaves, where the enzymes are in short supply. This permits substantial hydrolysis of protein providing a nitrogen source for the initiation of rapid growth by the fungus and simultaneously shortening the lag phase, during which time metabolic preparations for rapid growth are occurring. Any reduction in the lag phase would also shorten the time period in which the substrate would be susceptible to colonization by competitor saprotrophs. They were unable to detect any trace of digestive proteolytic enzymes produced by the ants. Thus the chief biochemical adap-tation of the Attine ant to the fungus-growing habit has been the cessation of the production of any digestive proteases, since they would degrade any fungal proteases as they passed through their guts.

Much of the ants' culturing and maintenance activities can be considered as enhancing the competitive ability of the fungus so that it emerges in the nest as the dominant micro-organism excluding competitors. The scraping and pulping will liberate cell contents so that they are immediately available to be absorbed. Rapid and extensive colonization is enhanced by planting mycelial inocula. This gives a better competitive advantage over contami-nants starting from spores. In many ways, in this association, the ants are dominated by the fungus. They forage for substrate for it, prepare the substrate for inoculation, inoculate and fertilize it, extract and transplant enzymes from one part to another, weed the gardens and remove old expended material. The fungus is cellulolytic and it indirectly contributes its cellulose degrading ability to the ants. It converts cellulose into fungal carbohydrates, especially glycogen and trehalose, which the ants can metabolize and so, by eating the fungus, the ants can tap the virtually inexhaustible cellulose supply in its environment. This they certainly could not do without the fungus. The fungus provides the ants with an excellent, rich and complete diet. More than 50% of the dry mass of the fungus is easily soluble. Carbohydrates make up 27%, free amino acids about 5%, protein 13% and lipids 0.2%. The lipid fraction also contains the vital sterol, ergosterol, which the ant larvae may require for normal development. The lack of digestive proteases makes the amino acids particularly important to the ants.

One can attribute the success of the Attine ant in the New World tropics to the ecological soundness of its leaf-cutting, fungus-growing strategy. Much of this success is due to the ability of the ant fungi to grow on almost any kind of leaf material. By constructing their nests in the soil rather than above ground, the ants ensure that the fungi have a suitably moist environment in which to grow and so may be found in a wide diversity of habitats, semi-desert, grassland, tropical rain forest and even in suburban gardens where the ants plunder the leaves on man's crop and garden plants to build their fungus gardens.

Termites and their fungi

Enzymes of fungal origin have also been found in the digestive fluids of fungus-growing termites. Termites are best known as destroyers of wood. The lower termites possess symbiotic protozoa and bacteria which digest cellulose in the paunch, an enlarged region of the hindgut. But the higher termites, the Termitidae, which make up about 75% of the existing termite species, lack these and afford an interesting parallel with the Attine ants in that they maintain fungus gardens or 'combs' in their nests. Whereas species of *Microtermes* construct a series of chambers containing fungus combs below ground in a diffuse pattern, species of *Macrotermes* build steeple-like mounds up to 10 m tall, especially in tropical savannahs. The workers collect and eat all sorts of fallen plant debris but especially wood and they use their faecal pellets to make combs in chambers within the mound. Undigested or just chewed wood is not used. Each comb becomes covered by mycelium and is sponge-like in appearance but corky in texture. Numerous small spheres or nodules, 0.5–2.0 mm diameter, develop over the surface. In *Macrotermes natalensis*, these nodules are the conidia of the fungus *Termitomyces*, a pink-spored member of the Agaricales. Several species of *Termitomyces* are known and all have an obligate association with termites (Fig. 9.6). Basidiocarps are produced early in the rainy season but only after the colony has abandoned its nest. All possess a stalk with a very distinctive long and tapering root-like base or pseudorhiza. The basidiocarp initials arise in the combs and this structure enables the remainder of the basidiocarp to be

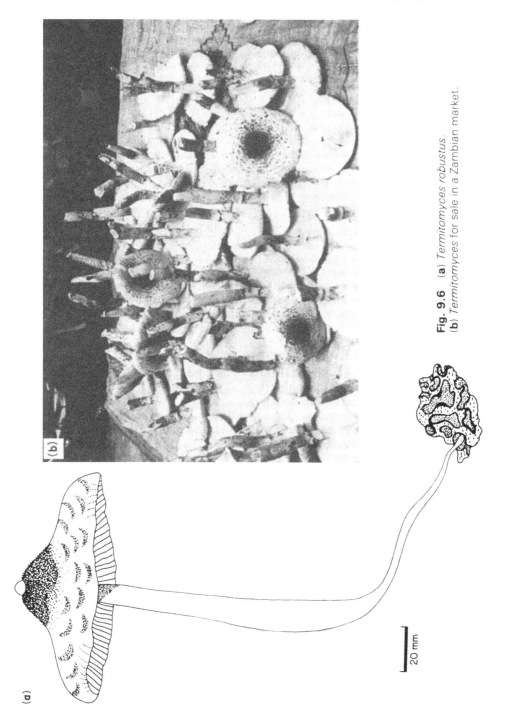

Fig. 9.6 (a) *Termitomyces robustus.*
(b) *Termitomyces* for sale in a Zambian market.

20 mm

pushed up and expand above the soil on top of a nest or around the base of a mound. The basidiocarps are esteemed as delicacies by man in all parts of the world wherever they are found. For example, *T. robustus*, which grows in large numbers in contact with termites' nests on forest soils in Nigeria, is called 'ewe' by the Yoruba. The word means 'to expand' and is given because of the way in which the globose cap expands and flattens into a plate-like disc, some 200 mm diameter. The rains, which occur at that time, are called 'eji ewe'. 'Eji' means rains and hence they are the '*T. robustus* rains'. This and five or six other species are collected to eat and can be purchased at the roadside or local markets. Other species of *Termitomyces* are smaller except *T. titanicus*, common in Zambia. It currently holds the record for size in the fungi, with caps up to 1 m diameter.

Most termites culture one fungus in their nests but the same species may associate with different species of *Termitomyces* in different areas. Other fungi known to associate with termites include the Gasteromycete *Podaxis pistillaris* which is found in the nests of some species of *Trinervitermes* and a number of species of the Ascomycete genus *Xylaria* (Fig. 9.7). *X. nigripes* is usually associated with *Odontotermes* sp. and *X. furcata* may also associate with *Macrotermes natalensis*. Like *Termitomyces* and the ant fungi, they produce their reproductive structures only when the nests are abandoned.

Numerous suggestions have been made as to why the termites grow their fungi. It has been argued that the fungi by their metabolic activities, in what are virtually miniature compost heaps, produce a controlled high humidity and an equable temperature regime in the mounds and thus help to maintain a favourable micro-climate and that this would be of an obvious advantage in particular to termites inhabiting hot, dry savannahs. The insides of the mounds of *M. natalensis* are certainly maintained at 98–99% RH even in the annual very hot dry season. The mounds are architecturally very complex with walls up to 600 mm thick, with an elaborate air duct system in their walls. It is more likely that it is the design and structure which maintain the internal microclimate.

The acquisition of fungal enzymes

The fungus comb with its mycelium and nodules is an important part of the termites' diet. The combs are eaten away from below when they reach a certain stage of maturity and are built up at the upper surface or are replaced by a new comb in the space beneath. Martin and Martin (1978, 1979) have investigated the *M. natalensis/Termitomyces* association. In this termite, ingested plant cellulosic wall material is digested in the gut. Any ingested material is finely chewed so that the cellulase enzymes can get at the cellulose, especially in the lignified walls. But the termites starve if fed only cellulosic materials such as wood with no access to the fungus.

Cellulose is a β–1,4 glucan and more than one enzyme is involved in its hydrolysis. Cellulose is first cleaved by an enzyme designated C_1 into linear glucan chains so that these can be hydrolysed by a second enzyme C_x to low molecular weight compounds such as glucose and cellobiose. The cellobiose is then hydrolysed by a β–glucosidase (see Fig. 3.6). The C_1/C_x story is a gross

(a)   (b)

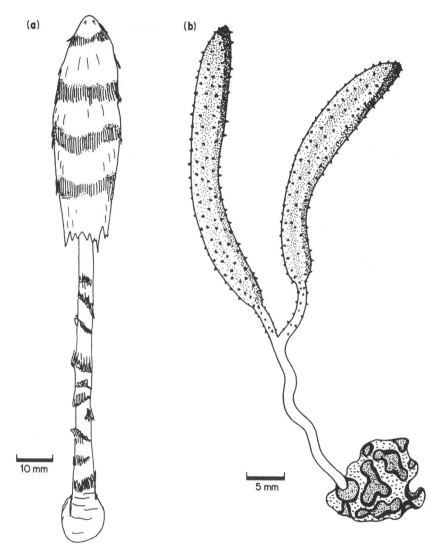

10 mm

5 mm

Fig. 9.7   Other termite fungi. (**a**) *Podaxis pistillaris*, a Gasteromycete with a stalked basidiocarp. (**b**) *Xylaria termitum*; ascocarps arising from an old piece of comb.

oversimplification and perhaps outdated. C_1 is now regarded as an enzyme complex consisting of an exo-β-glucanase, which removes cellobiose units from the end of the glucan chains, and an endo-β-glucanase with very special properties acting within the chains, but acting only on areas where the chains are highly ordered and strictly parallel, by disrupting H bonds between the chains. C_x is an endo-β-glucanase. The point that should be stressed is that the prior action of the C_1 component is required for ultimate hydrolysis by C_x.

Most celluloysis occurs in the midgut rather than the paunch of these termites and the entire set of enzymes required for cellulolysis are present there but the salivary glands and midgut epithelium only produce the C_x component and a β-glucosidase but not the vital C_1 component. Martin and Martin argue that the fungus is the source of the C_1. The termites feed on the nodules which contain high C_1 activity and they acquire the C_1 when they digest the fungus. Cellulose digestion in the midgut can thus occur only if the acquired fungal enzymes are present – thus the nutritional dependence of the termites on their fungus garden.

Such evidence as this has led Martin (1979) to put forward the very plausible hypothesis that many of the enzymes present in the guts of wood- and detritus-feeding Arthropods may be fungal enzymes acquired while feeding. The source of these may be the fungal tissue itself or simply the substrate into which the enzymes have been secreted. It also seems likely that the gut enzymes of many litter feeding invertebrates may be derived from fungal tissues consumed along with the detritus in their diet. He derived this hypothesis from the fact that wood- and detritus-feeding invertebrates routinely encounter fungal tissues during their feeding activities, that fungal tissues have been identified in the gut contents of many species and that the enzymes present in the guts of such species are common fungal enzymes. This acquisition of external sources of digestive enzymes can be seen as windfall of mycophagy and as an alternative to the maintenance of a culture of extra-cellular endosymbionts in the gut to produce the necessary enzymes. As a group, the termites have adopted both strategies.

The faecal pellets which the termites use to culture the fungus contain proportionally more lignin than the ingested wood because much, but not all, of the cellulose has been digested as it passes through the gut. It is assumed that the fungus is breaking down predominantly lignin in the comb material. It appears that the termites do not acquire fungal ligninolytic enzymes. As lignin degradation is an oxidation rather than an hydrolysis these enzymes in any case may not function in the gut because of the low oxygen tension. In addition some of the phenolic degradation products may be harmful or even poisonous to the termite.

ENDOSYMBIOTIC ASSOCIATIONS

Many insects, especially Coleoptera and Homoptera, possess intracellular endosymbiotic fungi. These are housed in specialized cells, mycetocytes, aggregated into special organs, mycetomes. The fungi are all yeasts or yeast-like and referred to such genera as *Candida* and *Torulopsis*; they appear to provide insects with essential nutrients which are lacking from their diet. The fungus probably benefits by being provided with a suitable environment, free of competition, for its development and by being dispersed by special transmission procedures without the fungus having to undergo a free-living

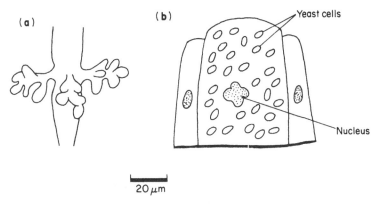

Fig. 9.8 *Sitodrepa panicea*, the drug store beetle. (**a**) Lobed, blind mycetomes at the anterior of the mid-gut. (**b**) A specialized cell, a mycetocyte, containing yeast cells from the lining of the mycetome.

phase. Many insects of the wood-eating group have become nutritional specialists. *Sitodrepa panicea,* the drug store beetle, feeds on carbohydrate-rich substances such as grain and in particular flour. It houses a yeast in very much enlarged cells, lining four voluminous and lobed blind sacs situated at the anterior end of the mid-gut (Fig. 9.8). It is generally believed that the beetles rely upon the yeast to provide essential amino acids, vitamins, especially those of the B group, and sterols which are deficient in their diet. Larvae can be grown free of the yeast. They show a marked retardation of growth. Few even reach the adult stage. They require B vitamins, sterols and all the essential amino acids whereas normal larvae require only arginine, leucine and threonine. Transmission of the fungus is ensured by other fungus-containing organs, intersegmental tubules which connect up with the ovipositor sheath and fungus-filled sacs at the base of the ovipositor. As the eggs are being deposited they are smeared with fungal cells from these. The larvae become infected as they hatch by eating the contaminated egg cases.

10

Fungi as parasitic symbionts of plants – an introduction

Parasitism is a mode of life adopted by many fungi. A number of examples have already been mentioned. In this mode as chemoheterotrophs, they derive their organic requirements from living plant or animal tissues. In so doing they come into an association with their host and share a common life. A symbiosis is thus established. In such situations the fungus is usually antagonistic to its host. It benefits and the host is harmed. The fungus is referred to as a parasite and the relationship is called a parasitic symbiosis.

Modes of nutrition

The degree of dependence of the fungus on the association with its host varies enormously. In some it is absolutely necessary. These fungi need to be associated with the living cells of the host in order to grow and reproduce under natural conditions. They are obligate parasites in the ecological sense. They are normally unable to grow as saprotrophs in nature because they lack the ability to compete successfully in mixed saprotrophic populations. Many cannot even be grown successfully or at all in culture. In others the association is not an absolute necessity. They can grow as saprotrophs away from their hosts. They are facultative parasites.

Parasitic fungi can also be characterized by their modes of nutrition. Some can derive their nutrients only from living cells of the host. They are called biotrophs – 'living feeders'. If the host or the part of it occupied by the fungus dies, the fungus also dies or goes into a resting state. Away from their hosts, biotrophs are encountered only as dispersal or dormant spores. Others derive their requirements from dead cells which they have killed. As parasites they continually create adequate substrate by killing host cells in advance of penetration. Since they derive their nutrients from dead host cells they are called necrotrophs – 'dead feeders'. In this respect they differ from saprotrophs only in that they kill host cells. This is also reflected in the fact that many necrotrophs are facultative parasites which can live as saprotrophs on the non-living organic components in their environment. Indeed the mode of nutrition of a fungus very much depends upon the conditions in which it finds itself. A single fungus at different times and under different circumstances may show all three modes.

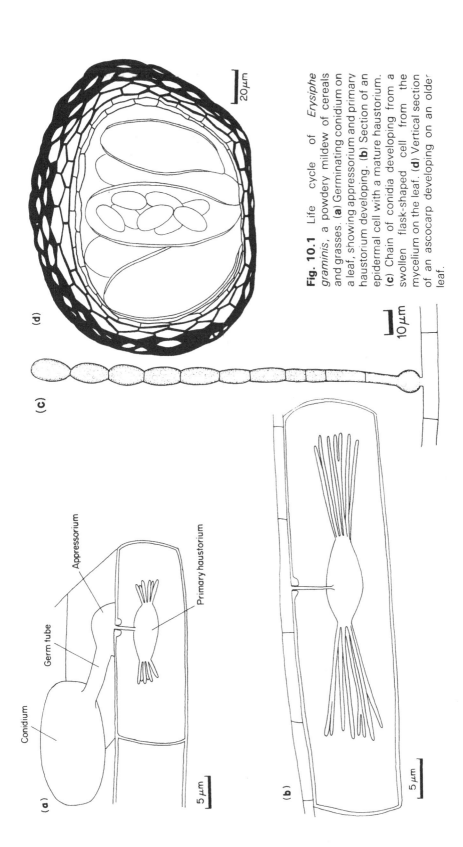

Fig. 10.1 Life cycle of *Erysiphe graminis*, a powdery mildew of cereals and grasses. (**a**) Germinating conidium on a leaf, showing appressorium and primary haustorium developing. (**b**) Section of an epidermal cell with a mature haustorium. (**c**) Chain of conidia developing from a swollen flask-shaped cell from the mycelium on the leaf. (**d**) Vertical section of an ascocarp developing on an older leaf.

Conidium

Germ tube

Appressorium

Primary haustorium

5 μm

(a)

5 μm

(b)

(c)

10 μm

(d)

20 μm

Obligate biotrophs

Obligate plant parasites occur in three orders each from different sub-divisions of the fungi. These are the Peronosporales or downy mildews in the Mastigomycotina, the Erysiphales or powdery mildews in the Ascomycotina and the Uredinales or rusts in the Basidiomycotina. From these three orders, only a few of the rusts have as yet been cultured in the absence of living host cells. They are nutritionally very exacting and are all biotrophs.

Infection

These arrive on the surfaces of susceptible hosts in the form of wind-blown spores. These germinate and penetrate either via natural openings such as stomata or directly through the cuticle and cell wall. The germ tubes of the urediospores of many rusts extend until they encounter stomata, each forms an appressorium and enters via the pore. The germ tubes of the conidia of *Erysiphe* similarly produce an appressorium by the tip swelling to form an ovoid structure which increases the area of contact with an attachment to the

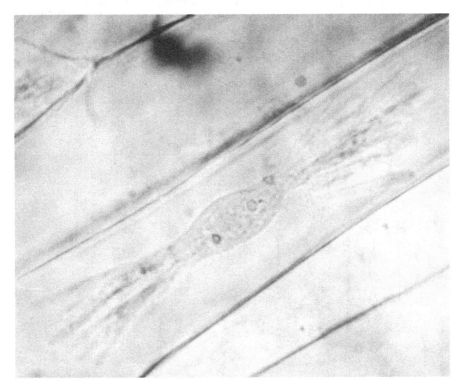

Fig. 10.2   Haustorium of *Erysiphe graminis*.

host surface. From the underside of the appressorium a very narrow hyphal infection peg penetrates the cuticle and cell wall (Figs. 10.1 and 10.2). Penetration in this case is partly enzymatic and partly mechanical. In *Erysiphe* differential staining techniques have revealed some dissolution of both the cuticle and cellulose of the wall immediately around infection pegs suggesting that hydrolytic enzymes are involved in penetration. In others, especially several facultative necrotrophs, such as species of *Botrytis*, penetration may be purely mechanical. The infection peg of such fungi can penetrate thin metal films applied to leaf surfaces. Appressorial formation further illustrates the versatility of fungal hyphae in terms of penetrating and permeating tissues. The firmly attached appressorium prevents lift off from the surface allowing the infection peg to apply considerable mechanical pressure as it bores through the cuticle and wall.

Haustoria

After infection a swelling or vesicle usually arises from the tip of the infection peg and in most biotrophs hyphal branches arise from this and grow between the host cells (Figs. 10.3 and 10.4). They are intercellular and do not penetrate the host cells. The cells are penetrated by specialized hyphal branches of determinate growth called haustoria. The walls are penetrated by very narrow hyphal branches which balloon out to form haustoria of a variety of shapes (Fig. 10.5). These invaginate but do not penetrate the plasmalemma. The host-fungus interface is thus fungal cell membrane, fungal cell wall and host plasmalemma. Very often, in addition, there may be a distinct layer of amorphous material, the extrahaustorial matrix between fungal wall and host plasmalemma (Fig. 10.6). The base of the neck of the haustorium may also be surrounded by a collar-like sheath of host-derived callose. Haustoria are nucleated and rich in organelles such as mitochondria, ribosomes and vesicles, indicating that they are sites of high metabolic activity. The presumption is that they are the organs responsible for the absorption of major nutrients from the host cells. They most certainly provide a surface for close contact between the host and fungus and increase the surface area available for any such absorption. Nevertheless data to support this view are not easy to find. There is ample evidence that substances do move from host to fungus but the very extensive intercellular hyphal system could be responsible for this. Some evidence for the nutritional functions of haustoria come from studies on wheat leaves infected with *Erysiphe graminis*. In the Erysiphales, unlike in the other two orders of obligate biotrophs, all the hyphae are outside on the surface of the host and only haustoria penetrate the cells and then only the epidermal cells. In infected wheat leaves, the surface hyphae start to grow only after the first haustorium arising from a germinating conidium has developed to a particular stage. Also labelled compounds supplied to the host move to the fungus only after the primary haustoria have also completed their development. In biotrophs with intercellular hyphae, the hyphae themselves provide a very large surface area and large quantities

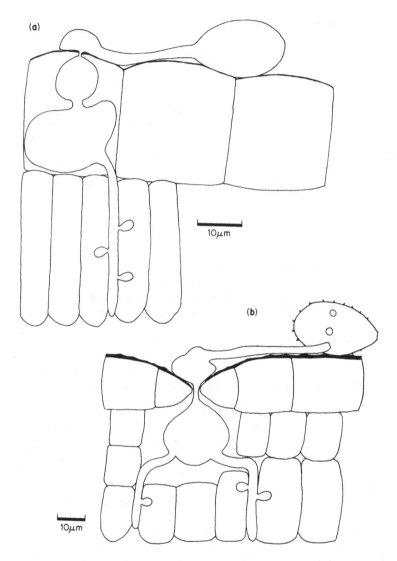

Fig. 10.3 Penetration structures in biotrophs. (**a**) *Bremia lactucae*. (**b**) *Puccinia graminis*.

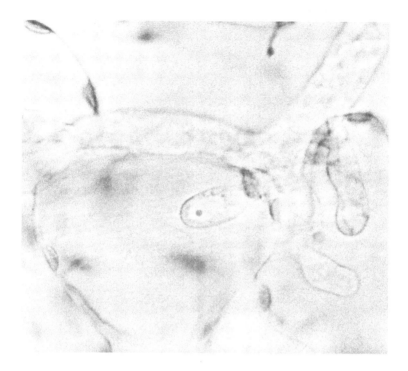

Fig. 10.4   Intercellular hypha and haustoria of *Bremia lactucae.*

of nutrients may move directly into these through the host cell walls as they do, for instance, in sheathing mycorrhizal associations (Chap. 7).

Features of biotrophy

As the fungi are dependent upon living host tissues to complete their development, they usually cause minimal tissue damage, initially at least. There is little or no immediate cell death and they do not produce large quantities of extracellular lytic enzymes or toxins. They do produce these but their production is controlled. There are a number of possibilities. It might be that their production is switched on when they are required and switched off when they are not. For example, wall-degrading enzymes are produced only when wall penetration occurs. It could also be that they are only ever produced in the smallest of quantities. If they were also hyphal wall-bound, they would not hydrolyse the walls beyond the sites of infection. Similarly, toxins might be produced in such small quantities that they increase membrane permeability to a degree without bringing about disruption.

The success of these fungi as parasites largely depends upon the extent to which they can extract nutrients for their own growth before the host is

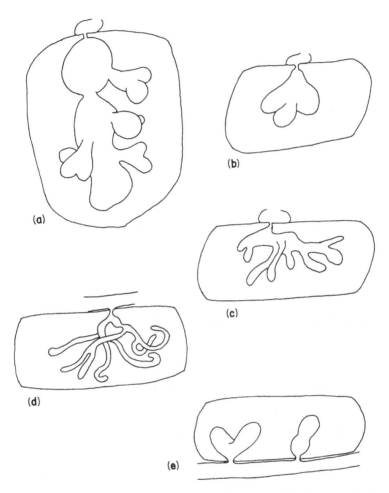

Fig. 10.5 Haustorial shapes. (**a**) *Hemileia holstii*. (**b**) *H. vastatrix*. (**c**) *Puccinia purpurea*. (**d**) *Peronospora calotheca*. (**e**) *Peronospora parasitica*.

sufficiently damaged to decrease the synthesis of these nutrients. Infection of part of the host creates a sink into which a variety of host-produced nutrients are rapidly removed. This induces the translocation of host metabolites into infected regions. It may well be that one function of haustoria is to introduce metabolites into the host which lead to such and other specific modifications of the host's metabolism in favour of the biotroph.

Perhaps the most marked feature of plants infected by obligate biotrophic fungi is a 2–4-fold increase in the respiration rate. By far the greater part of this is due to stimulation of the host's metabolism by the biotroph. The fungus itself directly contributes very little to this increase. In host tissues infected by rusts and powdery mildews, there is not only an increase in

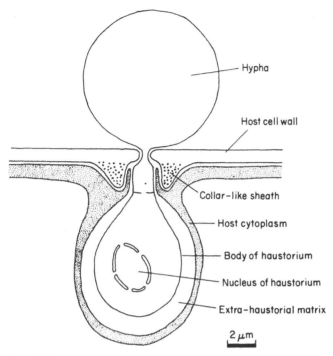

Host cell wall

Hypha

Collar-like sheath

Host cytoplasm

Body of haustorium

Nucleus of haustorium

Extra-haustorial matrix

2 μm

Fig. 10.6 Diagrammatic representation of a section of an haustorium of *Erysiphe graminis* to show the extrahaustorial matrix.

respiration rate but also a shift away from the major pathway of glycolysis and the tricarboxylic acid cycle to one involving the pentose phosphate pathway. This would not only provide more pentoses for the synthesis of nucleic acids, which in turn would affect protein synthesis, but also more NADPH which could be used in the synthesis of sugar alcohols and lipids which are all-important in the metabolism of the fungi.

Overall there is a tendency for a general increase in the biosynthetic processes of the host but the biosynthesis is of materials which the fungus requires. For instance there is increased synthesis of RNA. It has been suggested that once rust spores have germinated they may not be capable of further net protein synthesis and that for a successful infection to be established the biotroph must induce the host to produce proteins which it can utilize. In the early stages of infection of wheat leaves by rusts, the rate of RNA synthesis may be about three times that in healthy leaves suggesting enhanced protein synthesis.

Labelled amino acids are also incorporated about three times as rapidly into infected tissues as they are into healthy tissues and are rapidly built up into fungal proteins. Infection sites thus act as sinks for amino acids but the successful establishment and development of biotrophs within the host tissue also entails that there should be an adequate supply of carbohydrates from the host. Most biotrophs not only induce the diversion of host photosynthate towards infection sites but also rapidly convert it into substances which the host cannot utilize. This not only ensures a concentration gradient providing a continuous supply for the fungus but it also ensures that they are not re-utilized by the host. As in sheathing mycorrhizal associations, host photosynthate, especially sucrose, is rapidly converted to trehalose, glycogen, arabitol or mannitol. Starch may also be preferentially deposited in host chloroplasts in and around infection sites. It disappears as the fungus begins to reproduce. Both its deposition and disappearance may be fungus-induced. This diversion and conversion of host nutrients by the fungus is the major cause of reduced growth and yield of cereals infected by powdery mildews and rusts. Infected leaves no longer become effective exporters of photosynthate. Healthy leaves of wheat export over 80% of the photosynthate produced whereas rust-infected leaves export less than 5% and in very severe infection may even import photosynthate. In cereals infected by *Erysiphe graminis*, an early attack on the first formed leaves, which usually export to the roots, restricts root development and as a consequence reduces the number of fertile stems per plant. A later attack on the upper leaves which normally export to the inflorescences or ears causes poor filling and shrivelling of the grain. In some years, *E. graminis* reduces yield of barley in Great Britain by 10–25%. As infection by biotrophs progresses the rate of photosynthesis in infected leaves usually declines; the decline is often accompanied by a decrease in chlorophyll content although this may occur later than the decline in photosynthesis. Chloroplasts in infected leaves are also usually smaller and often show signs of membrane damage. The loss of chlorophyll results in chlorosis and premature senescence.

In infections caused by powdery mildews, the chlorophyll content of the leaves initially falls, resulting in chlorosis, but the pigments are then resynthesized in areas adjacent to infections. In many rust infections, tissues adjacent to lesions remain green and healthy, producing 'green islands' in otherwise chlorotic leaves. The retention of chlorophyll in these islands is another indication of the ability of the biotrophic fungus to control the metabolism of the host. The photosynthate produced in these islands may help to maintain the fungus until after it has reproduced but the 'green islands' may also be associated with the diversion of photosynthate. Cytokinins spotted onto uninfected senescing leaves often produce 'green islands' and the host also translocates towards these. Increased levels of growth substances are characteristic of biotrophic infections. High levels of auxin (IAA) are found in many rust-infected tissues. In rust-infected wheat plants ten days after infection there is a 24-fold increase in IAA content. It is not known whether the fungus produces IAA in the infected host or whether it stimulates the host to produce it. Another alternative would be for the fungus to interfere with the process by which IAA is degraded in the host by, for instance, bringing about a decrease in the activity of IAA oxidase. The precise role of these growth substances in influencing the metabolism of the host to favour and maintain the biotrophs is not clear but a considerable volume of circumstantial evidence indicates that their role is an all important one, especially in mediating the increases in metabolism and the diversion of nutrients.

Infection by biotrophs while causing minimal tissue damage, often, but not invariably, leads to morphological disturbances in the host. These vary from uniformly elongated internodes to markedly curled and distorted stems, as in the white blister rust of crucifers, to markedly contorted leaves, as in peach leaf curl, slightly swollen to grossly and grotesquely enlarged roots, as in club root, and the stimulation of the growth of dormant buds to produce masses of spindly twigs, as in witches' brooms. Such growth responses would be expected from the known levels of growth substances in infected tissues and are a direct result of these levels but are probably purely incidental. Higher levels of growth substances also lead to morphological changes in sheathing mycorrhizal associations. The branching pattern of the host root system is changed on infection.

In obligate biotrophic associations a relatively well-balanced relationship between the two partners is established initially since the fungi are physiologically highly specialized in being dependent upon living host tissues to complete their development. This is also reflected in the fact that most have limited host ranges. For example, the morphological species *Puccinia graminis*, the black stem rust of cereals and other grasses, can be divided into six pathogenic entities – formae speciales *tritici*, *secalis*, *avenae*, *agrostidis*, *poae* and *phlei-pratensis* according to their pathogenic adaptation to particular genera such as *Triticum* and *Secale*. They are named according to the host principally parasitized and crosses between them were usually infertile. Where man has developed genetically stable varieties or cultivars of the host as in wheat, it is possible to subdivide each forma specialis. For example, not all the strains within *P. graminis* f. sp. *tritici* are capable of infecting equally all the known cultivars of *Triticum*. By using a number of different cultivars of wheat with different degrees of resistance to the fungus as a so-called differential host series and assessing quantitatively the degree of infection by particular strains, physiologic races can be recognized. Well over 200 such races have been recognized in *P. graminis* f. sp. *tritici*. These are artificial entities in a sense that their recognition is quite arbitrary since the number depends upon the number of differential hosts used and the classes of pathenogenicity reactions recognized.

Necrotrophs

In marked contrast to these obligate biotrophs, most facultative necrotrophs, in addition to living parasitically, do have a free-living phase and can be grown on simple culture media. Many, the less specialized parasites, are not only able to grow on the host after they have killed it but they can also grow successfully and indefinitely as competitive saprotrophs on other dead substrates. This is not true of all necrotrophs as with increasing specialization towards parasitism there is a tendency towards a progressive loss of competitive saprotrophicability.

Secretion of extracellular enzymes

Necrotrophs usually kill host cells rapidly and before they penetrate them by the secretion of extracellular enzymes or toxins or both. It is these which are largely responsible for the disease symptoms seen. Their effects vary from localized necrosis to massive tissue destruction and disintegration. Many cause diseases in which the main symptom is a rapidly spreading soft, watery rot of parenchymatous tissues. This is well seen in such diseases as the soft rot of fruit, such as apples, caused by the conidial *Penicillium expansum*, or marrows, caused by the Zygomycete *Rhizopus stolonifer* and seedling diseases such as pre-emergence killing and 'damping-off', caused by the Oomycete *Pythium debaryanum*. On infection the hyphae again penetrate between the cells. No haustoria are formed but the hyphal tips secrete copious amounts of a wide variety of pectolytic enzymes which attack the anhydrogalacturonic acid units making up the pectic polymers. This brings about a maceration of the parenchymatous tissues as the cells separate along the middle lamella because of the breakdown of the pectic substances which cement them together. The tissues become very soft, watery and lack coherence.

In pectic substances a proportion of the anhydrogalacturonic units are esterified with methanol (Fig. 10.7). Apart from pectin methyl esterase

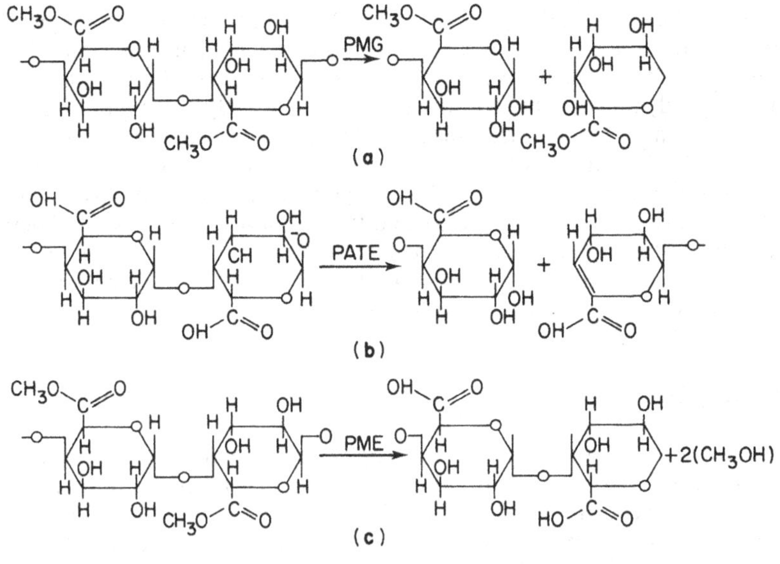

Fig. 10.7 Pectic substances and pectolytic enzymes. (a) Action of polymethylgalacturonase on pectin (a polymer of methylated galacturonic acid). (b) Action of pectic acid *trans*-eliminase on pectic acid (a polymer of galacturonic acid). (c) Action of pectin methylesterase on pectin. (After Goodman, Kiraly and Zaitlin, 1967.)

which acts upon these ester bonds and thus increases the solubility of the polymer by producing more free carboxyl groups, most other pectolytic enzymes are chain splitting ones which act upon the α-1,4 glycosidic linkages between adjacent residues. Some, such as polygalacturonases, rupture the bonds by hydrolysis. Others, such as pectin *trans*-eliminases, act by trans-elimination of a proton from C5 of one residue to give an unsaturated bond between C4 and C5 of the reaction product (Fig. 10.7). In both types of chain splitting enzymes, there are those which attack chains terminally (exo-) and those which attack chains at random (endo-), for example exo-polygalacturonase and endo-polygalacturonase. A further division can be made into those which preferentially attack esterified parts of the chain and those which preferentially attack non-esterified parts. Only pectolytic enzymes play a key role in the early and critical stages of soft-rotting. Enzymes which act upon other components of the cell wall such as cellulose and hemicelluloses are not produced initially. The pectolytic enzymes have another very important effect in that they bring about a marked increase in the permeability of the protoplasts. The cells leak so much that they lose turgor and they often die before they are separated. The enzymes diffuse out from the hyphal tips so that the cells are killed in advance of the hyphae. The fungi utilize both the residues from the hydrolysis of the pectic substances and the solutes which pour out of the affected cells.

Production of toxins

In other diseases caused by necrotrophs, the cells are killed but the tissues, although discoloured, remain dry rather than watery and there is no cell separation. Moreover the lesions, unlike rots, are slow in spreading and discrete. In these cases the fungi produce low molecular weight and heat-stable toxins which kill protoplasts at very low concentrations. These can be demonstrated in some leaf spot diseases such as the early blight of tomatoes and potatoes caused by the conidial *Alternaria solani*. This fungus causes small black necrotic lesions on the leaves between the veins. Infected leaves become chlorotic and senesce prematurely. When the lower leaves are heavily infected, the upper leaves soon appear chlorotic. The toxin moves about the plant. A crystalline dibasic acid with the formula $C_{21}H_{30}O_8$ has been obtained from culture filtrates of *A. solani* and has been called alternaric acid (Fig. 10.8). When healthy cuttings from potato shoots are placed with their stems in solutions containing as little as 20 p.p.m. of the acid, typical lesions occur in the leaves after 2–3 days. It is characteristic of toxins that they are active in very low concentrations and that they are mobile within the plant. Alternaric acid is non-specific in that it will affect a variety of plants that are not parasitized by the fungus. Other toxins are host specific showing the same specificity as the parasite itself. The conidial *Helminthosporium victoriae* causes a devastating seedling blight of oats in the USA. It is particularly virulent on one oat cultivar called Victoria where it causes a wilt, necrosis at the base of the stem and reddish necrotic lesions on the leaves. It produces a

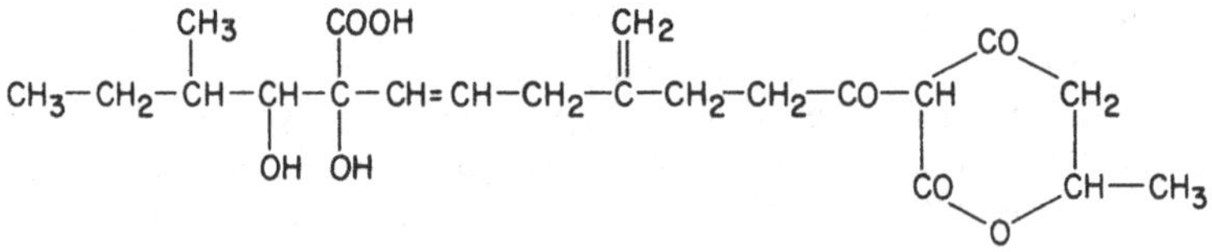

CH₃—CH₂—CH—CH—C—CH=CH—CH₂—C—CH₂-CH₂—CO—CH   CH₂

(structure with CH₃, COOH, CH₂, OH, OH, CO, CO, O, CH—CH₃)

Alternaric acid

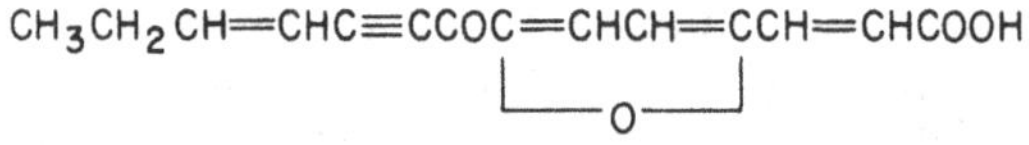

CH₃CH₂CH=CHC≡CCOC=CHCH=CCH=CHCOOH
└─O─┘

Wyerone acid

Fig. 10.8   The phytotoxin, alternaric acid, and the phytoalexin, wyerone acid.

toxin, a peptide linked to a tricyclic amine, named victorin. Resistant oat cultivars are not affected by it but susceptible ones, such as Victoria, are very sensitive to it. Not only does the toxin show the same specificity as the fungus itself but only toxin producing strains of the fungus cause disease symptoms. Resistance to a necrotrophic parasite is very often based upon insensitivity to a specific toxin or the ability to degrade it as soon as it is produced.

In toxin-producing necrotrophs the host tissues are again first killed and then exploited by the hyphae. Many toxins are also known to affect membrane permeability as do pectolytic enzymes. The rapid killing of host cells by the production of enzymes or toxins or both is a major characteristic of necrotrophs. There is a complete lack of any form of balanced relationship between host and parasite and further, unlike biotrophs, most necrotrophs have a wide host range. However in many cases it is not clear to what extent they can and do grow away from their hosts as competitive saprotrophs. Some necrotrophs are clearly obligate ones. This is particularly true of several representatives of a large group of white-rot wood-decay fungi, such as *Armillaria mellea*, which infect the roots of trees and shrubs and fungi, such as *Botrytis fabae*, which causes the chocolate spot disease of leaves and pods of broad beans (*Vicia faba*).

Armillaria mellea as an obligate necrotroph

A. mellea has been recorded as a parasite of a very wide range of dicotyledonous and coniferous trees. It is probably the most widespread and damaging root parasite of trees. It also attacks shrubs and occasionally monocotyledons, such as sugarcane, and herbaceous plants, such as potatoes, where it was formerly present on tree roots. It may rapidly kill its host by invading the cambium of the root system and spreading back to the base of the stem. The mycelium forms thick, creamy-white fan-like sheets under the bark and girdles the cambium. The tree then dies suddenly. After killing the tree, it colonizes the wood causing a white-rot with an intricate pattern of black lines, zone lines, running through it. Eventually the tree is wind-blown or falls and the fungus colonizes the whole trunk. Alternatively it may cause a butt rot without attacking the living tissues of the tree. A central conical zone of decay extends 1-2 metres upwards from the base of the stem. However the result is the same. The tree is wind-blown and the fungus colonizes the whole trunk. It persists for several years, even 10 or more, as an active saprotroph degrading the wood of the trunk. It has no free-living saprotrophic phase but is restricted to the dead tissues which it has killed. As these are usually very durable substrates it remains in the vegetative state. Eventually it produces basidiocarps and its air-borne basidiospores may infect recent wounds on other hosts at a distance. Alternatively healthy roots may come in contact with already colonized roots in the soil but once an infection site has been established the major means of local spread is by the initiation and growth of rhizomorphs from the mycelial fans. These grow out through the soil away from the dead host to infect another, perhaps metres away, obtaining the necessary nutrients directly from the dead host rather than from dead organic matter in the soil. In this case the massive tissue destruction characteristic of necrotrophy is not obvious immediately but occurs over a long period of time and is brought about by the production of cellulolytic and ligninolytic enzymes rather than pectolytic ones.

Botrytis fabae as an obligate necrotroph

Other obligate necrotrophs, such as *Botrytis fabae*, initially cause very local-ized tissue damage such as spots on leaves but the physiological state of the host and environmental conditions markedly influence the extent to which the parasite spreads further. When a wind-dispersed conidium of *B. fabae* lands on a broad bean leaf, it germinates, if moisture is present, by forming an appressorium from which a thin peg-like penetration hypha bores through the cuticle and cell wall. Once inside it swells and forms a short hypha often less than 100 μm long (Fig. 10.9). Within 24 h a discrete limited lesion – the chocolate spot – develops. This is without any further growth of the fungus. Some 20–50 host cells are killed around the fungus and, although there is little evidence of tissue breakdown, there is intense tissue browning. This is called the 'non-aggressive' phase of the disease. It appears that once inside the host the fungus begins to produce pectolytic enzymes. The fungus would rely upon these to provide nutrients for further growth from the hydrolysis of pectic substances and leakages of solutes from the cells. A host phenolase is activated by the action of the pectolytic enzymes and it oxidizes phenols which also leak from dying cells. The oxidized products are responsible for the brown coloration and they also inhibit the fungal pectolytic enzymes so that growth of the fungus is prevented. Other substances toxic to the fungus may also be produced. Many plants have been shown to produce a specific chemical on infection which prevents further growth of the fungus. These are called phytoalexins – 'plant warding off' compounds. When *B. fabae* infects broad beans a phytoalexin called wyerone acid (Fig. 10.8) is produced and it

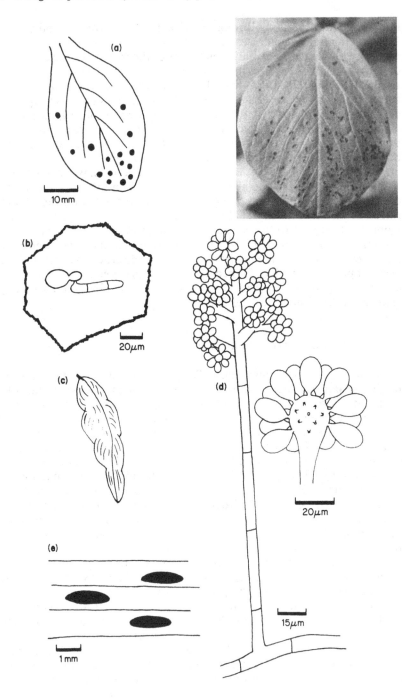

appears that in the 'non-aggressive' phase sufficient is produced to prevent further growth.

The 'aggressive' phase only develops either on older senescing leaves or under any environmental conditions which are unfavourable to the host such as poor phosphate supply, extensive waterlogging of the soil or, in particular, very high humidities which greatly favour the fungus. The fungus then behaves as a typical necrotroph. The main symptom is a rapid and progressive rot of the leaves followed by that of the parenchymatous parts of the stems. In this phase oxidation of the phenolics cannot keep pace with the production of pectolytic enzymes and the wyerone acid is inactivated or degraded. Conidia are soon produced and serve to spread the disease. Later other extracellular enzymes, such as cellulase, are produced which both facilitate growth of the fungus through the dead tissues and solubilize the insoluble components of the cells so that they can be absorbed by the fungus. When the supply of available carbon- and especially nitrogen-containing substances begins to fall, sclerotia are produced. These enable the fungus to survive in the absence of living host plants and when it has exhausted the nutrient supplies within any tissues which it has killed. Such survival structures are very important to necrotrophs where the host tissues, once killed, are relatively ephemeral and contain only a limited supply of nutrients and also where the host is seasonal in occurrence. In *B. fabae* the sclerotia overwinter on the dead stems and in the spring produce tufts of conidiophores; the conidia then re-infect the next season's crop of broad beans. *B. fabae*, like *A. mellea*, has no free-living saprotrophic phase, it again being restricted to host cells which it has killed.

Pythium debaryanum as a facultative necrotroph

Other necrotrophs are much less specialized than *B. fabae* and are facultative ones. *Pythium debaryanum* is a good example. It is an aggressive necrotroph in the parasitic phase but otherwise lives saprotrophically. It causes a number of seedling diseases such as pre-emergence killing and 'damping-off' and, like *B. fabae*, illustrates very well the importance of environmental factors in disease expression. *P. debaryanum* has a very wide host range. It has been recorded as attacking over 220 hosts. It is an exceedingly common soil saprotroph especially as a primary colonizer of plant debris. Under suitable conditions such as high density planting and in waterlogged or poorly drained soils, it becomes an aggressive parasite of seedlings. In pre-emergence killing, the seedlings fail to emerge and bare patches arise in seed beds. The seeds germinate but the radicles are killed. 'Damping-off' is a post-emergence

Fig. 10.9 (*Left*) Life cycle of *Botrytis fabae*, the cause of chocolate spot of broad beans. (**a**) Chocolate spots on a leaflet of broad bean, the non-aggressive phase. (**b**) Conidium penetrating an epidermal cell. (**c**) Shrivelled and blackened leaflet, the aggressive phase. (**d**) Conidiophore and blastoconidia. (**e**) Sclerotia on an old stem of broad bean.

condition, the term being used as it is usually associated with over-watering. Seedlings emerge and after a few days their hypocotyls or plumules lose turgor, become water-soaked and collapse. One or more hyphae penetrate at or around soil level and grow between the parenchymatous cortical cells producing copious amounts of pectolytic enzymes which again separate the cells and cause them to leak and die.

P. debaryanum, as an Oomycete, produces both zoospores and oospores but also intercalary chlamydospores (Fig. 10.10). Unlike *A. mellea*, its mycelium actually grows through the soil. Exudates, especially sugars and amino acids, from roots and hypocotyls are important in infection. They affect the fungus in a number of ways. They may stimulate resting oospores to germinate or they may induce directional growth of hyphae or directional movement of zoospores or their uptake and utilization may increase the capacity to infect. 'Damping-off' is most severe when environmental factors favour the growth of the fungus but not the host. External conditions may affect host and parasite individually or together.

The soil moisture regime is a most important factor. Disease severity usually increases with increase in soil moisture. High moisture levels favour zoospore production and free water is essential for their dispersal. High moisture levels usually mean poor gas exchange. *P. debaryanum* is unusually tolerant of low oxygen and high carbon dioxide concentrations and under such conditions it is at an advantage as its competitors are suppressed. The parasitic activity of *P. debaryanum* in very wet soils can be attributed as much to its high tolerance of poor gas exchange as to any positive need for water. Similarly if seedlings are overcrowded and over-watered, the fungus can grow over the surface of the soil, without being subjected to drying out, rather than grow through it. This is not only a path of less resistance but aeration is better and competitors are fewer. Waterlogging of roots, like extremes of temperature, causes more exudates to be produced and this again favours the fungus. The cortical tissues of the roots and hypocotyls also take longer to mature. As resistance to such fungi develops with maturity, they thus remain susceptible to attack for a longer time. The facts that such fungi are soil-borne, possess the ability to grow saprotrophically through the soil and have such a wide host range present problems in terms of control but disease can be minimized by careful manipulation of the environment in terms of seed bed management. As in many cases, environmental conditions may determine whether, at any one particular moment in time, a fungus is pathogenic or not.

Nutritional relationships of *Venturia inaequalis*

Biotrophy and necrotrophy have been considered as two quite distinct modes of nutrition. It is better to regard them as two extremes with a whole spectrum of nutritional relationships between. A number of parasitic fungi exhibit a biotrophic phase followed by a necrotrophic one as infection of the host develops. *Venturia inaequalis*, which causes apple scab, is an example. Apple

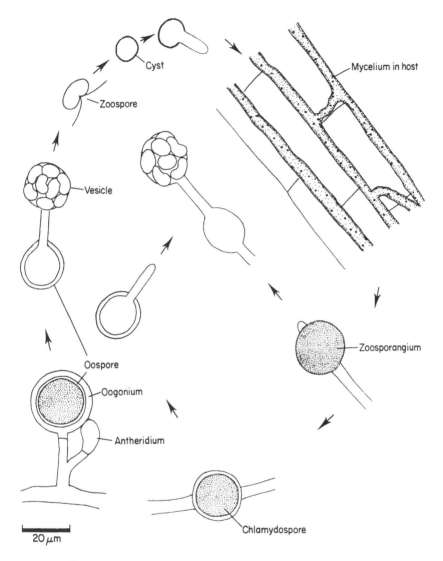

Fig. 10.10 Life cycle of *Pythium debaryanum*, the cause of seedling diseases such as pre-emergence killing and 'damping-off'

scab is one of the most serious diseases which apple growers have to contend with. In the spring, young leaves are infected by ascospores released from ascocarps which have developed over the winter in dead fallen leaves. The germ tube arising from a germinating ascospore produces an appressorium from which a hypha penetrates the cuticle. Hyphal branches from this then spread between the cuticle and the outer walls of the epidermal cells (Fig. 10.11). They fuse laterally to form flat, wide strands which develop in

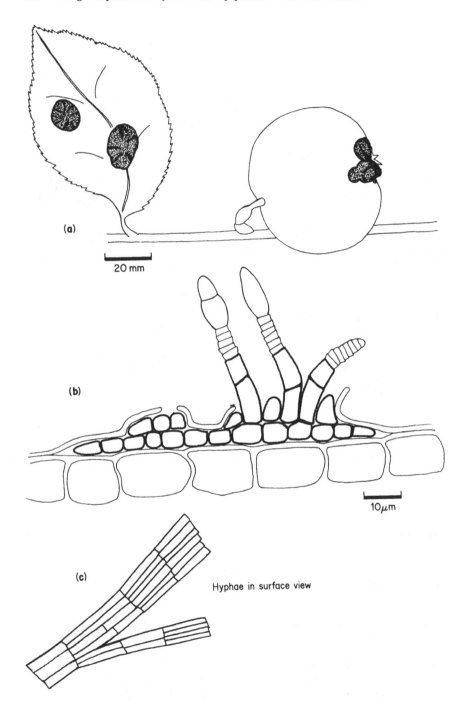

(a)

20 mm

(b)

10 μm

(c)

Hyphae in surface view

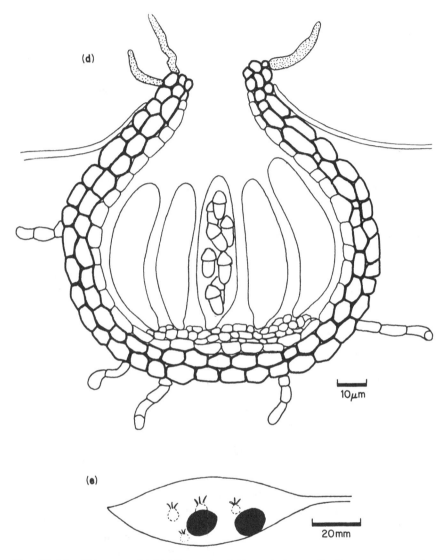

Fig. 10.11 (*Above and left*) Life cycle of *Venturia inequalis*, the cause of apple scab. (*Left*) (**a**) Scab lesions on a leaf and on an apple. (**b**) Conidiophores arising from hyphae growing between the cuticle and the epidermal cells. (**c**) Surface view of part of the strands of laterally fused hyphae beneath the cuticle. (*Above*) (**d**) Vertical section through a pseudothecium. (**e**) Ascocarps on an overwintering leaf.

the form of a plate under the cuticle. There may be no obvious signs of tissue damage for up to 2-3 weeks. Towards the end of this time the cuticle is raised and ruptured and conidia begin to be produced from short conidiophores arising from the sub-cuticular mycelium. These are rain-splashed to adjacent leaves and serve to spread the disease. There is no cellular penetration but up to 40% of the cuticle may be degraded in heavily infected leaves. Haustoria are not formed. Although the fungus is outside the host cells, it induces the host to translocate metabolites to infection sites. These are the major source of nutrients for the fungus. The export of photosynthate from infected leaves is also reduced. Thus the sub-cuticular phase is biotrophic. Eventually the epidermal cells under a lesion become more and more vacuolated and discoloured by the deposition of brown pigments. Eventually they collapse and die. When they are dead, conidial production ceases. At this stage the fungus is necrotrophic but its growth is still restricted.

On senescence prior to leaf-fall, the fungus grows out into the leaf meso-phyll and, once the leaves have died and fallen, the mycelium may permeate the whole leaf. This saprotrophic spread however is limited to the one infected leaf. It does not spread to others. The extent of its spread through the leaf is determined by the degree of competition which it encounters from the common primary saprotrophs, which would also be colonizing the leaves at that time, and leaf-inhabiting bacteria. Ascocarp initials develop in the scab lesions and in other colonized parts of the leaves in late autumn. They develop to maturity only if given a cold treatment such as they naturally experience in the winter months. In the laboratory pseudothecia, the ascocarps, are formed on leaf discs only after 4-5 months at 4-8°C. By the spring the pseudothecia are mature. Ascospores begin to be released as the new season's leaves unfold and continue to do so for a few hours after periods of thorough wetting by rain for the next 2-3 months.

A number of parasitic Ascomycotina have this saprotrophic phase during which the perfect or teleomorphic state develops and which serves to ensure survival of the fungus during periods when the host itself is dormant. Unlike the strictly obligate biotroph, it is during this phase that the fungus is subjected to competition from other heterotrophs. The significance of competition from the common primary saprotrophs and other competitive or antagonistic micro-organisms, especially bacteria, is seen in the degree of control of apple scab achieved by spraying infected trees just before leaf-fall, to achieve a good cover, with a 5% solution of urea. Not only does the addit-ional nitrogen in itself inhibit ascocarp development but more particularly it stimulates the growth of common primary saprotrophs such as species of *Cladosporium* and *Alternaria*. These achieve two effects. They either so rapidly decompose the leaves that they disappear before the spring when pseudothecia normally mature or they remove so much of the nutrients avail-able in the leaves that insufficient remain for the pseudothecial initials to develop to maturity. Such treatment can reduce the ascospore inoculum in the spring by 97%.

References

Ainsworth, G.C. (1973). Introduction and keys to higher taxa. In *The Fungi: An Advanced Treatise*, IVB, (Ainsworth, G.C., Sparrow, F.K. and Sussman, A.S., eds), pp. 1-7. Academic Press, London and New York.

Azcon-G. de Aguilar, C., Azcon, R. and Barea, J.M. (1979). Endomycorrhizal fungi and *Rhizobium* as biological fertilisers for *Medicago sativa* in normal cultivation. *Nature*, **279**, 325-7.

Bandoni, R.J. (1972). The terrestrial occurrence of some aquatic Hyphomycetes. *Canadian Journal of Botany*, **50**, 2283-8.

Barghoorn, E.S. and Linder, D.H. (1944). Marine fungi: their taxonomy and biology. *Farlowia Journal of Cryptogamic Botany*, **1**, 395-467.

Bärlocher, F. and Kendrick, B. (1973a). Fungi in the diet of *Gammarus pseudolimnaeus* (Amphipoda). *Oikos*, **24**, 295-300.

Bärlocher, F. and Kendrick, B. (1973b). Fungi and food preference of *Gammarus pseudolimnaeus*. *Archiv fur Hydrobiologie*, **72**, 501-16.

Barnes, R.S.K. (ed) (1984). *A Synoptic Classification of Living Organisms*. 273 pp. Blackwell Scientific Publications, London.

Bartnicki-Garcia, S. (1968). Cell wall chemistry, morphogenesis, and taxonomy of fungi. *Annual Review of Microbiology*, **22**, 87-108.

Bartnicki-Garcia, S. (1973). Fundamental aspects of hyphal morphogenesis. *Symposium of the Society for General Microbiology*, **23**, 245-67.

Bashi, E. and Fokkema, N.J. (1977). Environmental factors limiting growth of *Sporobolomyces roseus*, an antagonist of *Cochliobolus sativus*, on wheat leaves. *Transactions of the British Mycological Society*, **68**, 17-25.

Baylis, G.T.S. (1959). Effect of vesicular-arbuscular mycorrhizas on growth of *Griselinia littoralis* (Cornaceae). *New Phytologist*, **58**, 274-80.

Baylis, G.T.S. (1967). Experiments on the ecological significance of phycomycetous mycorrhizas. *New Phytologist*, **66**, 231-43.

Birchill, R.T. and Cook, R.T.A. (1971). The interaction of urea and micro-organisms in suppressing the development of perithecia of *Venturia inaequalis* (Cke.) Wint. In *Ecology of leaf-surface micro-organisms* (Preece, T.F. and Dickinson, C.H., eds), pp. 471-83. Academic Press, London and New York.

Björkman, E. (1960). *Monotropa hypopitys* L., an epiparasite on tree roots. *Physiologia Plantarum*, **13**, 308-27.

Bradley, R., Burt, A.J. and Read, D.J. (1982). The biology of mycorrhiza in the Ericaceae. VIII. The role of mycorrhizal infection in heavy metal resistance. *New Phytologist*, **91**, 197-210.

Brock, T.D. (1975). The effect of water potential on photosynthesis in whole lichens and in their liberated algae components. *Planta*, **124**, 13-23.

Burnett, J.H. (1976). *Fundamentals of Mycology*, 2nd edition, 673 pp. Edward Arnold, London.

Chang, Y. and Hudson, H.J. (1967). The fungi of wheat straw compost. I. Ecological studies. *Transactions of the British Mycological Society*, **50**, 649-66.

Christensen, C.M. (1965). *The Molds and Man*, 3rd edition. 284 pp. University of Minnesota Press, Minneapolis.

Christensen, C.M. (1975). *Molds, Mushrooms and Mycotoxins*. 264 pp. University of Minnesota Press, Minneapolis.

Christensen, C.M. and Kaufmann, H.H. (1965). Deterioration of stored grain by fungi. *Annual Review of Phytopathology*, **3**, 69-84.

Cooney, D.G. and Emerson, R. (1964). *Thermophilic Fungi*. 188 pp. W.H. Freeman and Co., London and San Francisco.

Crisan, E.V. (1973). Current concepts of thermophilism and the thermophilic fungi. *Mycologia*, **65**, 1171-98.

De Wildeman, E. (1893). Notes Mycologiques. Fascicle 2. *Annales de la Société belge de microscopie*, **17**, 35-68.

Dick, M.W. (1970). Saprolegniaceae on insect exuviae. *Transactions of the British Mycological Society*. **55**, 449-58.

Diem, H.G. (1971). Effect of low humidity on the survival of germinated spores commonly found in the phyllosphere. In *Ecology of leaf-surface micro-organisms* (Preece, T.F. and Dickinson, C.H., eds), pp. 211-19. Academic Press, London and New York.

Duddridge, J. and Read, D.J. (1982). An ultrastructural analysis of the development of mycorrhizas in *Monotropa hypopitys* L. *New Phytologist*, **92**, 203-14.

Farrar, J.F. (1973). Lichen physiology: progress and pitfalls. In *Air Pollution and Lichens* (Ferry, B.W., Baddeley, M.S. and Hawksworth, D.L., eds), pp. 238-82. University of Toronto Press, Toronto.

Farrar, J.F. (1976a). Ecological physiology of the lichen *Hypogymnia physodes*. I. Some effects of constant water saturation. *New Phytologist*, **77**, 99-103.

Farrar, J.F. (1976b). Ecological physiology of the lichen *Hypogymnia physodes*. II. Effects of wetting and drying cycles, and the concept of 'physiological buffering'. *New Phytologist*, **77**, 105-13.

Farrar, J.F. (1978). Symbiosis between algae and fungi. In *Handbook of Nutrition and Food*, B, (Rechcigl, M. ed.), CRC Press, Cleveland, Ohio.

Ferry, B.W., Baddeley, M.S. and Hawksworth, D.L. (eds) (1973). *Air Pollution and Lichens*. 389 pp. University of Toronto Press, Toronto.

Fokkema, N.J. (1971). The influence of pollen on saprophytic and pathogenic fungi on rye leaves. In *Ecology of leaf-surface micro-organisms* (Preece, T.F. and Dickinson, C.H., eds), pp. 277-82. Academic Press, London and New York.

Francis, R. and Read, D.J. (1984). Direct transfer of carbon between plants connected by vesicular-arbuscular mycorrhizal mycelium. *Nature*, **307**, 53-6.

Gay, P.E., Grubb, P.J. and Hudson, H.J. (1982). Seasonal changes in the concentration of nitrogen, phosphorus and potassium, and the density of mycorrhiza, in biennial and matrix-forming perennial species of closed chalkland turf. *Journal of Ecology*, **70**, 571-93.

Garrett, S.D. (1963). *Soil Fungi and Soil Fertility*. 165 pp. Pergamon Press, London.

Goodman, R.N., Király, Z. and Zaitlin, L. (1967). *The Biochemistry and Physiology of Infectious Plant Disease*. 354 pp. Van Nostrand Company, Inc., New Jersey.

Griffin, D.H. (1981). *Fungal Physiology*. 383 pp. John Wiley and Sons, New York.

Grove, S.N. and Bracker, C.E. (1970). Protoplasmic organisation of the hyphal tips among fungi; vesicles and Spitzenkörper. *Journal of Bacteriology*, **104**, 989-1009.

Gunderson, M.F. (1962). *Low Temperature Microbiology.* 322 pp. Campbell Soup Co., New Jersey.
Hadley, G. and Williamson, B. (1971). Analysis of post-infection growth stimulus in orchid mycorrhiza. *New Phytologist*, **70**, 445–55.
Hale, M.E. (1983). *The Biology of Lichens*, 3rd edition. 190 pp. Edward Arnold, London.
Harley, J.L. (1975). Problems of mycotrophy. In *Endomycorrhizas* (Sanders, F.E., Mosse, B. and Tinker, P.B., eds), pp. 1–24. Academic Press, London and New York.
Harley, J.L. and McCready, C.C. (1952). Uptake of phosphate by excised mycorrhiza of the beech. II. Distribution of phosphate between host and fungus. *New Phytologist*, **51**, 56–64.
Harley, J.L. and Smith, S.E. (1983). *Mycorrhizal Symbiosis.* 483 pp. Academic Press, London and New York.
Harper, J.E. and Webster, J. (1964). An experimental analysis of the coprophilous fungus succession. *Transactions of the British Mycological Society*, **47**, 511–30.
Hartig, T. (1844). Ambrosia de *Bostrichus* dispar. *Allgemeine Forstzeitung und Jagdzeitung*, **13**, 73–6.
Hatch, A.B. (1937). The physical basis of mycotrophy in the genus *Pinus*. *Black Rock Forestry Bulletin*, **6**, 168 pp.
Hawksworth, D.L. and Rose, F. (1970). Qualitative scale for estimating sulphur dioxide air pollution in England and Wales using epiphytic lichens. *Nature*, **227**, 145–8.
Hawksworth, D.L. and Rose, F. (1976). *Lichens as Pollution Monitors.* Studies in Biology. No. 66. 60 pp. Edward Arnold, London.
Hawksworth, D.L., Sutton, R.C. and Ainsworth, G.C. (1983). *Ainsworth and Bisby's Dictionary of the Fungi*, 7th edition. 412 pp. Commonwealth Mycological Institute, Kew.
Hayman, D.S. and Mosse, B. (1971). Plant growth responses to vesicular-arbuscular mycorrhiza. I. Growth of *Endogone* – inoculated plants in phosphate deficient soils. *New Phytologist*, **70**, 19–27.
Hedger, J.N. (1975). The ecology of thermophilic fungi in Indonesia. In *Biodegradation et Humification* (Kilbertus, G., Reisinger, O., Mourey, A. and Cancela da Fonseca, J.A., eds), pp. 59–65. Pierron, Nancy.
Hering, T.F. (1967). Fungal decomposition of oak leaf litter. *Transactions of the British Mycological Society*, **50**, 267–73.
Hill, D.J. (1972). The movement of carbohydrate from alga to the fungus in the lichen *Peltigera polydactyla*. *New Phytologist*, **71**, 31–9.
Hudson, H.J. (1968). The ecology of fungi on plant remains above the soil. *New Phytologist*, **67**, 837–74.
Hunsley, D. and Burnett, J.H. (1970). The ultrastructural architecture of the walls of some hyphal fungi. *Journal of General Microbiology*, **62**, 203–18.
Hunsley, D. and Kay, D. (1976). Wall structure of the *Neurospora* hyphal apex: Immunofluorescent localization of wall surface antigens. *Journal of General Microbiology*, **95**, 233–48.
Ikediugwu, F.E.O. and Webster, J. (1970). Antagonism between *Coprinus heptemerus* and other coprophilous fungi. *Transactions of the British Mycological Society*, **54**, 181–204.
Ingold, C.T. (1942). Aquatic hyphomycetes of decaying alder leaves. *Transactions of the British Mycological Society*, **25**, 339–417.
Ingold, C.T. (1966). The tetraradiate aquatic fungal spore. *Mycologia*, **58**, 43–56.
Ingold, C.T. (1971). *Fungal Spores, Their liberation and dispersal.* 302 pp. Clarendon Press, Oxford.

Ingold, C.T. (1975). *An illustrated guide to Aquatic and Water-borne Hyphomycetes (Fungi imperfecti) with Notes on their Biology.* 96 pp. Scientific Publication No. 30. Freshwater Biological Association, Ambleside.

Ingold, C.T. (1976). The morphology and biology of freshwater fungi, excluding Phycomycetes. In *Recent Advances in Aquatic Mycology* (Jones, E.B.G., ed), pp. 335-57. Elek Science, London.

Iqbal, S.H. and Webster, J. (1973a). The trapping of aquatic hyphomycete spores by air bubbles. *Transactions of the British Mycological Society*, **60**, 37-48.

Iqbal, S.H. and Webster, J. (1973b). Aquatic Hyphomycete spora of the River Exe and its tributaries. *Transactions of the British Mycological Society*, **61**, 331-46.

James, P.W. and Henssen, A. (1976). The morphological and taxonomic significance of cephalodia. In *Lichenology: Progress and Problems* (Brown, D.H., Hawksworth, D.L. and Bailey, R.H., eds), pp. 27-77 Academic Press, London and New York.

Jaquelinet-Jeanmougin, S. and Gianinazzi-Pearson, V. (1983). Endomycorrhizas in the Gentianaceae. I. The fungi associated with *Gentiana lutea* L. *New Phytologist*, **95**, 663-6.

Jones, E.B.G. (1971). The ecology and rotting abilities of marine fungi. In *Marine Borers, Fungi and Fouling Organisms of Wood* (Jones, E.B.G. and Eltringham, S.K., eds), pp. 237-58. O.E.C.D., Paris.

Jones, E.B.G. (1982). Observations on the ecology of lignicolous aquatic hyphomycetes. In *The Fungal Community: its organisation and role in the ecosystem* (Wicklow, D.T. and Carroll, G.C., eds), pp. 731-42. Marcel Dekker, New York.

Käärik, A.A. (1974). Decomposition of wood. In *Biology of Plant Litter Decomposition. I.* (Dickinson, C.H. and Pugh, G.J.F., eds), pp. 129-74. Academic Press, London and New York.

Kaushik, N.K. and Hynes, H.B.N. (1968). Experimental study on the role of autumn-shed leaves in aquatic environments. *Journal of Ecology*, **56**, 229-43.

Kaushik, N.K. and Hynes, H.B.N. (1971). The fate of dead leaves that fall into streams. *Archiv für Hydrobiologie*, **68**, 465-515.

Kidston, R. and Lang, W.H. (1921). On old red sandstone plants showing structure, from the Rhynie chert bed, Aberdeenshire. Part V. The thallophyta occurring in the Peat-bed; the succession of plants through a vertical section of the bed and the conditions of accumulation and preservation of the deposit. *Transactions of the Royal Society of Edinburgh*, **52**, 855-902.

Khan, A.G. (1975). Growth effects of VA mycorrhiza on crops in the field. In *Endomycorrhizas* (Sanders, F.E., Mosse, B. and Tinker, P.B., eds), pp. 419-35. Academic Press, London and New York.

Kirk, T.K. (1971). Effects of micro-organisms on lignin. *Annual Review of Phytopathology*, **9**, 185-210.

Kirk, T.K. and Fenn, P. (1982). Formation and action of the ligninolytic systems in basidiomycetes. In *Decomposer Basidiomycetes: their biology and ecology* (Frankland, J.C., Hedger, J.N. and Swift, M.J., eds), pp. 67-90. Cambridge University Press, Cambridge.

Kohlmeyer, J. and Kohlmeyer, E. (1979). *Marine Mycology, The Higher Fungi*, 670 pp. Academic Press, London and New York.

Lange, O.L., Schulze, E-D. and Koch, W. (1970). Experimentell-ökologische Untersuchungen an Flechten der Negev-Wuste. *Flora*, **159**, 38-62.

Lenne, J.M. and Parberry, D.G. (1976). Phyllosphere antagonists and appressorium formation in *Colletotrichum gloeosporioides. Transactions of the British Mycological Society*, **66**, 334-6.

Levi, M.P. and Cowling, E.B. (1969). Role of nitrogen in wood deterioration. VII. Physiological adaptation of wood-destroying and other fungi to substrates deficient in nitrogen. *Phytopathology,* **59**, 460-8.

Martin, M.M. (1979). Biochemical implications of insect mycophagy. *Biological Reviews,* **54**, 1-21.

Martin, M.M. and Martin, J.S. (1970). The biochemical basis for the symbiosis between the ant, *Atta colombica tonsipes* and its fungus food. *Journal of Insect Physiology,* **16**, 109-19.

Martin, M.M. and Martin, J.S. (1978). Cellulose digestion in the midgut of the fungus-growing termite, *Macrotermes natalensis:* the role of acquired digestive enzymes. *Science, Washington,* **199**, 1453-5.

Martin, M.M. and Martin, J.S. (1979). The distribution and origins of the cellulolytic enzymes in the higher termite *Macrotermes natalensis. Physiological Zoology,* **52**, 11-21.

Massee, G. and Salmon, E.S. (1901). Researches on coprophilous fungi. *Annals of Botany,* **15**, 313-57.

Massee, G. and Salmon, E.S. (1902). Researches on coprophilous fungi. II. *Annals of Botany,* **16**, 57-63.

Miehe, H. (1907). *Die Selbsterhitzung des Heus. Eine biologische Studie.* 127 pp. Gustav Fisher, Jena.

Mosse, B. and Phillips, J.M. (1971). The influence of phosphate and other nutrients on the development of vesicular-arbuscular mycorrhiza in culture. *Journal of General Microbiology,* **69**, 157-66.

Newton, J.A. (1971). *A mycological study of decay in the leaves of deciduous trees on a river bed.* Ph.D. thesis, University of Salford.

Nicholson, P.B., Bocock, K.L. and Heal, O.W. (1966). Studies on the decomposition of faecal pellets of a millipede (*Glomeris marginata* (Villers)). *Journal of Ecology,* **54**, 755-66.

Pace, M.A. and Campbell, R. (1974). The effect of saprophytes on infection of leaves of *Brassica* spp. by *Alternaria brassicicola. Transactions of the British Mycological Society,* **63**, 193-196.

Rayner, A.D.M. and Todd, N.K. (1979). Population and community structure and dynamics of fungi in decaying wood. *Advances in Botanical Research,* **7**, 333-420.

Read, D.J. and Stribley, D.P. (1973). Effect of mycorrhizal infection on nitrogen and phosphorus nutrition of ericaceous plants. *Nature (New Biology), London,* **244**, 81-2.

Richardson, M.J. (1970). Studies on *Russula emetica* and other agarics in a Scots pine plantation. *Transactions of the British Mycological Society,* **55**, 217-29.

Richardson, M.J. (1972). Coprophilous ascomycetes on different dung types. *Transactions of the British Mycological Society,* **58**, 37-48.

Richardson, M.J. and Watling, R. (1968). Keys to fungi on dung. *Bulletin of the British Mycological Society,* **2**, 18-43.

Richardson, M.J. and Watling, R. (1969). Keys to fungi on dung. *Bulletin of the British Mycological Society,* **3**, 86-8 and 121-4.

Ross, E.W. and Marx, D.H. (1972). Susceptibility of sand pine to *Phytophthora cinnamomi. Phytopathology,* **62**, 1197-200.

Schmidberger, J. (1836). Naturgeschichte des apfelborkenkafers *Apate dispar. Beitrage Obstbaumzucht Naturgesellschaft Obstbaumen schadlichen Insekten,* **4**, 313-30.

Slankis, V. (1973). Hormonal relationships in mycorrhizal development. In *Ectomycorrhizae* (Marks, G.C. and Kozlowski, T.T., eds), pp. 231-98. Academic Press, London and New York.

Smith, D.C. (1978). What can lichens tell us about real fungi? *Mycologia,* **70**, 915-34.

Smith, D.C. and Drew, E.A. (1965). Studies in the physiology of lichens. V. Translocation from the algal layer to the medulla in *Peltigera polydactyla*. *New Phytologist*, **64**, 195-200.

Smith, S.E. (1966). Physiology and ecology of orchid mycorrhizal fungi with reference to seedling nutrition. *New Phytologist*, **65**, 488-99.

Smith, S.E. (1967). Carbohydrate translocation in orchid mycorrhizal fungi. *New Phytologist*, **66**, 371-8.

Stribley, D.P. and Read, D.J. (1974). The biology of mycorrhiza in the Ericaceae. IV. The effect of mycorrhizal infection on the uptake of N^{15} from labelled soil by *Vaccinium macrocarpon* Ait. *New Phytologist*, **73**, 1449-55.

Stribley, D.P. and Read, D.J. (1980). The biology of mycorrhiza of the Ericaceae. VII. The relationship between mycorrhizal infection and the capacity to utilize simple and complex organic nitrogen sources. *New Phytologist*, **86**, 365-71.

Suberkropp, K. and Klug, M.J. (1981). Degradation of leaf litter by aquatic hyphomycetes. In *The Fungal Community: its organisation and role in the ecosystem* (Wicklow, D.T. and Carroll, G.C., eds), pp. 761-6. Marcel Dekker, New York.

Sumner, J.L., Morgan, E.D. and Evans, H.C. (1969). The effect of growth temperature on fatty acid composition of fungi in the order Mucorales. *Canadian Journal of Microbiology*, **15**, 515-20.

Swift, M.J. (1977). The role of fungi and animals in the immobilisation and release of nutrient elements from decaying branch-wood. In *Soil Organisms and Components of Ecosystems* (Lohm, U. and Persson, T., eds). pp. 193-202. Swedish National Science Council, Stockholm.

Tansey, M.R. (1971). Isolation of thermophilic fungi from self-heating industrial wood chip piles. *Mycologia*, **58**, 537-47.

Tansey, M.R. (1973). Isolation of thermophilic fungi from alligator nesting material. *Mycologia*, **65**, 594-601.

Thornton, D.R. (1963). The physiology and nutrition of some aquatic hyphomycetes. *Journal of General Microbiology*, **33**, 23-31.

Tresner, H.D. and Hayes, J.A. (1971). Sodium chloride tolerance of terrestrial fungi. *Applied Microbiology*, **22**, 210-13.

Webster, J. (1959). Experiments with spores of aquatic Hyphomycetes. I. Sedimentation, and impaction on smooth surfaces. *Annals of Botany, London, N.S.*, **23**, 595-611.

Webster, J. (1970). Coprophilous fungi. *Transactions of the British Mycological Society*, **54**, 161-80.

Webster, J. (1975). Further studies of sporulation of aquatic hyphomycetes in relation to aeration. *Transactions of the British Mycological Society*, **64**, 119-27.

Webster, J. (1977). Seasonal observations on "aquatic hyphomycetes" on oak leaves on the ground. *Transactions of the British Mycological Society*, **68**, 108-11.

Webster, J. (1980). *Introduction to Fungi*, 2nd edition. 669 pp. Cambridge University Press, Cambridge.

Webster, J. and Dix, N.J. (1960). Succession of fungi on decaying cocksfoot culms. III. A comparison of the sporulation and growth of some primary saprophytes of stem. leaf blade and leaf sheath. *Transactions of the British Mycological Society*, **43**, 85-99.

Webster, J. and Towfik, F.H. (1972). Sporulation of some aquatic hyphomycetes in relation to aeration. *Transactions of the British Mycological Society*, **59**, 353-64.

Willoughby, L.G. (1961). The ecology of some lower fungi at Esthwaite Water. *Transactions of the British Mycological Society*, **44**, 305-32.

Willoughby, L.G. (1962). The occurrence and distribution of reproductive spores of Saprolegniales in fresh water. *Journal of Ecology*, **50**, 733-59.

Index

Acer 128, 136, 140
2-acetamido-2-deoxy-D-
    glucopyranose 5, 14
Achlya 120, 121, 143
A. sparrowii 121
Acromyrmex 254
Actinastrum hanzschii 131
Acinospora
    melagospora 128-9,
    132
aero-aquatic
    hyphomycetes 141-3
agaric 44, 46-52, 58, 72, 73,
    177
Agaricales 46, 76, 87, 150,
    237
    as mycorrhizal fungi 146
Agaricus bisporus 29, 166-71
    cultivation of 167-70
    preparation of
        compost 167-8
    composting
        process 168-9
    preparation of
        inoculum 169
    casing 169
    cropping 170
    pathogens of 171
    pests and competitors in
        compost 171
A. campestris, green rings in
    grass 23
Agrostis tenuis 233
Alatospora 126
aleurioconidia 37
    development of 37-9
    in aquatic
        hyphomycetes 126-7
Alnus 136
Alternaria 72, 87, 180
A. alternata 63, 70, 72, 164
A. brassicicola 67-8
A. solani 275
alternaric acid, as a
        toxin 275-6

Amanita 46
    as a mycorrhizal
        genus 184
A. muscaria 184
A. phalloides 46
ambrosia fungi 244-8
    nutrition of 247
Ambrosiella xylebori 246
Amylostereum areolatum,
    association with
        Sirex 248-52
    mycetangia in 248-9
    as a pathogen of
        Pinus 250-52
anamorphic 1, 128, 129, 135,
    142
Anguillospora 128, 140
A. crassa 128, 143
A. longissima 128
antheridium 34, 40
antibiotics, antibacterial 139
    antifungal 69, 139
Aphanomyces 121
A. astaci 122
A. laevis 121
Aphyllophorales 47, 87, 237
apical body 15, 16
apical vesicular complex 15
Apiognomonia errabunda 75
aplanospores 33
    in Mucorales 35-7
Apostemidium 124
A. guernisaci 125
appressorium 63, 69, 132,
    215, 217, 267, 277
Aqualinderella
    fermentans 143, 145
aquatic hyphomycetes 124-41
    conidial
        development 124-7
    teleomorphic
        states 128-9
    significance of tetra-
        radiate conidial
        form 132-3

in other
        environments 134-5
    seasonal abundance
        of 135-7
    as intermediaries in the
        food chain 138-9
    influence on food
        selection by micro-
        fauna 139-41
Arbutus 231
arbuscules 215, 217
Arctostaphylos 231
Armillaria mellea, rhizomorphs
    of 30-1, 99, 237
    as an obligate
        necrotroph 276-7
Artemia salina (brine
        shrimp) 117
Articulospora
        tetracladia 132, 133,
        134, 138, 143
asci 10, 11
    unitunicate 41-2
    bitunicate 41-2, 116, 125
Ascobolus 147
A. glaber 151, 152
A. immersus 151, 152
A. stictoideus 151, 152
A. viridulus 151, 152-5
Ascophyllum nodosum,
        infection by
        Mycosphaerella
        ascophylli 115-6
ascospores 41-3
    thread-like 124
    with appendages 112
    gelatinous sheaths
        of 112, 124, 125
aspergillosis 166-7
Aspergillus 75
A. amstelodami 180-1
A. chevalierii 181
A. fumigatus 160, 162, 164,
        166, 173
A. glaucus group 180-2

A. halophilicus 181
A. niger 180-81
A. repens 181
A. restrictus group 180-82
A. versicolor 109
Asphondylia sorothamni (gall midge) 246
Aspidiotus, association with *Septobasidium* 252-4
Astacus astacus (crayfish) 122
Asteroxylon 214
Atta colombica tonsipes 255
A. texana 254
Attine ants, and fungi 254-8, the colony and 'fungus gardens' 254-5
biochemical basis of association 256-8
Aureobasidium 38, 60-62, 67, 180
A. pullulans 60, 65, 67, 70, 72-6, 142, 160, 164, 177
Auricularia mesenterica 102
Auriscalpium vulgare 102
Azalea indica 227

Bacillus 106, 108
Ballia callitricha 115
B. scoparia 115
ballistospore 43, 58, 59
Basidiobolus ranarum 155-7
life cycle 156
two-phase dispersal in 156
basidiocarps 10, 12
agaric type 46-7
polypore type 47-52
response to displacement 49-52
basidiospores 40
mechanisms of discharge 43-5
Basidiomycotina, life cycle 10, 12-13
basidium 10, 12
spore discharge from 43-5
holobasidium 43-4
heterobasidium 45
Betula 87, 177, 184
binding hyphae 21, 47-9
biotrophs 264
features of 269-73
infection by 266-7
haustoria in 267-9
morphological disturbances in host 273
diversion and coversion of host photosynthate by 271-2

bitunicate asci 41-2, 116, 125
Blackstonia perfoliata, mycorrhiza of 233-4
Blastocladia 3
B. ramosa 143
Blastocladiales 2, 143
blastoconidia 37
development of 37-8
blue-stain fungi 96-8
Bolbitius 146, 154
Boletus 46
as a mycorrhizal genus 184, 192
as a mycorrhizal fungus of *Monotropa* 232
B. (Suillus) aeruginascens 184
B. (Suillus) elegans 184
B. subtomentosus 185
Botrytis 26, 267
B. cinerea 28, 62, 63, 64, 69, 74, 75
B. fabae 63, 276
as an obligate necrotroph 277-9
penetration of host 277
'non-aggressive' and 'aggressive' phases 277-8
life cycle 278
Brassica 67, 68
Bremia lactucae 35
penetration structures and haustoria of 268-9
bromatia 256
brown-rot fungi 85, 87, 89, 92, 95, 100-102, 104-6
Bullera 58, 77

calcium phytate 191
calcolfluor 13
Calluna (ling) 226, 228
C. vulgaris 227, 232
Candida 262
C. aquatica 130
C. gelida 176
Cannabis achenes 119
carotene 147-8
cellulase 7, 32, 92, 93, 94, 157
acquisition by termites 260-62
'cellulin' 143
plugs in Leptomitales 143
cellulose 1, 5-7
enzymes involved in hydrolysis 260-62
Cenococcum graniforme mycorrhiza formed by 184-5
Centaurium erythraea 233
cephalodia 201, 211

Cephalosporium 99
Ceratobasidium 235
Ceratocystis 87, 97, 98
C. ulmi 99, 102, 104, 245
Ceriosporopsis calyptrata 112, 131
C. cambrensis 114
C. circumvestita 115
C. halima 114
Cetraria islandica (Iceland moss) 203
Chaetocladium 146
Chaetomium 70, 87, 146, 158
C. brasiliense 152
C. caprinum 151, 152
C. globosum 72, 92
C. thermophile 160, 165, 172, 173
Chalara cylindrospora 76
Cheilymenia 146
chemo-organotrophs 61
chimeras, in lichens 210-13
chitin 1, 5, 6, 7, 8, 14, 15
chitinase 7, 16
chitinolytic enzymes 69
chitin synthase 16
chitosan 5, 7
chlamydospores 19, 25-7, 28, 60, 61, 75, 215-18
Chlorosplenium aeruginascens 96
Chytridiales 2, 122-3
chytrids 2, 122-3
Cladonia fimbriata 200
C. rangiferina (reindeer moss) 203
Cladosporium 38, 60, 61, 62, 67, 72-3, 96, 164, 180
C. herbarum 70, 72-6, 177, 180
clamp connection 5, 20, 21, 128
classification 52-6
Clathrosphaerina 141
C. zalewskii 141, 142
Clavaria argillacea 227
C. vermicularis 227
Clavariopsis aquatica 128, 129, 136, 137
Clupea harengus (herring) 117
Cochliobolus sativus 63, 65-7
Coccomyxa 194, 201
Coelomomyces, infections in mosquito larvae 242
C. pentangulatus 243
Coleosporium senecionis 78
Collema 198
Colletotrichum gloeosporioides 69
Collybia 78

C. dryophila 28
columella 35, 36
commensals 76
common primary
    saprotrophs 61-2
    distribution 70-71
    attributes of 71-6
    desiccation tolerance
        of 74-5
    survival by 75-6
conchiolin 117
conidium 11
coniferyl alcohol 85, 86
Coniochaeta 146
Conocybe 146
Cooksonia 215
Coprinus 146, 147, 161, 170
C. cinereus 160, 165, 166, 172
C. heptemerus 151, 152,
    154-5
C. patouillardii 151, 152, 153
C. radiatus 151, 152, 153
Coprobia 41, 42, 146
C. granulatus 155
coprogen 148
coprophilous fungi 146-57
    adaptations to
        habitat 147
    dung as a substrate
        for 148-50
    analysis of fungal
        succession on 150
    competition and
        antagonism
        between 153-4
    preference for particular
        dung types 154-5
Cora 194
cordycepin 244
Cordyceps 242-4
C. militaris, as a parasite of
    insects 244
Coriolus hirsutus 99
C. versicolor 48, 87, 89, 94,
    99, 102, 104, 109
Corollospora maritima 114
Corticium 235
Cortinarius, as a mycorrhizal
    genus 184
Corylus avellana 157
coumaryl alcohol 85, 86
Curvularia 63
C. lunata 70
cuticular waxes 69
Cytisus scoparius 248

Dacrymycetales 45
Dacrymyces 45
Dactylaria gallopava 167
Dactylis glomerata 71
Dactylorchis 235

D. fuchsii 240
D. purpurella 236
    mycorrhiza of 236-40
Daedaleopsis confragosa 101
Daldinia concentrica 102
D. vernicosa 102
dehydrogenative polymerizate
    (DHP) 88
Delitschia 146
Dendriscocaulon 211
    chimeras with
        Sticta 211-13
Dendrospora 128
D. fusca 130
Dendrosporomyces 128
depsides 89, 209-10
depsidones 89, 209-10
desiccation 19, 28, 61, 69,
    75-6, 178
    and polyols 182
    tolerance in
        lichens 205-9
Desmazierella acicola 78
Dictyuchus 120, 121
Digitatispora marina 110,
    111, 131
dikaryon 10
dikaryophase 5, 10, 11, 12,
    128
Dimargaris 155
2-4-dinitrophenol 95
diplophase 7, 9, 10, 11, 12,
    13
Discomycetes 41, 111, 118,
    142
dispersal, two phase 154-7
dolipore septum 20, 128
Dothideales 113
Dutch elm disease 99

Ectocarpus fasciculatus 115
Elaphomyces 244
encephalitis 167
Endophragmia
    hyalosperma 158
Eniocyla pusilla (terrestrial
    caddis fly) 158
Entomophthora 131, 242
E. muscae, infection of house
    flies by 242-3
Entomophthorales 156, 242
Epicoccum 186
E. purpurascens 63, 67-8,
    72-5, 142, 164
Erica (heather) 226, 228
Ericoid mycorrhiza 226-31
    occurrence 226
    structure 226-7
    benefits to host in
        nitrogen uptake 228-30

    utilisation of amino
        acids 229-30
    nutrition of fungi 230-31
    sequestration of heavy
        metals by 232-3
Erysiphales 63, 75, 267
Erysiphe graminis 265-71
    ascocarps and
        conidia 265
    haustoria of 266-7
    infection by 266-7, 271
Eucalyptus 70
Eu-mycota 1
extension zone 13, 15
exochthonous fungi 64

Fagus sylvatica 183
    structure of
        mycorrhiza 183-4,
        186-7
    mycorrhizal fungi
        of 184-5
'Fairy rings' 23-5
Festuca ovina 224-5, 229
fibre saturation point 30
'fibrous' hyphae 30
'field fungi' 180
Fistulina hepatica 101
Flagellospora 128
F. curvula 136
F. penicillioides 128
Flammulina velutipes 102,
    104
Flavobacterium 106
flavonoids 95, 108
Fraxinus 140
fulvic acid 23, 107
Fungi Imperfecti 1
fungitoxins 99
'fungus gardens', of
    Atta 254-5
    or 'combs' of
        termites 258-61
Fusarium 38, 70, 142, 177
Fusicoccum bacillare 70, 78

Galeola septentrionalis 237
gallic acid 69, 84
Gammarus
    pseudolimnaeus 140
Ganoderma 47, 51
G. adspersum 47, 48
Gasteromycetes 43, 44, 58
Gastrodia elata 237
generative hyphae 47-9
Gentiana lutea 235
Gentianaceae, mycorrhiza
    of 233-5
Gentianella amarella 237
Geotrichum 59
Gigaspora 217

G. margarita 233
Glomeris 76
G. marginata (millepede) 157
Glomus 6, 216, 217, 222-3, 233
G. mosseae 220-21, 225
glucanase, endo- 107, 261
    exo- 107, 261
glucans 5, 6, 7, 8, 13, 14, 15
β-glucosidase 106, 107
glycoproteins 7, 14
'green islands' 272
Guignardia fagi 76
Gyromitra esculenta 184

Haber–Weiss reaction 91
halophilous fungi 179-80
Halosphaeria appendiculata 114, 115
H. quadricornuta 121, 131
H. salina 112
Harpella melusinae 243
haustoria 63, 215
    of *Bremia lactucae* 268-9
    of *Erysiphe graminis* 265-6, 271
    of *Hemileia holstii* 270
    of *H. vastatrix* 270
    of *Peronospora calotheca* 270
    of *P. parasitica* 270
    of *Puccinia graminis* 268
    of *P. purpurella* 270
Helicobasidium purpureum 29
Helicodendron 141, 142
H. conglomeratum 141
Helicoma monospora 78, 80
Helicoon 141
Heliscus lugdunensis 128, 136, 138
Helminthosporium victoriae 275-6
Helotium caudatum 76
hemicellulases 32
Hemileia holstii, haustoria of 270
H. vastatrix, haustoria of 270
Hendersonia acicola 78
heterobasidia 45
Heterobasidion annosum 97, 99, 102, 103, 104
Heterococcus 194
heterokaryosis 13
heteromerous lichens 194, 195
Hirschoporus abietinus 102
holobasidium 43-4
holomorph 1
homoiomerous lichens 195, 198

homolactic fermentation 145
humic acid 23, 107-9
Humicola 106
H. grisea 173
H. insolens 165, 172
humin 107-9
humus 106-9
    decomposition in the soil 106-7
    nature of 107-8
    turnover in soil 108-9
Hyalella azteca 139
Hyaloscypha zalewskii 142
hydrocyanic acid 25
Hydnum 46
Hylastes 97
Hymenogaster 44
Hymenomycetes 43, 46, 47
hyphae 2
    wall composition of 5-6
    wall structure of 6-7, 8
    growth of 13-19
    branching of 18-19
    septation of 19-21
hyphal interference 104, 154, 166
Hyphochytrium catenoides 157
Hypholoma fasciculare 29, 101, 104, 109
Hypogymnia physodes 205, 208
Hysterangium 44

Ichthyophonus hoferi 117
Ingoldiella hamata 128, 130, 134
Itersonilia 58
I. perplexans 59

Kalotermes flavicallis 104
keto-adipic acid 90, 91
Kriegeriella mirabilis 78, 80, 81

Laboulbeniales 151, 242
laccase 89
Lachnella villosa 76
Lactarius 184
    as a mycorrhizal genus 184
    nutrition of 185, 192
Lactobacillus 145
Lagenidiales 117
lag phase 21
Laminaria 115
Lasiobolus 146
Lasioptera rubi (gall midge) 248
Lecanora conizaeoides 203
lecanoric acid 210

Lemanea 124
Lemonniera 126
L. aquatica 126
Lenzites betulina 99
Leontodon hispidus 233
Lepiota 255
L. procera, rings in grass 23
Leptographium 97-9
Leptolegnia 121
Leptomitales 142, 143
Leptomitus 143
L. lacteus 143, 145
Leptosphaeria 41, 42, 70, 113
L. contecta 113
L. lemaneae 124
L. microscopica 70
L. neomaritima 113
Leptosporomyces galzinii 128, 130, 135
Leucopaxillus cerealis 194
Lichens 1, 2
    fungi involved in 194
    forms of 194-5
    algae and cyanobacteria of 194
    internal structure of 195-6
    nitrogen fixation in 201
    accumulation of ions and other nutrients 202
    accumulation of strontium and caesium 203
    sensitivity to sulphur dioxide 203-4
    as biological indicators of pollution 203-4
    growth rates 205
    water relations of 205-9
    resistance to environmental extremes 209
    role of lichen substances 209-10,
    chimeras 210-13
lichen substances 201, 204, 209-10
Lichina 115
lignin 84-92
    structure of 84-5
    of angiosperms 85
    of conifers 85
    extracted 88
    synthetic 88
lignin degradation 88-92
    role of phenolases in 89
    role of agents other than enzymes in 91-92
    schema for 90-91
    in soil 106-7
linear growth phase 21
Listera ovata 241

Littorella uniflora (shore weed) 122
Loculoascomycetes 41, 42, 61, 65, 112, 128, 142
Lophodermella sulcigena 78
Lophodermium pinastri 78, 79
Lulworthia floridiana 114
L. purpurea 114
L. rufa 114
Lunulospora curvula 136, 137
Lycoperdon 44

Macrotermes 258
    mounds of 257-8
M. natalensis 258
mannan 5, 6
mannitol 27, 28, 188, 204, 206, 231, 239
Marasmius androsaceus 78, 80, 81
M. oreades, in 'fairy rings' 23-5
marine fungi 110-117
    diversity of 110-14
    substrates for 114-17
    on animals 117
    on seaweeds 115-16
    on wood 114
Massarina 128, 129
mechanical pressure 18
Medicago sativa (alfalfa) 222
melanin 19, 28, 96, 108
melanized 19, 27, 28, 35, 61, 69, 148
melezitose 61
Mercurialis perennis (Dog's Mercury) 218
mesophiles 100
Metschnikowia bicuspidata var *australis* 117, 118
Micronectriella nivalis 177
Microsporum 39
Microthyrium fagi 76
Miladina lechithina 128, 129
Mollisia 128
Monacrosporium 247
Monilia candida 246
Monotropa hypopitys (yellow Bird's nest or Pinesap) 231-2
    mycorrhiza of 231-2
    fungal pegs in 231
    three tier system with *Boletus* and *Pinus* 232
Mortierella 76
mor type soil 77
Mucor 9, 36, 37, 76, 145
M. miehe 161
M. mucedo 151, 152

M. plumbeus 153
M. pusillus 160, 161, 164, 165-6
M. ramannianus 158
M. rouxii 3, 6
Mucorales 27, 33, 37, 58, 71, 76, 105, 146-55, 158, 176, 218
Mycena 76
M. galopus 72
mycelial strands 29-32
    structure and development 29-30
    functions of 31-2
mycetangia, in Scolytid and Platypodid beetles 245-7
    in *Sirex* 248-50
mycetocyte 263
mycetomes 262-3
    in *Sitodrepa panicea* 263
Myelophilus 97
mycophagy 80
Mycosphaerella 113
M. ascophylli 115-16, 124
M. ligulicola 65
M. tassiana 76

Naemocyclus niveus 78
necrotrophs 264, 273-84
    secretion of extracellular enzymes by 274
    production of toxins by 275
Nectria 11, 128
N. lugdunensis 136
Neurospora 8
N. crassa 6, 7, 13, 15, 22
Nia vibrissa 110, 111, 114, 131, 132
Nigrospora 63, 70
N. sphaerica 70
Nostoc 194, 199, 201
nuclear phases 7-13
nutrition, ambrosia fungi 247
    aquatic hyphomycetes 138
    coprophilous fungi 151
    ericoid mycorrhizal fungi 230-31
    lichen fungi 200-1
    parasitic fungi 264
    sheathing mycorrhizal fungi 185-7
    vesicular-arbuscular mycorrhizal fungi 255

Odontotermes 260
Oedocephalum 38
Oidium 39
Olpidium brassicae 3

oogonium 34, 40, 143
    in *Saprolegnia* 34, 40
    in *Pythium* 280-1
Oomycetes 5, 6, 7, 13, 15, 16, 33, 40, 58, 113, 142, 143
oospore 40
    in *Saprolegnia* 34, 40
    in *Pythium* 280-1
Orbilia marina 111
orchidaceous mycorrhizas 235-41
    of heterotrophic seedling 235-6
    fungi involved 235-7
    features of infection 237
    carbon nutrition of fungi 237
    effects of infection on growth of seedling 237
    mode of transfer of nutrients from fungus to host 237-8
    translocation from fungus to orchid 239
    orchid as a parasite of fungus 239
    fungus and adult green orchid 239-41
osmophilous fungi 179-80
    aspergilli in stored grain 180-82
Ostracoblabe implexa 117
Ostrea edulis (oyster) 117

Paecilomyces crustaceus 162
Paneolus 146, 154
Panicum maximum 70
Papulospora byssina 171
parietin 209
Parthenium argentatum (guayule) 166
Paxillus atrotomentosus 102
P. involutus 102
pectin methyl esterase 274
pectolytic enzymes 63, 274, 279, 280
Peltigera 198-9
P. aphthosa 201
P. polydactyla 199
    carbon fixation by 199-202
    water relation of 206-7
Pelvetia canaliculata 115
Penicillium 75, 76, 79, 105
P. emersonii 162
P. expansum 274
P. frequentans 108
Peniophora gigantea 103, 104
pentachlorophenol 95
peripheral growth zone 13, 15
perithecia 19, 43, 97, 146

Peronospora calotheca,
    haustoria of   270
P. parasitica   35
    haustoria of   270
Peronosporales   33, 58, 117,
    266
peroxidase   89
Pezizella ericae   227-8, 231
Phacidium infestans   177
*Phaeotrichum
    hystricinum*   155
Phallus impudicus   104
*Phanerochaete
    chrysosporium*   91
Pharcidia laminariicola   115
Phellinus pomaceus   47
P. igniarius   104
phenolases   89, 96, 277
phenols   69, 209, 251
phialide   37, 127
phialoconidia   37, 38
    development in aquatic
        hyphomycetes   126-7
    in thermophilous
        fungi   162-3
Phialophora   87, 103, 106
Phialophoropsis   247
Phlebia merismoides   104
phloroglucinol   107
phototropic responses   147,
    156
phosphatases   191, 230
Phragmites australis   124
Phycomyces blakesleeanus   6
phylloplane fungi   57-70
    leaf as a spore trap   57-8
    phylloplane
        inhabitants   58-62
    nutrient sources for   61-2
    pollen as a nutrient
        source   65-7
    competition with
        pathogens for   65-8
    antagonistic reactions
        of   67-9
physodic acid   210
phytoalexins   276-8
Phytophthora   5, 58, 143
P. cinnamomi   193, 194
P. parasitica   7, 13
Picea abies   184
Piggotia stellata   70
Pilaira   147
P. anomala   151
Pilobolus   147-155, 156
P. crystallinus   151, 152-5
P. kleinii   149, 153
Piptocephalis   147, 158
pine needles   77-83
    decomposition of   77-9
    role of micro-fauna in

decomposition   79,
    82-3
pinoresinol   87
pinosylvin   95, 251
Pinus   88
    structure of mycorrhiza
        of   183-4
    fungi involved   184-5,
        186, 232
P. clausa   193
P. echinata   194
P. radiata   251
P. strobus   189
P. sylvestris   71, 77, 79, 184
Piptoporus betulinus   49, 52,
    87, 94, 99-101, 104
Pisonia grandis   232
Pistillaria pusillus   76
plant galls and fungi   248
Plantago lanceolata   224
Platypoecilus maculatus
    (platyfish)   122
pleomorphism   6
Pleospora   41, 113
P. infectoria   42
P. scirpicola   124, 125
Pleurotus cornucopiae   102
Podaxis   44
P. pistillaris   260, 261
podetia   200
Podosphaera leucotricha   70
Podospora   146
P. appendiculata   155
P. curvula   155
P. minuta   151, 152
polygalacturonase   138, 274-5
polygalacturonate
    transeliminase   138,
        274-5
polyhydric alcohols (polyols),
    and desiccation   182,
        206
    in carbon fixation in
        lichens   199-201
    on wetting and drying in
        lichens   208
polypore   46-52, 58
Polyporus squamosus   101-2
Polyscytalum   158
P. fecundissimum   76
powdery mildews   63, 266
proline   182
    as a compatible solute in
        xerophiles   182
proteases, in faecal droplets of
    Atta   257
protocatechuic acid   90, 91
Psathyrella   146
Pseudomonas   108
P. putida, in mushroom
    compost   169-70

pseudothecia   19, 43, 146
Pseudotsuga menziesii   173
psychrophilous fungi   175-8
    basis of
        psychrophily   175-6
    distribution of   176-7
    as 'snow moulds'   177
    in frozen food   177-8
Puccinia   45
P. graminis   268
    formae speciales in   273
P. menthae   62
P. purpurea, haustoria of   270
Pythium   143
P. debaryanum   274
    as a facultative
        necrotroph   279-81
    disease severity and
        moisture regime   280
    life cycle of   281

Quercus   128, 136, 146
quinones   108

Rafflaelea   247
Ramalina maciformis   207
Ramaria   47
Readeriella mirabilis   70
resting sporangia   3
Rhipidium   143
R. americanum   143, 144
Rhizobium   222-3
Rhizoctonia   235, 237-40
R. solani   28
rhizoids   2, 3
rhizomorphs   19, 28, 29
    structure and
        development   30-31
    functions of   31-2
Rhizophlyctis rosea   122, 157
Rhizophydium   122
R. eleyensis   123
R. globosum   123
Rhizopogon   184
Rhizopus   33, 35, 36, 40, 41,
    70
R. nigricans   160
R. stolonifer   274
Rhododendron ponticum   232
Rhynia   214
Rhyparobius   146
R. dubius   146
Ruppia maritima   110
Russula, as a mycorrhizal
    genus   184-5, 192

Saccharomyces   13
S. cerevisiae   37
Saccobolus   146
salicylaldehyde   109
salicyl alcohol   109

Salix 177
Salmo salar (salmon),
ulcerative dermal
necrosis in 123
Sanguisorba minor 233
Saprolegnia 33, 34, 40, 120,
143
S. diclina 121
S. litoralis 121
S. parasitica 121
S. terrestris 121
Saprolegniales 33, 117,
118–23, 135, 143
as 'water
moulds' 119–21
on insect exuviae 121–2
as parasites 122–3
Sapromyces elongatus 145
Schizosaccharomyces 41, 42
Schizophyllum commune 7
Sclerocystis rubiformis 216,
218
sporocarps of 216
Scleroderma 184
Sclerophoma pithiophila 70,
77, 78
sclerotia 19, 25–8, 75
structure and
development 27–8
Sclerotinia 27, 175
S. borealis 175, 176, 177
Sclerotium 27, 177
S. cepivorum 27
S. rolfsii 28
Scoliolegnia asterophora 121
Scolytus scolytus 99, 104
Scytalidium 105
Scytonema 194
septal pore cap 20
septation 19–21
Septobasidium 252–4
association with
Aspidiotus 252–4
infection and parasitism
of *Aspidiotus* 252–3
Septoria nodorum 69
Sericomyrmex 254
Serpula lacrimans 29, 32, 95,
100
mycelial strands of 29–30
'shadow yeasts' 60
sheathing mycorrhiza 183–94
structure 183–4, 186–7
fungi involved 184–5
nutrition of fungi 185–7
dependence of fungi on
host for carbon
sources 188
enhanced mineral ion
uptake 188–9

sheath as a reservoir for
mineral ions 190
dual role of sheath 191–2
production of growth
substances by
fungi 192–3
and susceptibility of roots
to pathogens 193
sinapyl alcohol 85, 86
Sirex noctilio 248–52
association with
*Amylostereum
areolatum* 248–50
benefit to fungus and
wasp 250
problems for the plant
pathologist 250–52
Sirodrepa panicea (drug store
beetle) 263
and endosymbiotic
yeasts 263
skeletal hyphae 21, 30, 47–9
'snow mould' 177
soft-rot fungi 85–8, 92
on wood in sea 114
sooty moulds 61
Sordaria 41
S. fimicola 151, 152
soredia 202
Sphaceleria 131
Sphaeriales 112, 113
*Spathulospora
phycophila* 115
Spiranthes spiralis 235
spitzenkörper 15
sporangiospores 33–7
sporangium 9, 35–7, 148,
149, 156
Sporobolomyces 61, 62, 65,
68, 69
S. roseus 58, 59, 65, 77
Sporormia 146
S. intermedia 151, 152
Sporotrichum carnis 177
staling phase 21
stalked spore drop 37, 97, 98
staphyla 256
Stemphylium 38
S. botryosum 63
Stereum 104
S. hirsutum 102, 104
S. sanguinolentum 102
Sticta 211–13
S. felix 212
Stilbella erythrocephala 154
stilbenes 95
'storage fungi' 180
Stropharia 146
succession on dung 146–7,
150–54

in wheat straw
compost 164–5
on wood 103–4
on pine needles 77–9
'sugar fungi' 71
Sympodiella acicola 78
syringaldehyde 90, 106
syringic acid 107

tannic acid 89
tannins 96, 100, 102, 194
Talaromyces crustaceus 162
teleomorphic 1, 76
states in aquatic
hyphomycetes 128–9,
135, 136
'tendril' hyphae 30
termites, and fungi 258–62
acquisition of fungal
enzymes by 260–62
Termitomyces 258
association with
termites 258–60
T. robustus 259–60
T. titanicus 260
terpenoids 95, 102
Tetrachaetum 126
Tetracladium 140
T. marchalianum 138
Tetraploa aristata 74
tetra-radiate spores, in aquatic
hyphomycetes 124–7
in other fungi 110–11,
131–2
biological significance
of 132–3
thalloconidia, development
of 37–9
in aquatic
hyphomycetes 126–7
Thelebolus 146
T. stercoreus 155
thermophilous fungi 159–64
basis of thermophily
in 159–61
variety and distribution
of 161–4
beneficial and detrimental
activities of 166–7
in garden compost 172
in wood chip piles 172–3
in soils 173–4
in bird's nests 174–5
in alligator nests 175
Thermoascus crustaceus 162
*Thermomyces
lanuginosus* 160, 162,
164, 165, 166
Thielavia thermophile 171
Thraustochytriales 113–14,
117

Thraustochytrium
   proliferum 113
Thraustotheca 120
thujaplicin 95
Tilia 61
Tilletiopsis 58
Torula herbarum 74
Torulopsis 262
Trachymyrmex 254
Trametes suaveolens 109
Trebouxia 194, 199, 201
trehalose 28, 188, 237, 239,
   258, 272
Tremellales 45
Trentepohlia 194, 199
Trichoderma 38, 76, 79, 103,
   105
Tricholoma, as a mycorrhizal
   genus 184
Tricholomopsis rutilans 106
Tricladium 126, 128, 130
T. angulatum 140
T. chaetocladium 136, 137
T. splendens 132
Trifolium parviflorum 225
Triscelophorus
   monosporus 134
tropolones 95
Tuber melanosporum 184
Tulasnella 235
Turdus merula
   (blackbird) 157
turgor 14
turgor pressure 14, 17
Typha latifolia 124
Typhula 27, 177
T. idahoensis 175, 176
T. trifolii 177
tyrosinase 89

Ulmus 139, 140
unitunicate asci 41–2
urea, application to
   leaves 72–3
Uredinales 45, 267
urediospores 57
Urtica dioica 71
usnic acid 210
Ustulina deusta 87, 102

Vaccinium (bilberry) 226
V. macrocarpon 228–30, 232
vanillic acid 90, 106, 107
vanillin 90
Venturia inaequalis 69, 72,

   75, 280
   nutritional relationships
     of 280–84
   life cycle of 282–3
   as a saprotroph 284
Verrucaria 115
Verticillium dahliae 27
V. malthousei 171
vesicles 215–17
vesicular-arbuscular
   mycorrhizas 214–26
   occurrence 214–15
   features of
     infection 215–18
   arbuscules 215, 217
   vesicles 215–17
   chlamydospores 215, 216
   fungi involved 218
   spores in soil 218–19
   as obligate biotrophs 219
   enhanced host
     growth 219–23
   increased phosphate
     uptake by 219–23
   source of phosphate taken
     up 223–4
   possible use as biological
     fertilisers 221–3
   inter plant transfer of
     carbon via
     fungus 224–5
   benefits to fungus 225–6
'vessel' hyphae 30
Vibrio marinum 176
Vicia faba (broad bean),
   chocolate spot disease
   of 277–9
victorin 276
Volvariella volvacea,
   cultivation of 171–2

water activity (a_w) and wood
   decay 94–5
   and fungal growth 178–9
   and xerophilous and xero-
     tolerant fungi 179–80
   and 'osmophilous'
     aspergilli 180–82
'water moulds' 119–21
   on insect exuviae 121
wheat straw compost 164–6
   succession of fungi
     in 164–5
   nutrition of fungi
     involved 165

white-rot fungi 85–91, 99,
   101, 103, 104, 105, 106
   specificity of 100–102
wood, structure and
   components 84–8
   resistance to fungal
     decay 93–6
   nitrogen content and
     decay 93–4
   moisture content and
     decay 94–5
   toxic substances in 95–6
   environmental factors and
     decay 99–100
wood decay, types 85–8
   see also white-, brown-
     and soft-rot fungi
wyerone acid 276, 277–9

Xanthoria aureola 201
X. parietina, and
   pigmentation 209
xerophilous fungi 178–82
   basis of thermophily 182
   'osmophilous'
     aspergilli 180–82
xero-tolerant fungi 179–80
Xylaria hypoxylon 87, 102
X. polymorpha 87, 102
X. nigripes 260
X. termitum 261
Xyleborus 245
   mycetangia in 247

yeasts 2, 57
yeast-like 2, 6, 58

Zalerion maritima 114
zoosporangium 3
   in *Saprolegnia* 33–4,
     120, 123, 143
   in *Pythium* 281
zoospores 3, 6, 33–4, 117,
   120–21, 135
   in *Pythium* 280–81
Zygomycetes 15, 16, 33, 40,
   41, 58, 71, 76, 146, 165
Zygomycotina, life cycle 7,
   9–10
zygospore 9, 10
   in *Rhizopus* 40–41

Printed in the United States
By Bookmasters